T0297707

REFRIGERATION, AIR CONDITIONING AND HEAT PUMPS

Refrigeration: The process of removing heat.

Air-conditioning: A form of air treatment whereby temperature, humidity, ventilation, and air cleanliness are all controlled within limits determined by the requirements of the air conditioned enclosure.

BS 5643: 1984

Heat pump: A device which extracts energy from a source at low temperature and makes it available as useful heat energy at a higher temperature.

R. D. Heap, 1982

REFRIGERATION, AIR CONDITIONING AND HEAT PUMPS

Fifth Edition

G F HUNDY

A R TROTT

T C WELCH

Amsterdam • Boston • Heidelberg • London
New York • Oxford • Paris • San Diego
San Francisco • Singapore • Sydney • Tokyo
Butterworth-Heinemann is an imprint of Elsevier

Butterworth-Heinemann is an imprint of Elsevier
The Boulevard, Langford Lane, Kidlington, Oxford OX5 1GB, UK
50 Hampshire Street, 5th Floor, Cambridge, MA 02139, USA

First published by McGraw-Hill Book Company (UK) ltd 1981
Second edition published by Butterworth-Heinemann 1989
Third edition 2000
Transferred to digital printing 2002
Fourth edition 2008
Fifth edition 2016

British Library Cataloguing-in-Publication Data
A catalogue record for this book is available from the British Library

Library of Congress Cataloging-in-Publication Data
A catalog record for this book is available from the Library of Congress

ISBN: 978-0-08-100647-4

For information on all Butterworth-Heinemann publications
visit our website at http://elsevier.com/

CONTENTS

PREFACE

In this fifth revision I have included further information about developments and innovations whilst at the same time retaining the fundamentals and depth of technical detail. The addition of heat pumps to the title reflects the fact that heating provision using refrigeration technology is now a serious contender in many situations. A further area of innovation is integrated heating and cooling whereby previously discarded heat is utilised or raised to a temperature where it can be utilised. Refrigeration is all about moving or 'pumping' heat from a lower to a higher temperature.

The task of refrigeration engineers is to achieve this movement of heat with the lowest power consumption in a cost effective manner. This is always a trade-off, but innovative technology is delivering advances. This has to be set against a background of environmental impact. Many effective refrigerants contribute to global warming if released to atmosphere, and improved containment together with legislation regarding the usage of these substances has impelled developments. The industry has paved the way for these constraints by developing new, less-damaging chemicals, and showing how other so-called 'natural' products can be used in applications where they were previously considered unsafe or just impractical. Standards, good practice guides, and training, all supported by industry, are vital components of this story, which is continuing today.

A.R. Trott designed this book to give an appreciation of the subject, building on the unchanging fundamentals in a logical way. This was further developed by T.C. Welch and I am indebted to both these authors for developing this approach. I would also like to thank friends and colleagues in the industry who have helped with information, proofread drafts, and provided suitable illustrations.

Guy Hundy
October 2015

PREFACE TO THE PREVIOUS EDITION

Refrigeration and air conditioning absorb about 15% of the United Kingdom's electrical generation capacity and it is not always appreciated that refrigeration technology is essential to our modern way of life. Without it, distribution of food to urban areas may not be possible. In a typical office, air conditioning can account for over 30% of annual electricity consumption, yet who cares about checking the system to find out if it is working efficiently?

Reducing the environmental impact of cooling whilst maintaining and expanding expectations is the driver of many of the developments which have been made since the last edition of this book. Aimed at students, and professionals in other disciplines, not too theoretical but with sufficient depth to give an understanding of the issues, this book takes the reader from the fundamentals, through to system design, applications, contract specifications and maintenance. Almost every chapter could be expanded into a book in itself and references are provided to assist those wishing to delve deeper. Standards and legislation are subject to change and readers are recommended to consult the Institute of Refrigeration web site for the latest developments.

This edition gives an up-to-date appreciation of the issues involved in refrigerant choice, efficiency, load reduction and effective air conditioning. Managing heat energy is going to be crucial in the quest of the United Kingdom to reduce carbon emissions – and managing heat rather than burning fuel to generate more of it, is what heat pumps do. Refrigeration technology has a potentially huge role to play in heating, which is where a very large proportion of the energy in United Kingdom is spent.

In navigating this book you should be guided by the context of your interest, but at the same time develop an awareness of related topics. Most real problems cross boundaries, which are in any case difficult to define, and some of the most exciting developments have occurred when taking concepts from various branches to other applications in innovative ways.

I am much indebted to friends and colleagues in the industry who have helped with information, proofread drafts, and given guidance on many of the topics. Thanks are due in particular to individuals who have gone out of their way to provide suitable illustrations and to their organisations for supporting them.

Guy Hundy
July 2008

PREFACE TO THE PREVIOUS EDITION

ACKNOWLEDGEMENTS

Front cover pictures (Clockwise from top):

Industrial water chiller	J & E Hall International;
Refrigerant Circuit Illustration	Business Edge Ltd
Refrigerated container (Reefer)	Cambridge Refrigeration Technology

Mollier diagrams drawn with the aid of CoolPack software:
Department of Mechanical Engineering, University of Denmark

Pictures and diagrams within the text are reproduced by courtesy of the following organizations:

Advanced Engineering Ltd
Airedale International Air Conditioning Ltd
Alfa Laval
Baltimore Aircoil
Bitzer Kühlmaschinenbau GmbH
Michael Boast Engineering Consultancy
Business Edge Ltd
Cambridge Refrigeration Technology
Carrier Corporation
CIBSE – Chartered Institution of Building Services Engineers
FMA – Fan Manufacturers Association
Climacheck Sweden AB
Climate Center
M Conde Engineering (Switzerland)
Danfoss A/S
Emerson Climate Technologies GmbH
RD&T – Refrigeration Development and Testing Ltd
Glasgow University Archive Services
Gram Equipment A/S
GEA Refrigeration
Heatking
Henry Technologies
IOR – Institute of Refrigeration
Howden Compressors Ltd
Hubbard Products Ltd
J & E Hall International

Jackstone Froster Ltd
Johnson Controls
Kensa Heat Pumps Ltd
Searle
Star Instruments Ltd
Star Refrigeration Ltd
Thermo King
Titan Engineering Ltd
XL Refrigerators Ltd
Harry Yearsley Ltd

LIST OF ABBREVIATIONS

IOR	Institute of Refrigeration, London
IIR	International Institute of Refrigeration, Paris
IPCC	International Panel on Climate Change
ASHRAE	American Society of Heating, Refrigerating and Air-Conditioning Engineers, Atlanta
IMechE	Institution of Mechanical Engineers, London
CIBSE	Chartered Institution of Building Services Engineers, London
ACRIB	Air Conditioning and Refrigeration Industry Board (United Kingdom)
DEFRA	Department for Environment Food & Rural Affairs (United Kingdom)
ASERCOM	Association of European Refrigeration Compressor and Controls Manufacturers
IEA	International Energy Agency
VDMA	Verband Deutscher Maschinen- und Anlagenbau e.V. (German engineering association)
LSBU	London South Bank University
ODP	Ozone Depletion Potential
GWP	Global Warming Potential
TEWI	Total Equivalent Warming Impact

CHAPTER 1

Fundamentals

1.1 INTRODUCTION

Refrigeration is the action of cooling, and in practice this requires removal of heat and discarding it at a higher temperature. Refrigeration is therefore the science of moving heat from low temperature to high temperature. In addition to chilling and freezing applications, refrigeration technology is applied in air conditioning and heat pumps, which therefore fall within the scope of this book. The fundamental principles are those of physics and thermodynamics, and these principles, which are relevant to all applications, are outlined in this opening chapter.

1.2 TEMPERATURE, WORK AND HEAT

The temperature scale now in general use is the *Celsius scale*, based nominally on the melting point of ice at 0°C and the boiling point of water at atmospheric pressure at 100°C (by strict definition, the triple point of ice is 0.01°C at a pressure of 6.1 mbar).

The law of conservation of energy tells us that when work and heat energy are exchanged there is no net gain or loss of energy. However, the amount of heat energy that can be converted into work is limited. As the heat flows from hot to cold, a certain amount of energy may be converted into work and extracted. For example, it can be used to drive a generator.

The minimum amount of work to drive a refrigerator can be defined in terms of the absolute temperature scale. Fig. 1.1 shows a reversible engine E driving a reversible heat pump P; *Q* and *W* represent the flow of heat and work. They are called reversible machines because they have the highest efficiency that can be visualised, and because there are no losses, E and P are identical machines.

The arrangement shown results in zero external effect because the reservoirs experience no net gain or loss of heat. If the efficiency of P were to be higher, that is, if the work input required for P to lift an identical quantity of heat Q_2 from the cold reservoir were to be less than *W*, the remaining part of *W* could power another heat pump. This could lift an additional amount of heat. The result would be a net flow of heat from the low temperature

Refrigeration, Air Conditioning and Heat Pumps
http://dx.doi.org/10.1016/B978-0-08-100647-4.00001-2

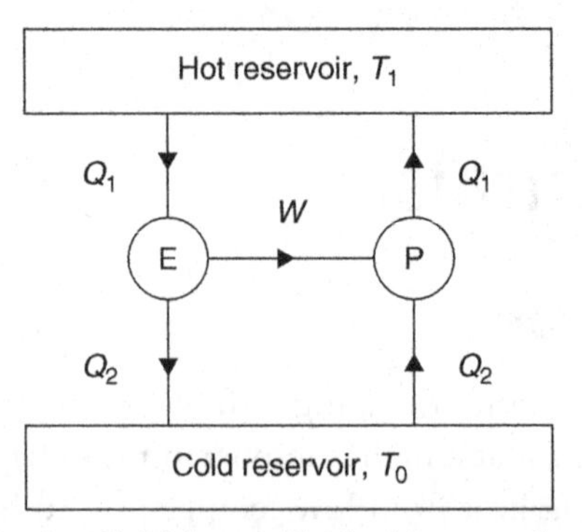

Figure 1.1 ***Ideal heat engine, E, driving an ideal refrigerator (heat pump), P.***

to the high temperature without any external work input, which is impossible. The relationship between Q_1, Q_2 and W depends only on the temperatures of the hot and cold reservoirs. The French physicist Sadi Carnot (1796–1832) was the first to predict that the relationship between work and heat is temperature dependent, and the ideal refrigeration process is known as the Carnot cycle. In order to find this relationship, temperature must be defined in a more fundamental way. The degrees on the thermometer are only an arbitrary scale. William Thomson, 1824–1907, (Fig. 1.2) together with other leading physicists of the period concluded that an absolute temperature scale can be defined in term sof the efficiency of reversible engines. William Thomson was appointed to the chair of natural philosophy at Glasgow University aged 22, and became Lord Kelvin in 1892.

Kelvin (1824–1907), together with other leading physicists of the period, concluded that an absolute temperature scale can be defined in terms of the efficiency of reversible engines (Fig. 1.2).

The ideal 'never-attainable-in-practice' ratio of work output to heat input (W/Q_1) of the reversible engine E equals: Temperature Difference ($T_1 - T_0$) divided by the Hot Reservoir Temperature (T_1).

In Fig. 1.1 the device P can be any refrigeration device we care to invent, and the work of Kelvin tells us that the minimum work, W necessary to lift a quantity of heat Q_2 from temperature T_0 to temperature T_1 is given by:

$$W = \frac{Q_2(T_1 - T_0)}{T_0}$$

The temperatures must be measured on an absolute scale, that is one that starts at absolute zero. The Kelvin scale has the same degree intervals as the Celsius scale, so that ice melts at +273.16 K, and water at atmospheric

Figure 1.2 ***William Thomson (Glasgow University).***

pressure boils at +373.15 K. On the Celsius scale, absolute zero is −273.15°C. Refrigeration 'efficiency' is usually defined as the heat extracted divided by the work input. This is called COP, *coefficient of performance.* The ideal or Carnot COP takes its name from Sadi Carnot and is given by:

$$\mathrm{COP} = \frac{Q_2}{W} = \frac{T_0}{(T_1 - T_0)}$$

Example 1.1

Heat is to be removed at a temperature of -5°C and rejected at a temperature of 35°C. What is the Carnot or Ideal COP?

Convert the temperatures to absolute:

-5°C becomes 268 K and 35°C becomes 308 K (to the nearest K)

$$\mathrm{Carnot\,COP} = \frac{268}{(308 - 268)} = 6.7$$

1.3 HEAT AND ENTHALPY

Heat is one of the many forms of energy and is commonly generated from chemical sources. The heat of a body is its thermal or internal energy, and a change in this energy may show as a change of temperature or a change between the solid, liquid and gaseous states.

Matter may also have other forms of energy, potential or kinetic, depending on pressure, position and movement. Enthalpy is the sum of its internal energy and flow work and is given by:

$$H = u + Pv$$

In the process where there is steady flow, the factor Pv will not change appreciably and the difference in enthalpy will be the quantity of heat gained or lost.

Enthalpy may be expressed as a total above absolute zero, or any other base which is convenient. Tabulated enthalpies found in reference works are often shown above a base temperature of -40°C, since this is also -40°F on the old Fahrenheit scale. In any calculation, this base condition should always be checked to avoid the errors which will arise if two different bases are used.

If a change of enthalpy can be sensed as a change of temperature, it is called *sensible heat*. This is expressed as specific heat capacity, that is, the change in enthalpy per degree of temperature change, in kJ/(kg K). If there is no change of temperature but a change of state (solid to liquid, liquid to gas, or vice versa) it is called *latent heat*. This is expressed as kJ/kg but it varies with the boiling temperature, and so is usually qualified by this condition. The resulting total changes can be shown on a temperature–enthalpy diagram (Fig. 1.3).

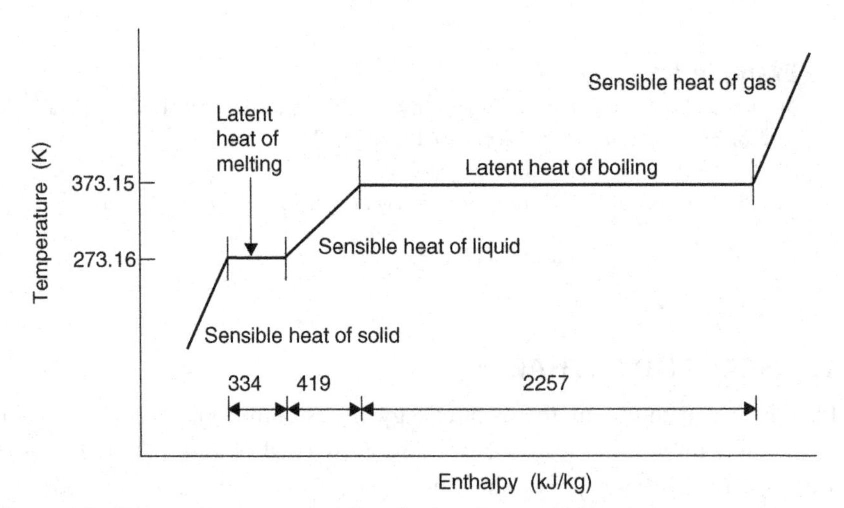

Figure 1.3 ***Change of temperature (K) and state of water with enthalpy.***

Example 1.2

The specific enthalpy of water at 80°C, taken from 0°C base, is 334.91 kJ/kg. What is the average specific heat capacity through the range 0–80°C?

$$\frac{334.91}{(80-0)} = 4.186\,\text{kJ/(kg K)}$$

Example 1.3

If the latent heat of boiling water at 1.013 bar is 2257 kJ/kg, the quantity of heat which must be added to 1 kg of water at 30°C in order to boil it is:

$$4.19(100-30)+2257=2550.3\ \text{kJ}$$

1.4 BOILING POINT

The temperature at which a liquid boils is not constant, but varies with the pressure. Thus, whilst the boiling point of water is commonly taken as 100°C, this is only true at a pressure of one standard atmosphere (1.013 bar) and, by varying the pressure, the boiling point can be changed (Table 1.1). This pressure–temperature property can be shown graphically (see Fig. 1.4).

The boiling point of a substance is limited by the *critical temperature* at the upper end, beyond which it cannot exist as a liquid, and by the *triple point* at the lower end, which is at the freezing temperature. Between these two limits, if the liquid is at a pressure higher than its boiling pressure, it will remain a liquid and will be sub-cooled below the saturation condition, whereas if the temperature is higher than saturation, it will be a gas and superheated. If both liquid and vapour are at rest in the same enclosure,

Table 1.1 Boiling point of water at different pressures

Pressure (bar)	Boiling point (°C)
0.006	0
0.04	29
0.08	41.5
0.2	60.1
0.5	81.4
1.013	100.0

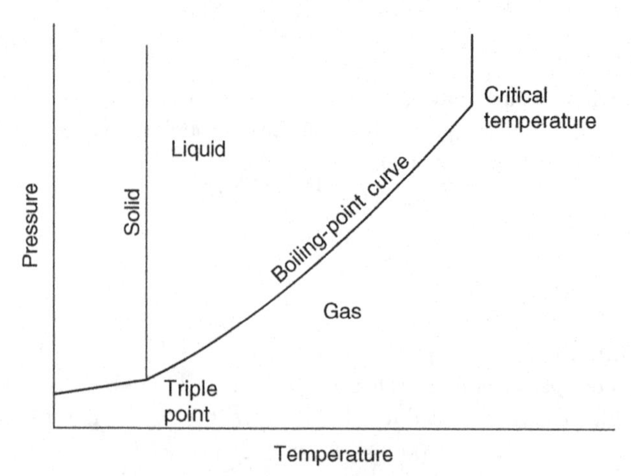

Figure 1.4 ***Change of state with pressure and temperature.***

and no other volatile substance is present, the condition must lie on the saturation line.

At a pressure below the triple point pressure, the solid can change directly to a gas (sublimation) and the gas can change directly to a solid, as in the formation of carbon dioxide snow from the released gas.

The liquid zone to the left of the boiling point line is sub-cooled liquid. In refrigeration the term *saturation* is used to describe the liquid/vapour boundary, saturated vapour being represented by a condition on the line and superheated vapour below the line. For more information on saturated properties for commonly used refrigerants see Chapter 3: Refrigerants.

1.5 GENERAL GAS LAWS

Many gases at low pressure, that is atmospheric pressure and below for water vapour and up to several bar for gases such as nitrogen, oxygen and argon, obey simple relations between their pressure, volume and temperature, with sufficient accuracy for engineering purposes. Such gases are called 'ideal'.

Boyle's law states that, for an ideal gas, the product of pressure and volume at constant temperature is a constant:

$$pV = \text{constant}$$

Example 1.4

A volume of an ideal gas in a cylinder and at atmospheric pressure is compressed to half the volume at constant temperature. What is the new pressure?

$$p_1V_1 = \text{constant}$$
$$= p_2V_2$$
$$\frac{V_1}{V_2} = 2$$

$$\text{so} \quad p_2 = 2 \times p_1$$
$$= 2 \times 1.01325\,\text{bar}\,(101\ 325\,\text{Pa})$$
$$= 2.0265\,\text{bar}\,(\text{abs.})$$

Charles' law states that, for an ideal gas, the volume at constant pressure is proportional to the absolute temperature:

$$\frac{V}{T} = \text{constant}$$

Example 1.5

A mass of an ideal gas occupies 0.75 m^3 at 20°C and is heated at constant pressure to 90°C. What is the final volume?

$$V_2 = V_1 \times \frac{T_2}{T_1}$$
$$= 0.75 \times \frac{273 + 90}{273 + 20}\,(\text{temperatures to the nearest K})$$
$$= 0.93\,\text{m}^3$$

Boyle's and Charles' laws can be combined into the ideal gas equation:

$$pV = (\text{a constant}) \times T$$

The constant is mass × R, where R is the specific gas constant, so:

$$pV = mRT$$

Example 1.6

What is the volume of 5 kg of an ideal gas, having a specific gas constant of 287 J/(kg K), at a pressure of one standard atmosphere and at 25°C?

$$pV = mRT$$

$$V = \frac{mRT}{p}$$

$$= \frac{5 \times 287(273 + 25)}{101\ 325}$$

$$= 4.22\ m^3$$

1.6 DALTON'S LAW

Dalton's law of partial pressures considers a mixture of two or more gases, and states that the total pressure of the mixture is equal to the sum of the individual pressures, if each gas separately occupied the space.

Example 1.7

A cubic metre of air contains 0.906 kg of nitrogen of specific gas constant 297 J/(kg K), 0.278 kg of oxygen of specific gas constant 260 J/(kg K) and 0.015 kg of argon of specific gas constant 208 J/(kg K). What will be the total pressure at 20°C?

$$pV = mRT$$

$$V = 1m^3$$

$$so \quad p = mRT$$

For the nitrogen $p_N = 0.906 \times 297 \times 293.15 = 78{,}881$ Pa

For the oxygen $p_O = 0.278 \times 260 \times 293.15 = 21{,}189$ Pa

For the argon $p_A = 0.015 \times 208 \times 293.15 = \underline{915\ \text{Pa}}$

Total pressure = 100,985 Pa

(1.009 85 bar)

The properties of refrigerant fluids at the pressures and temperatures of interest to refrigeration engineers exhibit considerable deviation from the ideal gas laws. It is therefore necessary to use tabulated or computer-based information for thermodynamic calculations.

1.7 THE PRESSURE – ENTHALPY CHART

In order to study the properties of a fluid in the region of interest, refrigeration engineers use a pressure–enthalpy or *P–h Diagram* (Fig. 1.5a). This diagram is a useful way of describing the liquid and gas phase of a substance. On the vertical axis is pressure, P, usually on a logarithmic scale, and on the horizontal, h, enthalpy. The *saturation curve* defines the boundary of pure liquid and pure gas, or vapour. At pressures above the top of the curve, there is no distinction between liquid and vapour. Above this pressure the gas cannot be liquefied. This is called the *critical pressure*. In the two-phase region beneath the curve, there is a mixture of liquid and vapour.

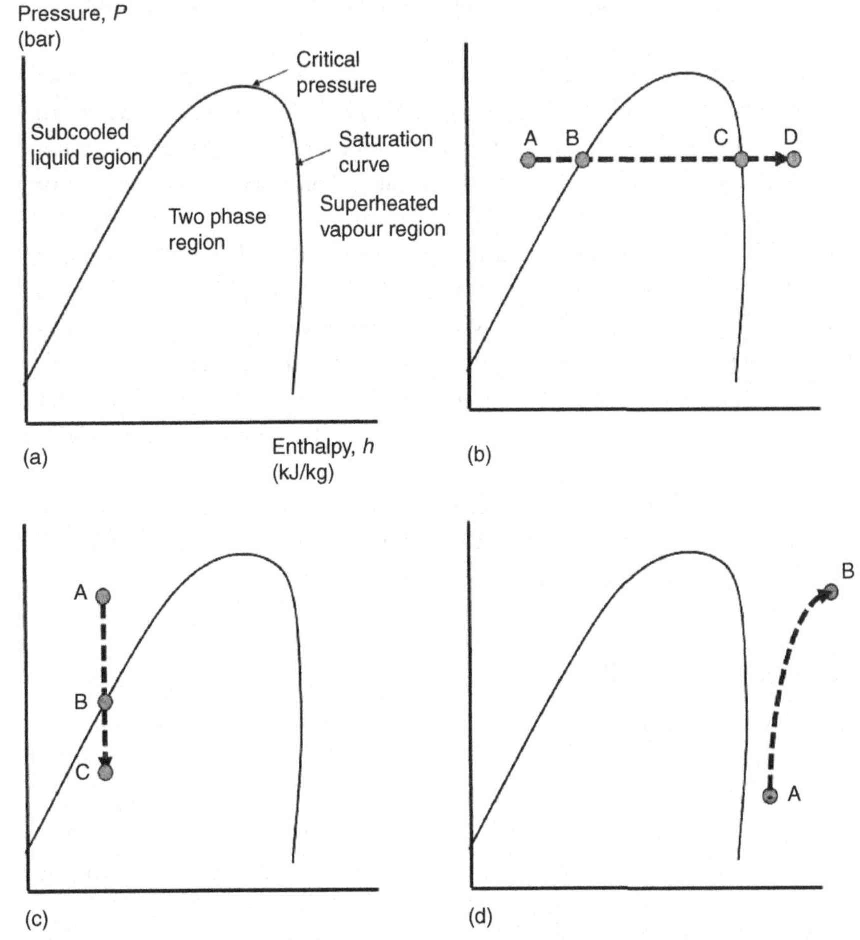

Figure 1.5 ***The P–h diagram.*** (a) Definitions, (b) constant pressure process, (c) constant enthalpy process, (d) compression process.

This chart is probably the most recognisable tool in the industry, and with good reason. It is essential to understand how it can be used to give quantitative property values. It is sometimes referred to as a *Mollier diagram*. Richard Mollier created the first chart using enthalpy as an axis in 1904. Fig. 1.5b shows a constant pressure process in which heat is added to a liquid causing its enthalpy to increase. Starting at A the liquid becomes warmer until at B it starts to boil. More heat causes more of the liquid to turn into a gas, or vaporise, until at C there is 100% gas. Further addition of heat causes temperature rise or increase in superheat. Fig. 1.5c is a constant enthalpy process; no heat is added or removed, but the pressure of a liquid is reduced until vapour starts to form and the fluid enters the two phase region. In Fig. 1.5d a vapour is compressed. The pressure rises and enthalpy (from compression work) also increases.

Turning now to Fig. 1.6 a process can be plotted on the chart itself. Starting from A, which represents liquid R717 at 40 bar (*y*-axis) where the enthalpy is 500 kJ/kg (*x*-axis), the pressure is reduced to 20 bar (B), where the temperature is 50°C (marked on the saturation curve). The temperature change during this reduction of liquid pressure is very small and the near-vertical lines of constant temperature are not shown on this side of the curve. More heat is now added until the enthalpy is 950 kJ/kg (C). At this point the fraction of gas/liquid is 50% (follow the steep 50% line down to the *x*-axis). The entropy and specific volume may also be read from interpolation of the sloping lines. At point D all the liquid has turned to vapour, still at 50°C. Further heat addition to 1650 kJ/kg results in a temperature rise to 100°C, pressure still unchanged.

It is difficult to read the chart accurately without a very large printed version, but computer generated data is almost universally used to calculate the refrigeration cycle discussed in the next chapter.

1.8 HEAT TRANSFER

Heat will move from a hot body to a colder one, and can do so by the following methods:

1. *Conduction*: Directly from one body touching the other, or through a continuous mass
2. *Convection*. By means of a heat-carrying fluid moving between one and the other
3. *Radiation*. Mainly by infrared waves (but also in the visible band, eg solar radiation), which are independent of contact or an intermediate fluid.

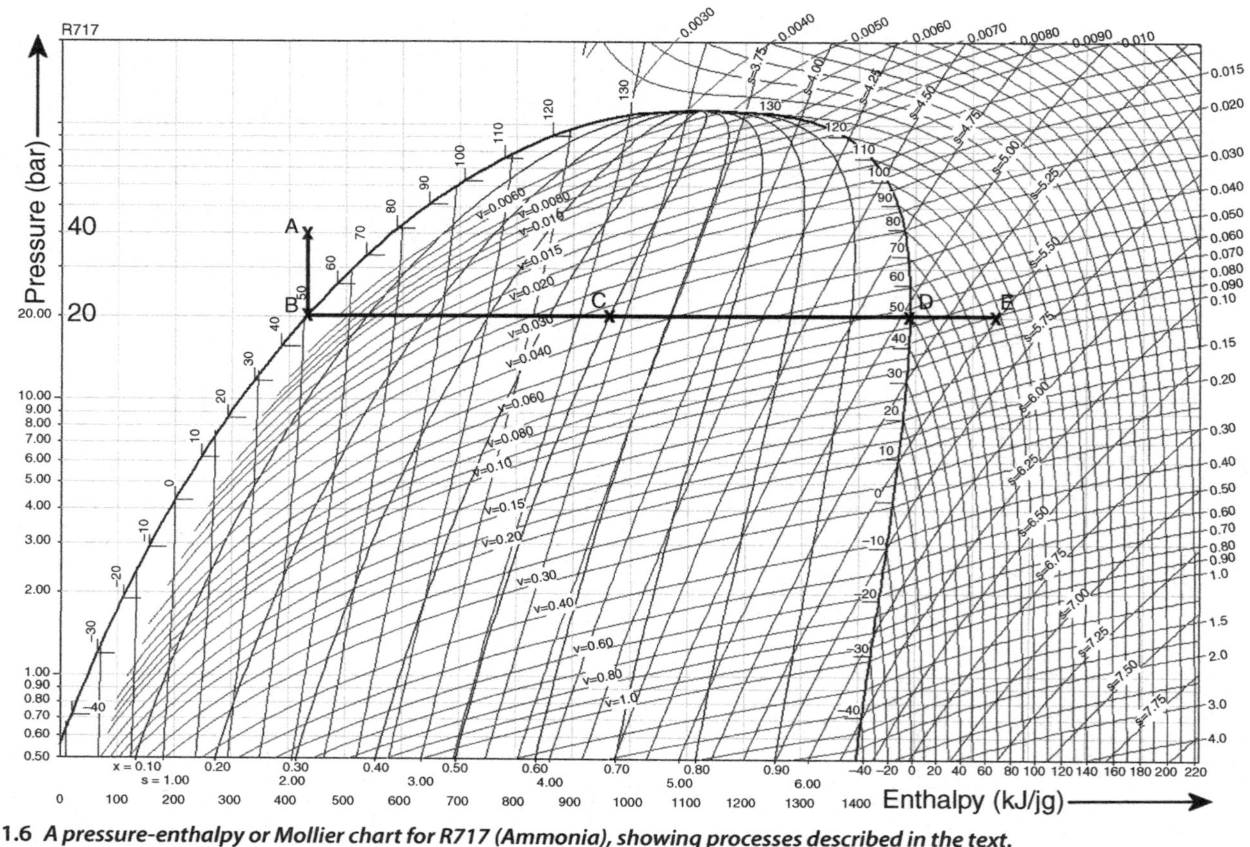

Figure 1.6 ***A pressure-enthalpy or Mollier chart for R717 (Ammonia), showing processes described in the text.***

Conduction through a homogeneous material is expressed directly by its area, thickness and a conduction coefficient. For a large plane surface, ignoring heat transfer near the edges:

$$\text{Conductance} = \frac{\text{area} \times \text{thermal conductivity}}{\text{thickness}}$$
$$= \frac{A \times k}{L}$$

and the heat conducted is

$$Q_f = \text{conductance} \times (T_1 - T_2)$$

Example 1.8

A brick wall, 225 mm thick and having a thermal conductivity of 0.60 W/(m K), measures 10 m long by 3 m high, and has a temperature difference between the inside and outside faces of 25 K. What is the rate of heat conduction?

$$Q_f = \frac{10 \times 3 \times 0.60 \times 25}{0.225}$$
$$= 2000\ \text{W} (\text{or } 2\ \text{kW})$$

Thermal conductivities, in watts per metre Kelvin, for various common materials are as in Table 1.2. Conductivities for other materials can be found from standard reference works.

Convection requires a fluid, either liquid or gaseous, which is free to move between the hot and cold bodies. This mode of heat transfer is complex and depends firstly on whether the flow of fluid is 'natural', that is caused by thermal currents set up in the fluid as it expands, or 'forced' by fans or pumps. Other parameters are the density, specific heat capacity and viscosity of the fluid and the shape of the interacting surface.

With so many variables, expressions for convective heat flow cannot be as simple as those for conduction. The interpretation of observed data has been

Table 1.2 Thermal conductivity of various materials

Material	Thermal conductivity (W/m K)
Copper	200
Mild steel	50
Concrete	1.5
Water	0.62
Cork	0.040
Expanded polystyrene	0.034
Polyurethane foam	0.026
Still air	0.026

Table 1.3 Dimensionless groups

Number	Symbol	Group	Parameters	Typical relevance
Reynolds	Re	$\frac{\rho v x}{\mu}$	Velocity of fluid, v Density of fluid, ρ Viscosity of fluid, μ Dimension of surface, x	Forced flow in pipes
Nusselt	Nu	$\frac{hx}{k}$	Thermal conductivity of fluid, k Dimension of surface, x Heat transfer coefficient, h	Convection heat transfer rate
Prandtl	Pr	$\frac{C_p \mu}{k}$	Specific heat capacity of fluid, C_p Viscosity of fluid, μ Thermal conductivity of fluid, k	Fluid properties
Grashof	Gr	$\frac{\beta g \rho^2 x^3 \theta}{\mu^2}$	Coefficient of expansion of fluid, β Density of fluid, ρ Viscosity of fluid, μ Force of gravity, g Temperature difference, θ Dimension of surface, x	Natural convection

made possible by the use of a number of dimensionless groups which combine the variables and which can then be used to estimate convective heat flow.

The main groups used in such estimates are as shown in Table 1.3. A typical combination of these numbers is that for turbulent flow in pipes expressing the heat transfer rate in terms of the flow characteristic and fluid properties:

$$\mathrm{Nu} = 0.023(\mathrm{Re})^{0.8}(\mathrm{Pr})^{0.4}$$

The calculation of every heat transfer coefficient for a refrigeration or air-conditioning system would be a very time-consuming process, even with modern methods of calculation. Formulas based on these factors will be found in standard reference works, expressed in terms of heat transfer coefficients under different conditions of fluid flow.

Where heat is conducted through a plane solid which is between two fluids, there will be the convective resistances at the surfaces. The overall heat transfer must take all of these resistances into account, and the unit transmittance, or '*U*' value is given by the following:

$$R_t = R_i + R_c + R_o$$
$$U = 1/R_t$$

where R_t = total thermal resistance
R_i = inside convective resistance
R_c = conductive resistance
R_o = outside convective resistance

Example 1.9

A brick wall, plastered on one face, has a thermal conductance of 2.8/ (m² K), an inside surface resistance of 0.3 (m² K)/W, and an outside surface resistance of 0.05 (m² K)/W. What is the overall transmittance?

$$R_t = R_i + R_c + R_o$$

$$= 0.3 + \frac{1}{2.8} + 0.05$$

$$= 0.707$$

$$U = 1.414\ \text{W/(m}^2\text{K)}$$

Typical values for thermal transmittance ae shown in Table 1.4.

Special note should be taken of the influence of geometrical shape, where other than plain surfaces are involved. The overall thermal transmittance, U, is used to calculate the total heat flow. For a plane surface of area A and a steady temperature difference ΔT, it is

$$Q_f = A \times U \times \Delta T$$

If a non-volatile fluid is being heated or cooled, the sensible heat will change and therefore the temperature, so that the ΔT across the heat exchanger wall will not be constant. Since the rate of temperature change (heat flow) will be proportional to the ΔT at any one point, the space–temperature curve will be exponential. In a case where the cooling medium is an evaporating liquid, the temperature of this liquid will remain substantially constant throughout the process, since it is absorbing latent heat, and the cooling curve will be as shown in Fig. 1.7.

Table 1.4 Typical thermal transmittance of heat flow barriers

Material	Thermal transmittance, W/(m² K)
Insulated cavity brick wall, 260 mm thick, sheltered exposure on outside	0.69
Chilled water inside copper tube, forced draught air flow outside	15–28
Condensing ammonia gas inside steel tube, thin W/(m² K) film of water outside	450–470

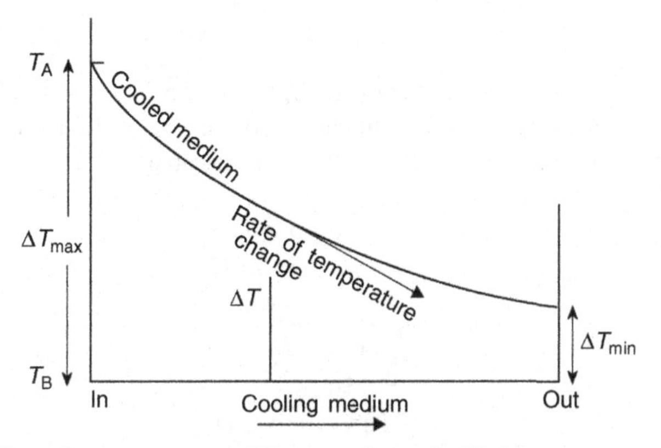

Figure 1.7 ***Changing temperature difference of a cooled fluid.***

Provided that the flow rates are steady, the heat transfer coefficients do not vary and the specific heat capacities are constant throughout the working range, the average temperature difference over the length of the curve is given by

$$\Delta T = \frac{\Delta T_{max} - \Delta T_{min}}{\ln(\Delta T_{max} / \Delta T_{min})}$$

This is applicable to any heat transfer where either or both the media change in temperature (see Fig. 1.8). This derived term is the *logarithmic mean temperature difference* (LMTD) and can be used as ΔT in the general equation, provided that U is constant throughout the cooling range, or an average figure is known, giving

$$Q_f = A \times U \times \text{LMTD}$$

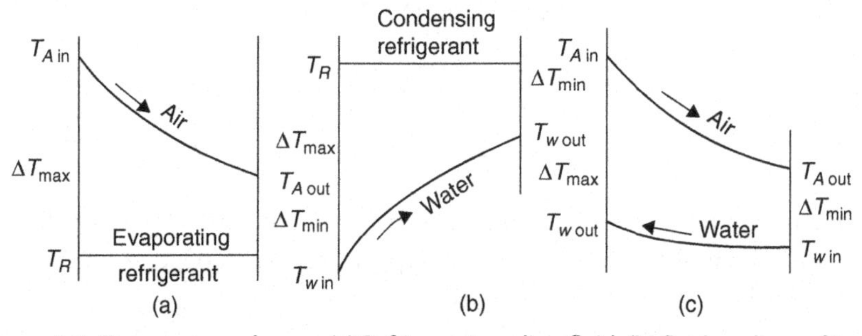

Figure 1.8 ***Temperature change.*** (a) Refrigerant cooling fluid; (b) fluid cooling refrigerant; (c) two fluids.

Example 1.10

A fluid evaporates at 3°C and cools water from 11.5 to 6.4°C. What is the logarithmic mean temperature difference and what is the heat transfer if it has a surface area of 420 m² and the thermal transmittance is 110 W/(m² K)?

$$\begin{aligned}
\Delta T_{\max} &= 11.5 - 3 = 8.5\,\mathrm{K} \\
\Delta T_{\min} &= 6.4 - 3 = 3.4\,\mathrm{K} \\
\mathrm{LMTD} &= \frac{8.5 = 3.4}{\ln(8.5/3.4)} \\
&= 5.566\,\mathrm{K} \\
Q_{\mathrm{f}} &= 420 \times 110 \times 5.566 \\
&= 257{,}000\,\mathrm{W} \text{ or } 257\,\mathrm{kW}
\end{aligned}$$

In practice, many of these values will vary. A pressure drop along a pipe carrying boiling or condensing fluid will cause a change in the saturation temperature. With some liquids, the heat transfer values will change with temperature. For these reasons, the LMTD formula does not apply accurately to all heat transfer applications.

If the heat exchanger was of infinite size, the space–temperature curves would eventually meet and no further heat could be transferred. The fluid in Example 1.10 would cool the water down to 3°C. The *effectiveness* of a heat exchanger can be expressed as the ratio of heat actually transferred to the ideal maximum:

$$\Sigma = \frac{T_{A\,\mathrm{in}} - T_{A\,\mathrm{out}}}{T_{A\,\mathrm{in}} - T_{B\mathrm{in}}}$$

Taking the heat exchanger in Example 1.10:

$$\begin{aligned}
\Sigma &= \frac{11.5 - 6.4}{11.5 - 3.0} \\
&= 0.6 \text{ or } 60\%
\end{aligned}$$

Radiation of heat was shown by Boltzman and Stefan to be proportional to the fourth power of the absolute temperature and to depend on the colour, material and texture of the surface:

$$Q_{\mathrm{f}} = \sigma \varepsilon T^4$$

where σ is Stefan's constant ($5.67 + 10^{-8}$ W/(m^2 K^4)) and ε is the surface *emissivity*.

Emissivity figures for common materials have been determined, and are expressed as the ratio to the radiation by a perfectly black body, namely

Rough surfaces such as brick, concrete or tile, regardless of colour	0.85–0.95
Metallic paints	0.40–0.60
Unpolished metals	0.20–0.30
Polished metals	0.02–0.28

The metals used in refrigeration and air-conditioning systems, such as steel, copper and aluminium, quickly oxidise or tarnish in air, and the emissivity figure will increase to a value nearer 0.50.

Surfaces will absorb radiant heat and this factor is expressed also as the ratio to the *absorptivity* of a perfectly black body. Within the range of temperatures in refrigeration systems, that is 270–150°C (203–323 K), the effect of radiation is small compared with the conductive and convective heat transfer, and the overall heat transfer factors in use include the radiation component. Within this temperature range, the emissivity and absorptivity factors are about equal.

The exception to this is the effect of solar radiation when considered as a cooling load, such as the air-conditioning of a building which is subject to the Sun's rays. At the wavelength of sunlight the absorptivity figures change and calculations for such loads use tabulated factors for the heating effect of sunlight. Glass, glazed tiles and clean white-painted surfaces have a lower absorptivity, whilst the metals are higher.

1.9 TRANSIENT HEAT FLOW

A special case of heat flow arises when the temperatures through the thickness of a solid body are changing as heat is added or removed. This *non-steady* or *transient* heat flow will occur, for example when a thick slab of meat is to be cooled, or when sunlight strikes on a roof and heats the surface. When this happens, some of the heat changes the temperature of the first layer of the solid, and the remaining heat passes on to the next layer, and so on. Calculations for heating or cooling times of thick solids consider the slab as a number of finite layers, each of which is both conducting and absorbing heat over successive periods of time. Original methods of solving transient heat flow were graphical, but could not easily take into account any change in the conductivity or specific heat capacity or any latent heat of the solid as the temperature changed.

Complicated problems of transient heat flow can be resolved by computer. Typical time–temperature curves for non-steady cooling are shown in Figs 16.1 and 16.3, and the subject is met again in Section 22.3.

1.10 TWO-PHASE HEAT TRANSFER

Where heat transfer is taking place at the saturation temperature of a fluid, evaporation or condensation (mass transfer) will occur at the interface, depending on the direction of heat flow. In such cases, the convective heat transfer of the fluid is accompanied by conduction at the surface to or from a thin layer in the liquid state. Since the latent heat and density of fluids are much greater than the sensible heat and density of the vapour, the rates of heat transfer are considerably higher. The process can be improved by shaping the heat exchanger face (where this is a solid) to improve the drainage of condensate or the escape of bubbles of vapour. The total heat transfer will be the sum of the two components.

Rates of two-phase heat transfer depend on properties of the volatile fluid, dimensions of the interface, velocities of flow and the extent to which the transfer interface is blanketed by fluid. The driving force for evaporation or condensation is the difference of vapour pressures at the saturation and interface temperatures. Equations for specific fluids are based on the interpretation of experimental data, as with convective heat transfer.

Mass transfer may take place from a mixture of gases, such as the condensation of water from moist air. In this instance, the water vapour has to diffuse through the air, and the rate of mass transfer will depend on the concentration of vapour in the air as well. In the air–water vapour mixture, the rate of mass transfer is roughly proportional to the rate of heat transfer at the interface and this simplifies predictions of the performance of air-conditioning coils.

CHAPTER 2

The Refrigeration Cycle

2.1 IDEAL CYCLE

An ideal reversible cycle based on the two temperatures of the system in Example 1.1 can be drawn on the basis of temperature–entropy relation (see Fig. 2.1).

In this cycle a unit mass of fluid is subjected to four processes after which it returns to its original state. The compression and expansion processes, shown as vertical lines, take place at constant entropy. A constant entropy (isentropic) process is a reversible or an ideal process. Ideal expansion and compression engines are defined in Section 1.2. The criterion of perfection is that no entropy is generated during the process, that is the quantity '*s*' remains constant. The addition and rejection of heat takes place at constant temperature and these processes are shown as horizontal lines. Work is transferred into the system during compression and out of the system during expansion. Heat is transferred across the boundaries of the system at constant temperatures during evaporation and condensation. In this cycle the net quantities of work and heat are in proportions which provide the maximum amount of cooling for the minimum amount of work. The ratio is the Carnot coefficient of performance (COP).

This cycle is sometimes referred to as a *reversed Carnot cycle* because the original concept was a heat engine and for power generation the cycle operates in a clockwise direction, generating net work.

2.2 SIMPLE VAPOUR COMPRESSION CYCLE

The *vapour compression cycle* is used for refrigeration in preference to gas cycles; making use of the latent heat enables a far larger quantity of heat to be extracted for a given refrigerant mass flow rate. This makes the equipment as compact as possible.

A liquid boils and condenses – the change between the liquid and the gaseous states – at a temperature which depends on its pressure, within the limits of its freezing point and critical temperature (see Fig. 2.2). In boiling it must obtain the latent heat of evaporation and in condensing the latent heat is given up.

Refrigeration, Air Conditioning and Heat Pumps
http://dx.doi.org/10.1016/B978-0-08-100647-4.00002-4

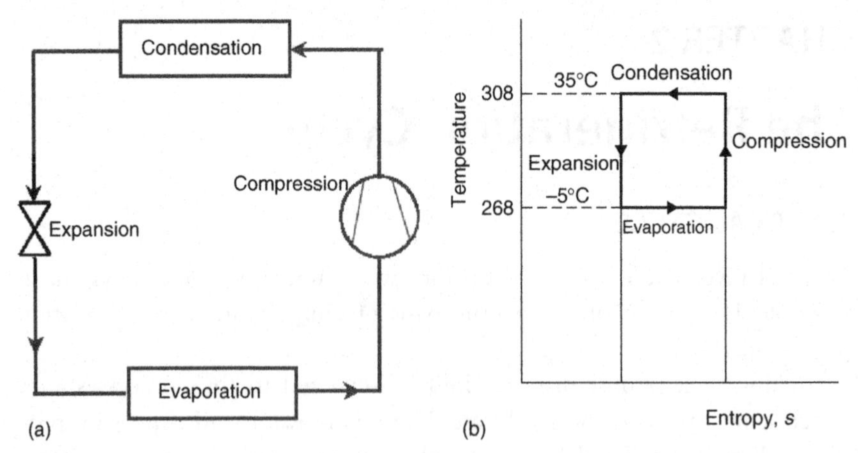

Figure 2.1 ***The ideal reversed Carnot cycle: (a) circuit and (b) temperature–entropy diagram.***

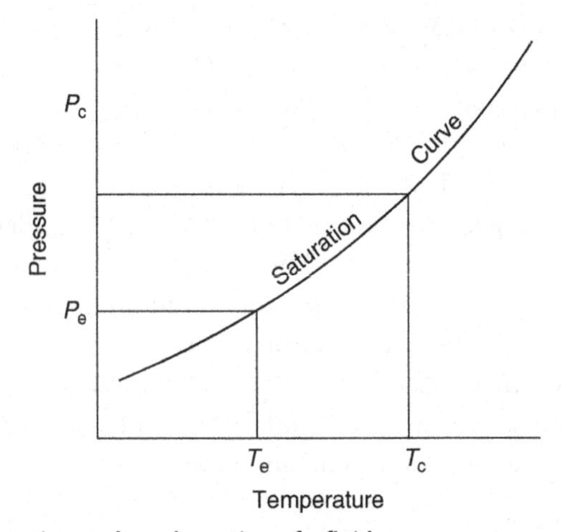

Figure 2.2 ***Evaporation and condensation of a fluid.***

Heat is put into the fluid at the lower temperature and pressure, thus providing the latent heat to make it vaporise. The vapour is then mechanically compressed to a higher pressure and a corresponding saturation temperature at which its latent heat can be rejected so that it changes back to a liquid. The cycle is shown in Fig. 2.3. The cooling effect is the heat transferred to the working fluid in the evaporation process, that is the change in enthalpy between the fluid entering and the vapour leaving the evaporator.

The simple vapour compression cycle is superimposed on the *P*–*h* diagram in Fig. 2.4. The evaporation process or vaporisation of refrigerant is a

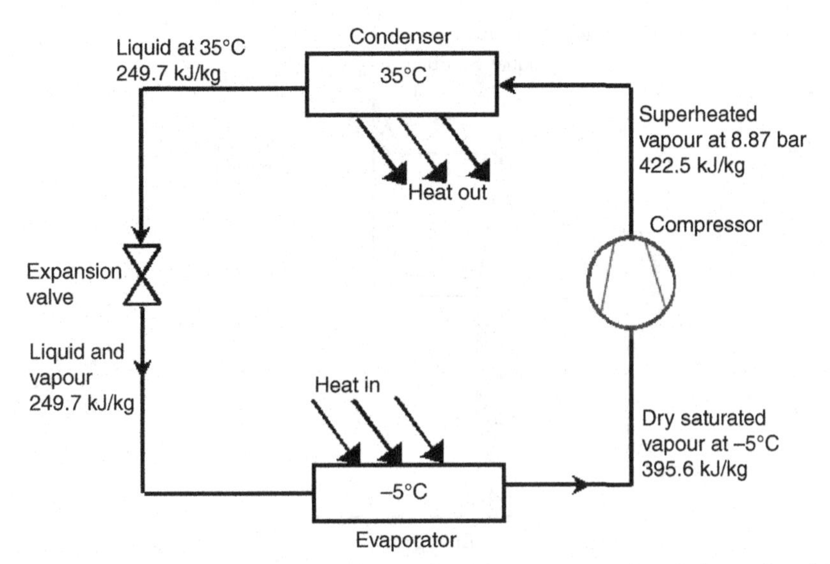

Figure 2.3 ***Simple vapour-compression cycle with pressure and enthalpy values for R134a.***

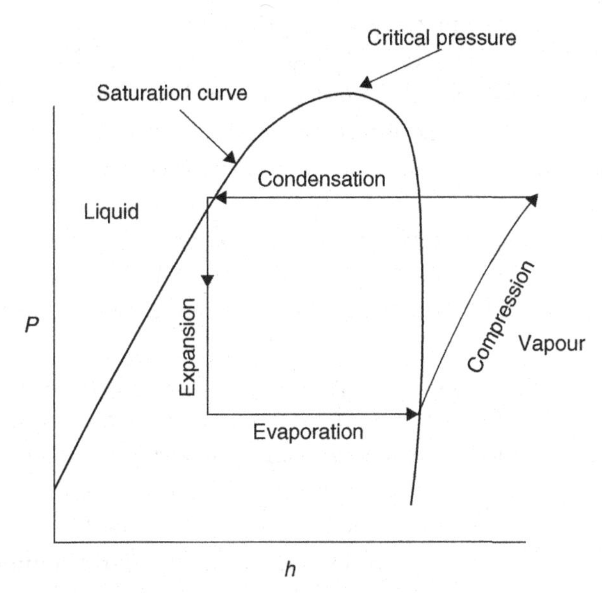

Figure 2.4 ***Pressure–enthalpy,* P–h *diagram, showing vapour compression cycle.***

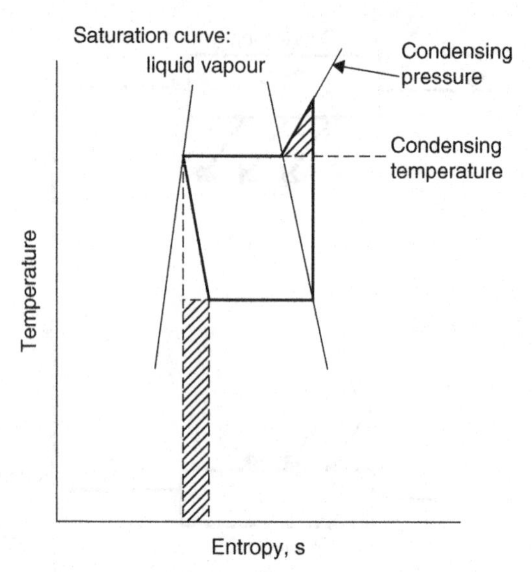

Figure 2.5 ***Temperature–entropy diagram for ideal vapour compression cycle.***

constant pressure process and therefore it is represented by a horizontal line. In the compression process the energy used to compress the vapour turns into heat and increases its temperature and enthalpy, so that at the end of compression the vapour state is in the superheated part of the diagram and outside the saturation curve. A process in which the heat of compression raises the enthalpy of the gas is termed *adiabatic compression*. Before condensation can start, the vapour must be cooled. The final compression temperature is almost always above the condensation temperature as shown, and so some heat is rejected at a temperature above the condensation temperature. This represents a deviation from the ideal cycle. The actual condensation process is represented by the part of the horizontal line within the saturation curve.

When the simple vapour compression cycle is shown on the temperature–entropy diagram (Fig. 2.5), the deviations from the reversed Carnot cycle can be identified by shaded areas. The adiabatic compression process continues beyond the point where the condensing temperature is reached. The shaded triangle represents the extra work that could be avoided if the compression process changed to isothermal (i.e. at constant temperature) at this point, whereas it carries on until the condensing pressure is attained.

Expansion is a constant enthalpy process, shown as a vertical line on the *P–h* diagram. No heat is absorbed or rejected during this expansion, the liquid just passes through a valve. Since the reduction in pressure at this valve must cause a corresponding drop in temperature, some of the fluid will flash off into vapour to remove the energy for this cooling. Therefore, volume of the working fluid increases at the valve by this amount of flash gas, and gives rise to its name, the expansion valve. No attempt is made to recover energy from the expansion process, for example by use of a turbine. This is a second deviation from the ideal cycle. The work that could potentially be recovered is represented by the shaded rectangle in Fig. 2.5.

2.3 PRACTICAL CONSIDERATIONS AND COP

For a simple circuit, using the working fluid Refrigerant R134a, evaporating at −5°C and condensing at 35°C, the pressures and enthalpies will be as shown in Fig. 2.3:

Enthalpy of fluid entering evaporator =249.7 kJ/kg

Enthalpy of saturated vapour leaving evaporator =395.6 kJ/kg

Cooling effect 395.6 − 249.7 =145.9 kJ/kg

Enthalpy of superheated vapour leaving compressor (isentropic compression) 422.5 kJ/kg

Since the vapour compression cycle uses energy to move energy, the ratio of these two quantities can be used directly as a measure of the performance of the system. As noted in chapter: Fundamentals, this ratio is termed the COP. The ideal or theoretical vapour compression cycle COP is less than the Carnot COP because of the deviations from ideal processes mentioned in Section 2.2. The ideal vapour compression cycle COP is dependent on the properties of the refrigerant, and in this respect some refrigerants are better than others.

Transfer of heat through the walls of the evaporator and condenser requires a temperature difference as illustrated in Fig. 2.6. The larger the heat exchangers are, the lower will be the temperature differences, and so the closer the fluid temperatures will be to those of the load and condensing medium. The COP of the cycle is dependent on the condenser and evaporator temperature differences (see Table 2.1).

Table 2.1 shows how the Carnot COP decreases as the cycle temperature lift increases due to larger heat exchanger temperature differences, ΔT.

The practical effects of heat exchanger size can be summarised as follows:

Larger evaporator: (1) Higher suction pressure to give denser gas entering the compressor and therefore a greater mass of gas for a given swept

volume, and so a higher refrigerating duty; (2) higher suction pressure, so a lower compression ratio and less power for a given duty.
Larger condenser: (1) Lower condensing temperature and colder liquid entering the expansion valve, giving more cooling effect; (2) lower discharge pressure, so a lower compression ratio and less power.

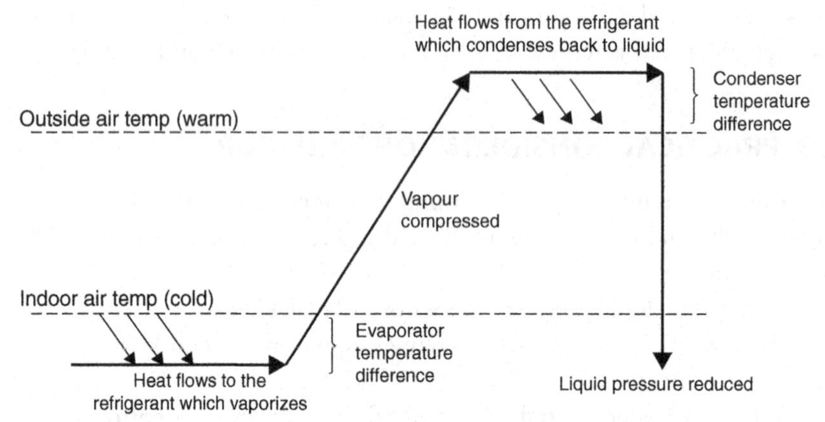

Figure 2.6 ***The temperature rise or 'lift' of the refrigeration cycle is increased by temperature differences in the evaporator and condenser.***

Table 2.1 COP values for cooling a load at −5°C, with an outside air temperature of 35°C (refrigerant R134a)

ΔT at evaporator and condenser (K)	0	5	10
Evaporating temperature (°C)	–5	–10	–15
Condensing temperature (°C)	35	40	45
Temperature lift (K)	40	50	60
Evaporating pressure, bar absolute	2.43	2.01	1.34
Condensing pressure, bar absolute	8.87	10.16	11.6
Pressure ratio	3.65	5.05	8.66
Carnot COP (refrigeration cycle)	6.70	5.26	4.30
COP, ideal vapour compression cycle★	5.45	4.03	3.07
COP with 70% efficient compression★★	3.82	2.82	2.15
System efficiency index, SEI†	0.57	0.54	0.50

★ The ideal vapour compression cycle with constant enthalpy expansion and isentropic adiabatic compression with refrigerant R134a.

★★ The vapour compression cycle as above and with 70% efficient compression with R134a and no other losses.

† SEI is the ratio between the actual COP and the Carnot COP with reference to the cooling load and outside air temperatures, that is when the heat exchanger temperature differences, ΔT, are zero. SEI decreases as ΔT increases due to less effective heat exchangers. Values are shown for the cycle with 70% efficient compression. Actual values will tend to be lower due to pressure drops and other losses.

Example 2.1

A refrigeration circuit is to cool a room at 0°C using outside air at 30°C to reject the heat. The refrigerant is R134a. The temperature difference at the evaporator and the condenser is 5 K. Find the Carnot COP for the process, the Carnot COP for the refrigeration cycle and the ideal vapour compression cycle COP when using R134a.

Carnot COP for 0°C (273 K) to 30°C (303 K)

$$= \frac{273}{(303 - 273)} = 9.1$$

Refrigeration cycle evaporating −5°C, condensing 35°C, Carnot COP

$$= \frac{268}{(308 - 268)} = 6.7$$

For R134a
Cooling effect = 395.6 − 249.7 = 145.9 kJ/kg
Compressor energy input = 422.5 − 395.6 = 26.9 kJ/kg
Ideal R134a vapour compression cycle COP

$$= \frac{145.9}{26.9} = 5.4$$

Since there are additional mechanical and thermal losses in a real circuit, the actual COP will be even lower. For practical purposes in working systems, the COP is the ratio of the cooling effect to the compressor input power.

System COP normally includes all the power inputs associated with the system, that is fans and pumps in addition to compressor power. A ratio of system COP to Carnot COP (for the process) is termed system efficiency index, SEI.

This example indicates that care must be taken with definitions when using the terms efficiency and COP. There is more information on this topic in chapter: Efficiency, Running Cost and Carbon Footprint.

A pressure–enthalpy chart, introduced in Section 1.7, is drawn in Fig. 2.7 with a vapour compression cycle superimposed. The lines of constant temperature (isotherms) are indicated.

A refrigeration cycle is represented by *A*, *A*1, *B*, *C*, *C*1, *D*. With a compression efficiency of 70% the final temperature at the end of compression is approximately 140°C. The value is dependent on the refrigerant and the compressor efficiency. This is a more practical cycle because the vapour leaving the evaporator is superheated (*A* to *A*1) and the liquid leaving the condenser sub-cooled (*C* to *C*1). Superheat and sub-cooling

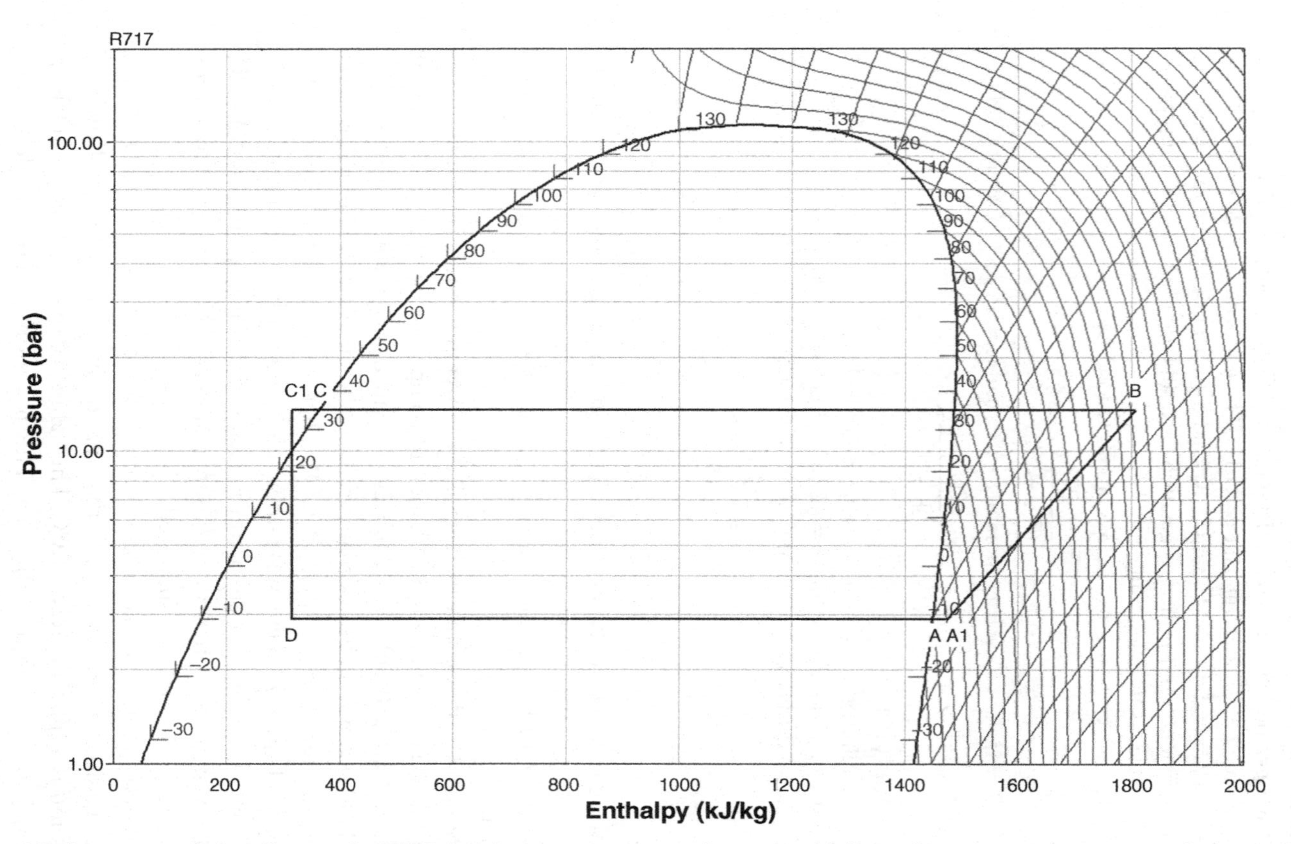

Figure 2.7 *Pressure–enthalpy diagram for R717 showing vapour compression cycle with lines of constant temperature and entropy in the vapour phase.*

occupy quite small sections of the diagram, but they are very important for the effective working of the system. Superheat ensures that no liquid arrives at the compressor with the vapour where it could cause damage. Sub-cooling ensures that the fluid flowing through the line from the condenser to the control or expansion valve is 100% liquid. If some vapour is present here, it can cause excessive pressure drop and reduction in performance of the system. Therefore in Fig. 2.7 the gas leaving the evaporator is superheated to point *A*1 and the liquid sub-cooled to *C*1. Taking these factors into account, the refrigerating effect per unit mass flow rate (*D* to *A*1) and the compressor energy (*A*1 to *B*) may be read off directly in terms of enthalpy of the fluid. In practice, pressure losses will occur across the compressor inlet and outlet, and there will be pressure drops through the heat exchangers and piping and these can also be plotted on the chart. There will also be some heat loss to atmosphere from the compressor and discharge piping.

The position of *D* inside the curve indicates the proportion of flash gas at that point. The condenser receives the high-pressure superheated gas, *B*, cools it down to saturation temperature, condenses it to liquid, *C*, and finally sub-cools it slightly, *C*1. The energy removed in the condenser, or *heat rejection* (*B* to *C*1) is seen to be the refrigerating effect plus the heat of compression. Computer software is available to make all these calculations and the usual reference for refrigerant property data is NIST Refprop.

A suction/liquid heat exchanger can be a beneficial addition to any vapour compression cycle. This device transfers heat from the relatively warm liquid to the suction vapour returning to the compressor. It is shown in Fig. 2.11 in the R744 booster circuit, but is frequently applied in single-stage systems. It can have a dual benefit if ensuring the liquid entering the expansion valve is sub-cooled to prevent premature gas formation, and that the vapour entering the compressor is superheated to prevent liquid return. This is particularly useful with flooded evaporators where there is no superheat control. The effect on cycle performance is small and refrigerant property dependent – the sub-cooling is increased but the refrigerant mass flow rate is reduced because the suction gas density is reduced. There is a positive benefit where long suction lines pick up non-useful heat. Then a suction/liquid heat exchanger at the evaporator outlet can raise the gas temperature and reduce this pick up. Then the heat transferred usefully cools the liquid.

2.4 MULTISTAGE CYCLES

Where the ratio of suction to discharge pressure is high enough to cause a serious drop in volumetric efficiency (see Chapter 4) or an unacceptably high discharge temperature, vapour compression can be carried out in two or more stages.

Two-stage systems use the same refrigerant throughout a common circuit, compressing in two stages. By using separate compressors for each stage, the second-stage displacement can be adjusted to accommodate an extra cooling load, *side load*, at the intermediate pressure. Compression in two stages within a single machine can be accomplished with multi-cylinder compressors. The first stage of compression takes place in, say, four cylinders and the second stage in two cylinders of a six-cylinder machine. Hot discharge gas from the first compression stage may be passed via an intercooler to the high-stage compressor. An intercooler may consist of a small evaporator supplied by refrigerant from the condenser. Alternatively a water-cooled heat exchanger could be used, or simple injection of a controlled amount of liquid refrigerant (from the condenser) to mix with the intermediate pressure gas.

A more energy-efficient alternative is the arrangement shown in Figs 2.8 and 2.9. Part of the refrigerant liquid from the condenser is taken

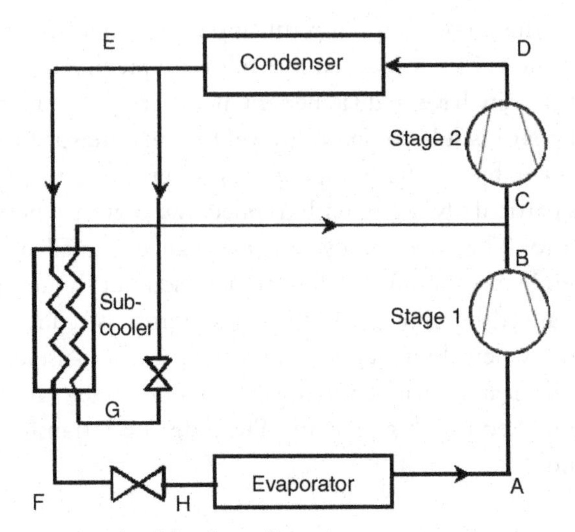

Figure 2.8 *Two-stage cycle with sub-cooler.*

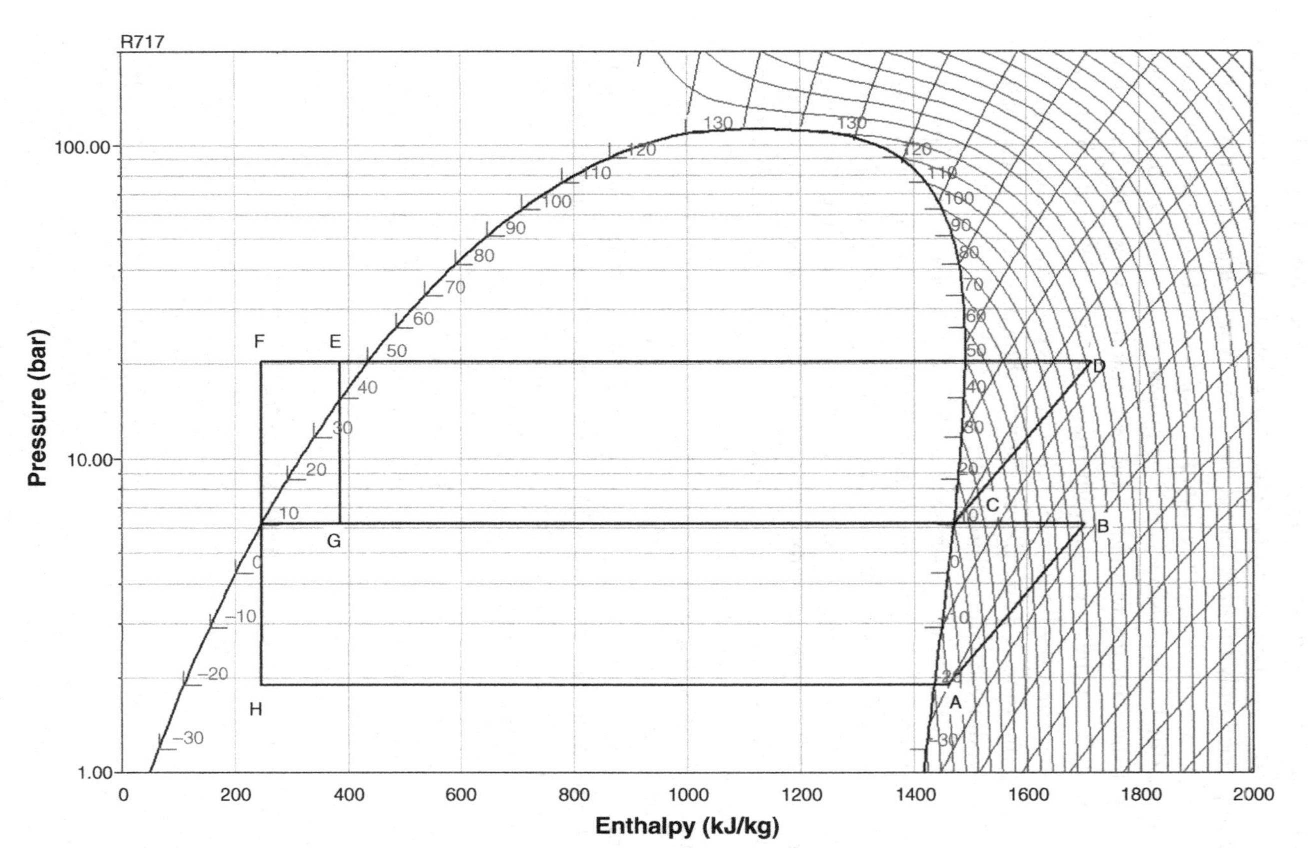

Figure 2.9 ***P-h diagram for R717 showing two-stage vapour compression cycle with sub-cooler.***

to a sub-cooler, where the main liquid flow to the expansion valve is cooled from E to F, and this increases the duty of the evaporator (H to A). This cycle is more efficient than the single-stage cycle because part of the mass flow is compressed only through the second stage. A *flash intercooler* may be used instead of a sub-cooler. All the liquid is then reduced to intermediate pressure via a suitable expansion valve. The intercooler acts as a separation vessel in which the flash gas formed in the expansion process is separated from the liquid. From the intercooler, the flash vapour is led to the high-stage compressor, whilst the liquid, which has been separated, is further expanded to the low pressure. A float valve of the type shown in Fig. 8.10 can be used to control admission to the intercooler.

A version of the two-stage cycle, called an *economiser cycle*, can be applied with scroll and screw compressors. With these machines, access to the intermediate pressure within the compression process via an additional port on the casing allows vapour from the sub-cooler to be injected part-way through the compression process. Only one compressor is needed, and it is almost identical to the single-stage version, requiring just the additional vapour injection port. The economiser cycle is a very cost-effective way of gaining improved performance.

The *cascade cycle* has two separate refrigeration systems, one acting as a condenser to the other. This arrangement permits the use of different refrigerants in the two systems and high-pressure refrigerants such as R23 are used in the low stage for very low temperature that would require conventional refrigerants to evaporate below atmospheric pressure.

The Mollier diagrams for compound systems indicate the enthalpy change per kilogram of circulated refrigerant, but it should be borne in mind that the mass flows rates in the low and high stages differ, and this must be accounted when calculating capacities.

2.5 TRANSCRITICAL CARBON DIOXIDE CYCLES

The low critical temperature for carbon dioxide can be seen in the pressure–enthalpy diagram (Fig. 2.10), and the shape of the constant temperature lines near the critical point results in unexpected system behaviour in that region. A cycle with heat rejection at 31°C would have a much lower refrigerating effect than one condensing at, say 27°C. Above the critical point, the gas cannot be condensed, and it is necessary to move into this region if the temperature at which heat can be rejected approaches 30°C. If the gas can be cooled, to say 40°C as shown in Fig. 2.10, the refrigerating

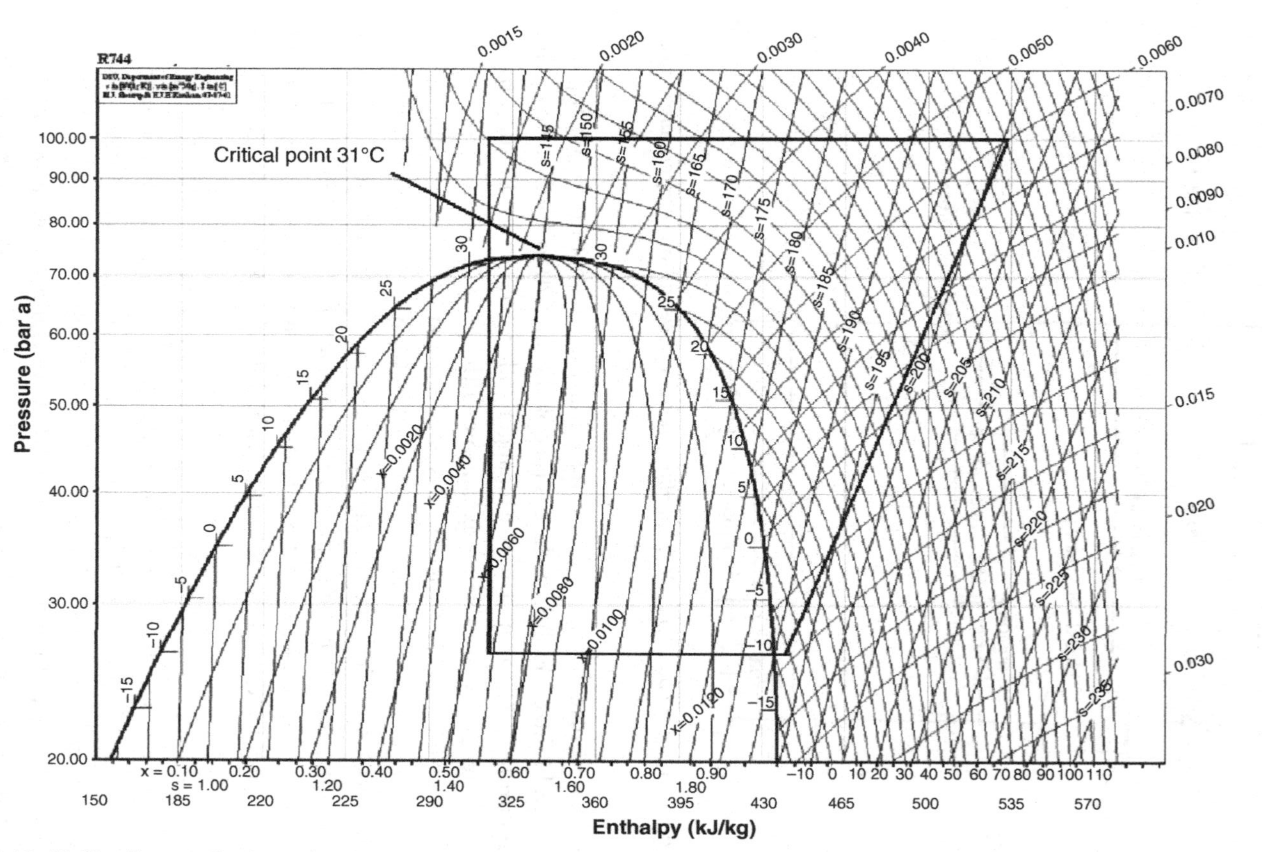

Figure 2.10 *Mollier diagram for R744 showing transcritical cycle with evaporation at −10°C, compression to 100 bar and gas cooling to 40°C.*

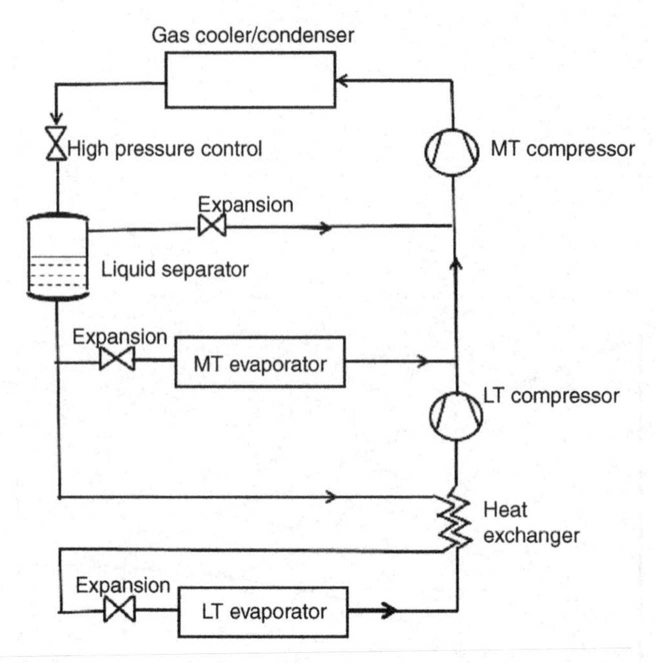

Figure 2.11 ***R744 booster cycle for medium- temperature and low-temperature loads.***

effect is similar to that with heat rejection at 30°C, but the power input is higher. In the cycle shown, the gas is cooled from 120 to 40°C at a constant pressure of 100 bar in a heat exchanger called a *gas cooler*. Liquid formation only takes place during expansion to the lower pressure level. It may be possible to operate a system designed for transcritical operation in the sub-critical mode, that is as a vapour compression cycle, under low ambient conditions in which case the gas cooler becomes a condenser.

Regulation of the high pressure is necessary for the transcritical cycle. The optimum pressure is determined as a function of the gas cooler outlet temperature and is a balance between the highest possible refrigerating effect and the smallest amount of compressor energy. A detailed insight into the optimisation of the gas cooler pressure is given by Cavallini (2004).

The transcritical cycle has a poor efficiency compared to a vapour compression cycle and normally reverts to a conventional vapour compression cycle when outdoor temperature is sufficiently low to enable heat rejection below the critical temperature. For low-temperature applications, in order to enhance efficiency and avoid excessive discharge temperatures it is necessary to use two stage or cascade systems. A two-stage, or booster schematic is shown in Fig. 2.11.

2.6 HEAT POWERED CYCLES

2.6.1 Absorption Cycle

The principle of the absorption cycle is given in Fig. 2.12. The refrigerant flowing through the condenser, expansion valve and evaporator performs in just the same way as in the vapour compression cycle (Fig. 2.3). The difference is that the compressor is replaced by a thermal compressor. The refrigerant leaving the evaporator is absorbed by a liquid absorbant; the strength of the solution increasing as the liquid absorbant passes through the absorber. This solution is then pumped up to condenser pressure and the vapour is driven off in the generator by direct heating. The high-pressure refrigerant, now a gas, can then be condensed in the usual way and passed back through the expansion valve into the evaporator. The weak solution from the generator is passed through another pressure-reducing valve and back to the absorber.

There are two main refrigerants in use, ammonia with water absorbant and water with lithium bromide absorbant. The water/lithium bromide systems are suited to air-conditioning water chiller temperatures, whereas ammonia systems are suited for evaporating temperatures below 0°C. With water as a refrigerant the whole system operates well below atmospheric pressure. Ingress of air due to leakages can cause problems.

The diagram in Fig. 2.12 is illustrative only. The low-pressure vessels (evaporator and absorber) and the high-pressure vessels (condenser and

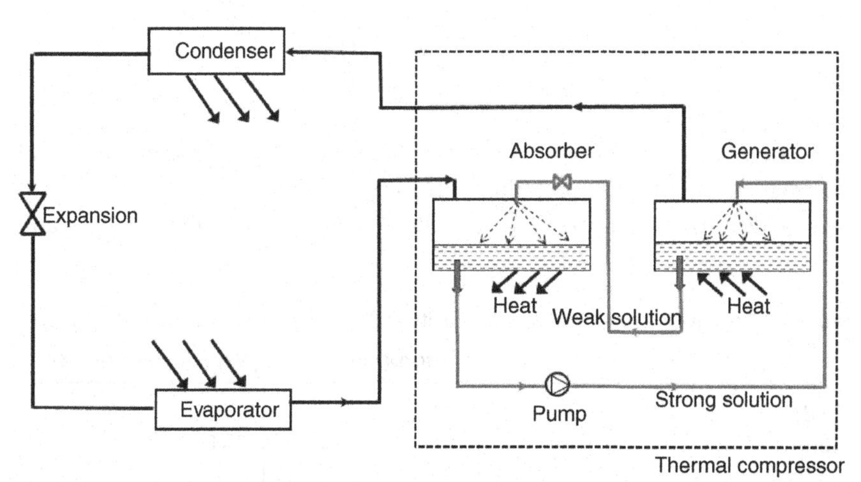

Figure 2.12 ***Absorption cycle – basic circuit.*** Refrigerant compression is replaced by a thermal compressor.

generator) are commonly combined within one shell each. Dual effect systems can be used. Overall thermal efficiency may also be improved by a heat exchanger between the two solution paths and a suction-to-liquid heat exchanger for the refrigerant.

Absorption system can be used to advantage where:

- a CHP unit has spare heat available.
- a low-cost supply of waste heat is available.
- heat from landfill gas or geothermal can be used.
- electrical load limits apply at the site.
- low noise and/or vibration are major considerations.
- solar energy can be harnessed.

Because of the need for additional processes, an absorption systems can be significantly more costly than a vapour compression alternative.

The energy input to an absorption cycle is higher than for a compression cycle, but the main driving energy is in the form of low-grade heat, not electricity. Typical figures are shown in Table 2.2. The ideal COP can be shown to be

$$\mathrm{COP} = \frac{\frac{1}{Ta} - \frac{1}{Tg}}{\frac{1}{Te} - \frac{1}{Tc}}$$

If the absorber and condenser temperatures are the same,

$$\mathrm{COP} = \frac{Te(Tg - Ta)}{Tg(Ta - Te)}$$

Where Te, Tg, Tc and Ta are the evaporating, generator, condenser and absorber respective temperatures. The ideal absorption COP is 1.4 for the conditions in Example 2.1, with Tg = 100°C (steam heat source) and

Table 2.2 Energy per 100 kW cooling capacity at 3°C evaporation, 42°C condensation

	Absorption	Vapour compression
Load	100.0	100.0
Pump/compressor (electricity)	0.1	30.0
Low-grade heat	165	–
Heat rejected	265.1	130.0

Ta = 25°C. Compare this with 6.7 for a Carnot COP, and 5.4 for an ideal vapour compression process with R134a. An analysis of the ideal cycle is given by Tozer and James (1997).

COP enhancement can be achieved by using higher heat supply temperature, higher refrigerant evaporating temperature, lower heat rejection temperature. A typical COP for an air-conditioning absorption cycle in would be about 0.7, compared to approximately 3.5 for a vapour compression system. With double and triple effect generators COP can be between 1.2 and 1.7.

2.6.2 Adsorption Cycle

Adsorbents such as active carbons, zeolites or silica gels can adsorb large quantities (c. 30% by weight) of many gases within their micropores. The most widely used combinations are active carbons with ammonia or methanol, and zeolites with water, but the choice of which adsorbent and which refrigerant to use depends on the application. The quantity of refrigerant adsorbed depends on the temperature of the adsorbent and the system pressure. Heat is required to drive out the refrigerant in desorption and heat is generated during adsorption.

A basic adsorption cycle is illustrated in Fig. 2.13. Initially (a) the whole assembly is at low pressure and temperature, the adsorbent contains a large concentration of refrigerant within it and the other vessel contains refrigerant gas. The adsorbent vessel (generator) is then heated, driving out the refrigerant and raising the system pressure. The desorbed refrigerant condenses as a liquid in the second vessel, rejecting heat (b). Then the generator is cooled back towards ambient temperature, re-adsorbing the refrigerant and reducing the pressure (c). The reduced pressure above the liquid in the second vessel causes it to boil, absorbing heat and producing the refrigeration effect (d). The cycle is discontinuous since useful cooling only occurs for one half of the cycle. Two such systems can be operated out of phase to provide continuous cooling and heat rejected by one sorption bed can be used to pre-heat the other, thereby improving efficiency. Adsorption technology is also being developed for hybrid heat pumps, see chapter: Heat Pumps and Integrated Systems. A review of thermally driven heat pumps is given by Kühn (2013).

2.6.3 Desiccant Cooling

Desiccant cooling is an open cycle which uses a desiccant wheel and thermal wheel to achieve both cooling and dehumidification. Solar thermal

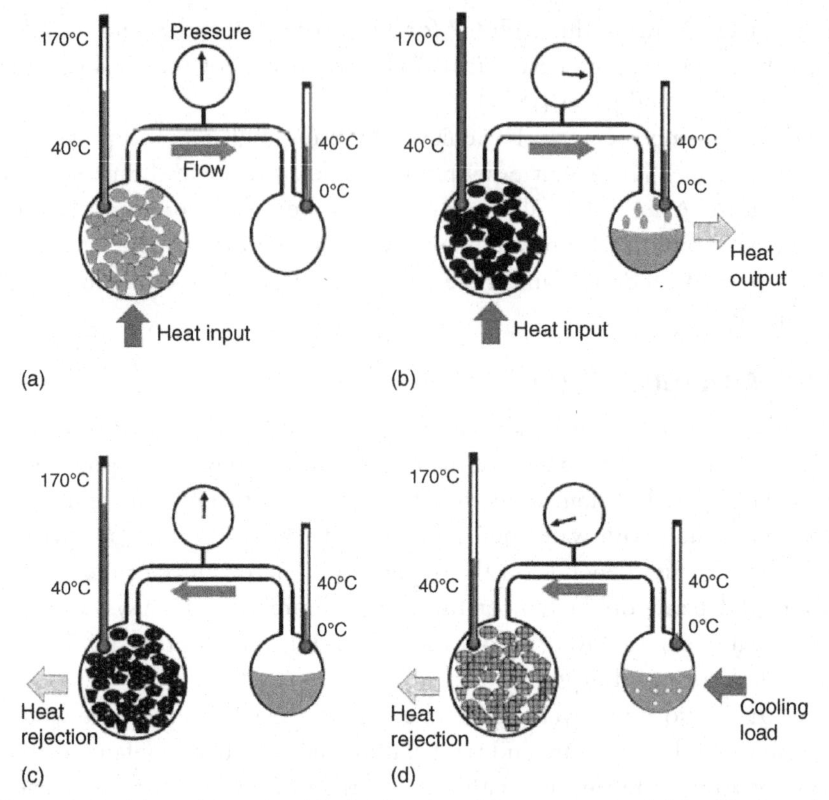

Figure 2.13 ***Adsorption cycle – idealised process (a) pressure raised, (b) refrigerant desorbed and moved to condenser, (c) pressure lowered, (d) refrigerant boiled in evaporator and adsorbed (*****University of Warwick*****).***

energy or waste heat can be used to re-activate the desiccant. The thermal wheel is a rotary heat exchanger positioned within the supply and exhaust air streams in order to recover the heat energy, and the desiccant wheel works in a similar way, additionally re-activating the desiccant. The principle of the cycle is shown in Fig. 2.14. There are a number of variations depending on the condition of ambient outdoor air, air change requirements and humidity control requirements. It offers high efficiency but requires large air flows and is limited in operational range, but it can be applied in conjunction with a conventional vapour compression cycle. Referring to Fig. 2.14, the outdoor air moisture content is reduced and its temperature increased as it passes through the desiccant wheel. It is sensibly cooled through the thermal wheel, further cooled to the required supply temperature, with some moisture gain as it passes through an evaporative cooler. The return air from

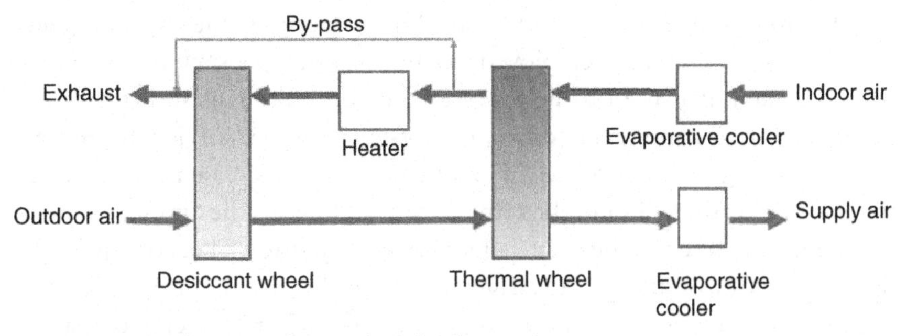

Figure 2.14 ***Desiccant cooling – principle of operation.***

indoors, at, say 25°C is passed through an evaporative cooler so that it enters the thermal wheel at a lower temperature and higher moisture content. As it passes through the thermal wheel, it is sensibly heated and further heated by the heater so that it can re-activate the desiccant before exhausting. The by-pass can be controlled so that unnecessary heat is not applied. The thermal wheel is illustrated in Fig. 25.1. More detail on desiccant cooling systems can be found in CIBSE Guide B3 (2016).

2.7 OTHER PROCESSES

2.7.1 Air Cycle

Air cycle refrigeration works on the reverse Brayton or Joule cycle. Air is compressed and then heat removed; this air is then expanded to a lower temperature than before it was compressed. Heat can then be extracted to provide useful cooling, returning the air to its original state (see Fig. 2.15). Work is taken out of the air during the expansion by an expansion turbine,

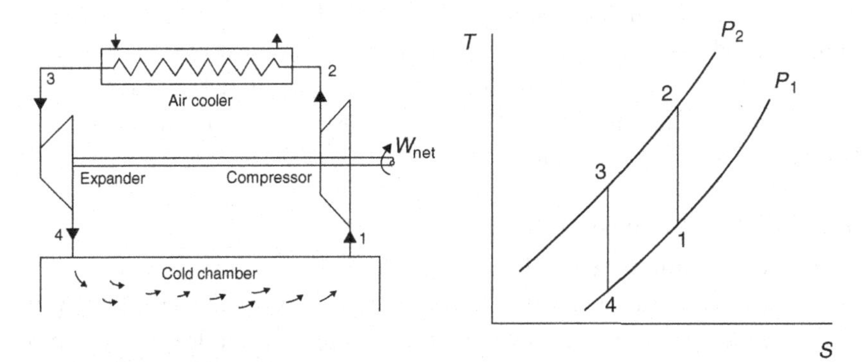

Figure 2.15 ***The air cycle – the work from the expander provides a portion of the work input to the compressor.***

which removes energy as the blades are driven round by the expanding air. This work can be usefully employed to run other devices, such as generators or fans. Often, it is used to help power the compressor, as shown. Sometimes a separate compressor, called a 'bootstrap' compressor, is powered by the expander, giving two stages of compression. The increase in pressure on the hot side further elevates the temperature and makes the air cycle system produce more useable heat (at a higher temperature). The cold air after the turbine can be used as a refrigerant either directly in an open system as shown or indirectly by means of a heat exchanger in a closed system. The efficiency of such systems is limited to a great extent by the efficiencies of compression and expansion, as well as those of the heat exchangers employed.

Originally, slow-speed reciprocating compressors and expanders were used. The poor efficiency and reliability of such machinery were major factors in the replacement of such systems with vapour compression equipment. However, the development of rotary compressors and expanders (such as in car turbochargers) greatly improved the isentropic efficiency and reliability of the air cycle. Advances in turbine technology together with the development of air bearings and ceramic components offer further efficiency improvements.

The main application for this cycle is the air conditioning and pressurisation of aircraft. The turbines used for compression and expansion turn at very high speeds to obtain the necessary pressure ratios and, consequently, are noisy. The COP is lower than with other systems.

2.7.2 Stirling Cycle

The Stirling cycle is an ingenious gas cycle which uses heat transferred from the gas falling in temperature to provide that for the gas rising in temperature. A detailed explanation of the cycle is beyond the scope of this book and readers are referred to Gosney (1982) and Hands (1993). The Stirling cycle has been successfully applied in specialist applications requiring low temperatures at very low duties.

2.7.3 Total Loss Refrigerants

Some volatile fluids are used once only and then escape into the atmosphere. Two of these are in general use: carbon dioxide and nitrogen. Both are stored as liquids under a combination of pressure and low temperature and then released when the cooling effect is required. Carbon dioxide is below its triple point at atmospheric pressure and can only exist as 'snow'

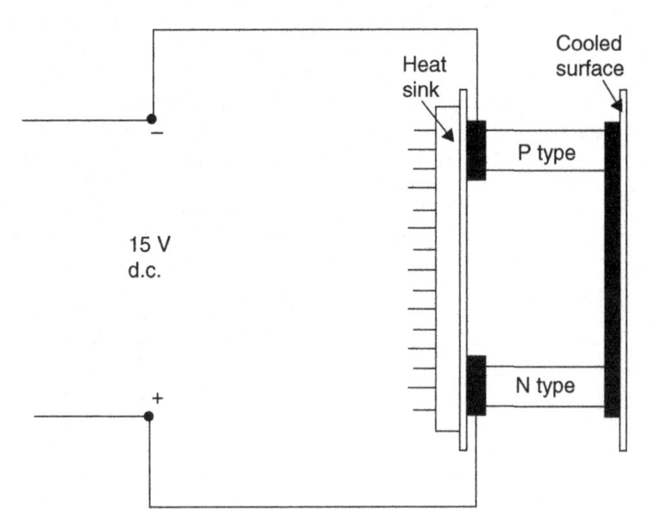

Figure 2.16 ***Thermoelectric cooling.***

or a gas. The *triple point* is where solid, liquid and vapour phases co-exist. Below this pressure, a solid sublimes directly to the gaseous state. Since both gases come from the atmosphere there is no pollution hazard. The temperature of carbon dioxide when released will be −78.4°C. Nitrogen will be at −198.8°C. Water ice can also be classified as a total loss refrigerant.

2.7.4 Thermoelectric Cooling

The passage of an electric current through junctions of dissimilar metals causes a fall in temperature at one junction and a rise at the other, the Peltier effect. Improvements in this method of cooling have been made possible in recent years by the production of suitable semiconductors. Applications are limited in size, owing to the high electric currents required, and practical uses are small cooling systems for military, aerospace and laboratory use (Fig. 2.16; Table 2.2).

2.7.5 Magnetic Refrigeration

Magnetic refrigeration depends on what is known as the *magnetocaloric effect*, which is the temperature change observed when certain magnetic materials are exposed to a change in magnetic field. Magnetic refrigeration is a research topic, and historically has been used at ultra-low temperatures. Only recently has it been seen as a possible means of cooling at near room temperatures. An overview of magnetic refrigeration is given by Wilson et al. (2007).

CHAPTER 3

Refrigerants

3.1 INTRODUCTION

The changes in the selection and use of refrigerants in response to environmental issues has accelerated during the last 30 years, a story which can be traced with the aid of a 'Refrigerant Time Line' (Fig. 3.1).

The earliest mechanical refrigeration used air as a working fluid. The introduction of the vapour compression cycle enabled more compact and effective systems. At first the only practical fluids were carbon dioxide and ammonia. One of the major requirements was preservation of meat on the long sea voyages from New Zealand and Australia to Europe, and for this ammonia was unsuitable owing to its toxic nature. Carbon dioxide, although requiring much higher pressures was used. Methyl chloride, although toxic and very unpleasant was used in some smaller systems.

A revolution came about with the invention of the chlorofluorocarbon R12 by Midgley in the early 1930s. This refrigerant and other members of the CFC family seemed to possess all the desirable properties. In particular they were non-toxic, non-flammable and with good thermodynamic properties and oil miscibility characteristics. The CFCs R12, R11, R114 and R502 together with HCFC R22 became the definitive refrigerants. They enabled the expansion of refrigeration into the commercial, domestic and air conditioning sectors. Carbon dioxide was eclipsed, but ammonia with its excellent thermodynamic properties and low cost continued to be applied in many industrial applications.

Environmental concerns of ozone depletion and global warming, explained in subsequent sections, have driven a major changes in refrigerants applied today and in the future. Figure 3.1 illustrates the rapid progress of the recent changes. Release into the atmosphere of substances that cause ozone depletion was the subject of the Montreal and Kyoto international protocols, and elimination of the chlorine containing CFC and HCFCs in new equipment has successfully countered ozone depletion.

The replacement chlorine-free HFCs were developed and have been successfully applied in almost all applications. However HFCs are powerful greenhouse gases, and as such have come under close scrutiny. Some of the most damaging HFCs, including R404A, are destined to be phased out and

Refrigeration, Air Conditioning and Heat Pumps
http://dx.doi.org/10.1016/B978-0-08-100647-4.00003-6

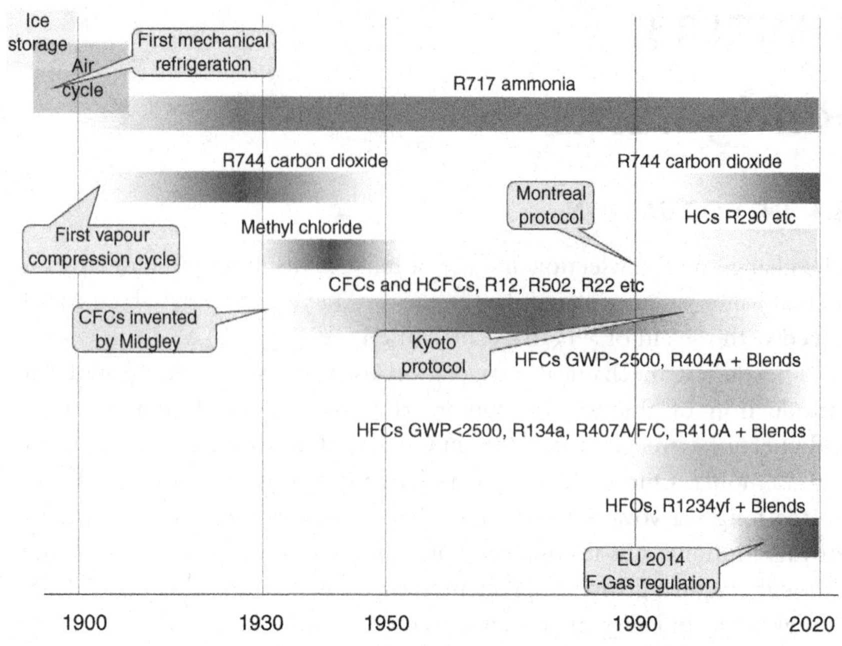

Figure 3.1 ***Time line for refrigerants illustrating the developments in HFCs and HFOs, the continued use of R717 and the resurgence of R744. 'Blends' are alternative refrigerant mixtures based on the same constituents.***

others regulated in various ways. This has speeded the development of a new class of hydrofluoroolefin (HFO) refrigerants. HFOs, with a short atmospheric life, have extremely low global warming impact but there are flammability and cost issues. R1234yf was initially developed as a replacement for R134a in automobile or mobile air conditioning (MAC) systems. Targeted regulation of MAC systems requires R134a phase out in car air conditioners in Europe.

The impact of global warming is being addressed by European F-Gas regulations and these were significantly strengthened in 2014. The regulations are complex, but essentially intended to substantially reduce greenhouse gas emissions. For refrigerants this means certain chemicals will be phased out and others restricted. Similar restrictions are appearing globally. The choice of refrigerant is becoming more complex and this chapter can only be considered as an introduction.

Key properties of the main refrigerants are given in Table 3.1 together with their typical application ranges; low, (−25 to −40°C) medium, (−5 to −25°C) and high temperature (+10 to −5°C). More information about refrigerants applied in specific applications can be found in the relevant chapters.

Table 3.1 Properties of refrigerants

Refrigerant	Composition	Application	GWP (CO_2 = 1)	Safety class	Boiling point °C	Vapour pressure at 50°C (dew) bar (abs)
HCFC						
R22	$CHClF_2$	HT, MT	1810	A1	−41	19.4
HFCs chlorine free						
R134a	CF_3CH_2F	HT, MT	1430	A1	−26	13.2
R125	CF_3CHF_2	Blends	3500	A1	−48	25.5
R143a	CF_3CHF_3	Blends	4470	A2	−48	23.2
R32	CH_2F_2	HT	675	A2L	−52	31.5
R404A	R143a/125/134a	LT	3922	A1	−47	23.0
R407C	R32/125/134a	HT	1774	A1	−44	19.8
R410A	R32/125	HT	2088	A1	−51	30.5
Other R32 Blends	R32 + HFCs	LT	1770–2280	A1	−46 to −48	21 to 23
Other R125 Blends	R125 + HFCs	HT, MT, LT	1830–3300	A1	−43 to −48	18 to 25
HFOs						
R1234yf	$CH_2 = CFCF_3$.	MAC, HT	4	A2L	−29	13.0
R1234ze[E]	CHF = CHCF3	HT	6	A2L	−19	10.0
HFO/HFC Blends	R1234yf/134a, R1234ze[E]/R134a	Various	600–1500	A1	−20 to −50	Various

(*Continued*)

Table 3.1 Properties of refrigerants (*cont.*)

Refrigerant	Composition	Application	GWP (CO_2 = 1)	Safety class	Boiling point °C	Vapour pressure at 50°C (dew) bar (abs)
HCs halogen free						
R290	C_3H_8 Propane	HT, MT	3	A3	−42	17.1
R1270	C_3H_6 Propylene	LT	3	A3	−48	20.6
R600a	C_4H_{10} IsoButane	MT	3	A3	−12	6.8
R290 Blends	R290 + HCs	HT, LT, MT	3	A3	−30 to −48	10 to 18
Other halogen free						
R717	NH_3 Ammonia	LT (MT, HT)	0	B2	−33	20.3
R744	CO_2 Carbon Dioxide	HT, MT, LT	1	A1	−57*	74**

GWP according to IPCC IV, time horizon 100 y
Safety limit classification according to EN378-1
* Triple point (5.2 bar a)
** At critical temperature 31°C

3.2 IDEAL PROPERTIES FOR A REFRIGERANT

These can be listed as follows:

- High latent heat of vaporisation
- High suction gas density
- Positive but not excessive pressures at evaporating and condensing conditions
- Critical temperature and triple point well outside the working range
- Chemically stable, compatible with construction materials and miscible with lubricants
- Non-corrosive, non-toxic and non-flammable
- High dielectric strength
- Environmentally friendly
- Low cost

Needless to say, no single fluid has all these properties, and the choice of fluid for any particular application will always be a compromise.

3.3 OZONE DEPLETION POTENTIAL

The ozone layer in our upper atmosphere provides a filter for ultraviolet radiation, which can be harmful to our health. Researchers found that the ozone layer was thinning, due to emissions into the atmosphere of chlorofluorocarbons (CFCs), halons and bromides. The ozone depletion potential (ODP) of a refrigerant represents its effect on atmospheric ozone, and the reference point usually adopted is ODP = 1 for the CFC R11.

After a series of rigorous meetings and negotiations, the Montreal Protocol on Substances that Deplete the Ozone Layer was finally agreed in 1987. Signatories agreed to phase out the production of these chemicals by 1995. Refrigerant emissions were only about 10% of the total, the remainder being made up of aerosol sprays, solvents and foam insulation.

The refrigeration industry was required to move rapidly away from CFCs to less ozone depleting HCFCs, namely R22 and HCFC replacement blends. At subsequent revisions of the Protocol, a phase-out schedule for HCFCs was also introduced as it was considered necessary to phase out all ozone depleting substances. Under the Protocol, HCFCs will be eliminated by 2030. This signalled the end of R22. Moreover the European Union drew up a far more stringent Regulation 2037/2000, which banned all new HCFC equipment in 2004, banned the sale of new HCFC refrigerant for service in Jan. 2010 and recycled refrigerant in 2015. The shift from CFCs to R22 or R22 blends resulted in an extensive estate of HCFC containing equipment, and then came phase out of R22 in many countries, combined with legislation, resulting in intensive replacement programmes with some owners experiencing difficulty in obtaining qualified people to change the refrigerant or replace equipment.

3.4 GLOBAL WARMING POTENTIAL

Global warming is possibly the most severe environmental issue faced by civilisation today. The risk posed by its effects has been described in terms of environmental disaster due to huge future climate changes. Global warming is the increasing of the world's temperatures, which results in melting of the polar ice caps and rising sea levels. A 'greenhouse' effect is caused by a blanket of gases in the upper atmosphere that reflect heat back to the Earth's surface, or hold heat in the atmosphere. This effect is intensified by the release into the atmosphere of so-called 'greenhouse' gases. The most infamous greenhouse gas is carbon dioxide (CO_2), which once released remains in the atmosphere for 500 years, so there is a constant build-up as time progresses. The exact extent of the contribution arising from man's activities may be uncertain, but in any case it is vital to keep it to a minimum and conserve fossil fuel reserves, that is minimise greenhouse gas emissions.

A major contributor of CO_2 emissions is in the generation of electricity at power stations. The CO_2 emission factor (kg of CO_2 emitted per kWh of electricity supplied) is dependent on the mix for electricity generation. For coal fired power stations the figure is relatively high, for gas-fired stations it is lower, and for hydroelectric, wind power or nuclear stations it is zero. Electricity suppliers may claim various mixes of generation type and hence differing emission factors. The DEFRA Greenhouse Gas Conversion Factor Repository, gives a 2014 UK emission factor of 0.49 kg CO_2 per kWh for generated electricity excluding transmission and distribution losses.

The global warming potential (GWP) of a gas may be defined as an index comparing the climate impact of its emission to that of emitting the same amount of carbon dioxide. The integrated effect over a fixed time allows for time decay of the substance. A time horizon of 100 years is usually adopted, although this is much less than the lifetime of CO_2 in the atmosphere. The refrigerant only affects global warming directly if it is released into the atmosphere. For example in Table 3.1, R134a has a GWP of 1430, which means that the emission of 1 kg of R134a is equivalent to 1430 kg of CO_2. GWP values are dependent on the assumed atmospheric life, and revisions by the IPCC are reported from time to time.

The choice of refrigerant affects the lifetime warming impact of a system and the term *total equivalent warming impact* (TEWI) is used to describe the overall impact. It includes the effects of refrigerant leakage, refrigerant recovery losses and energy consumption. TEWI should be calculated when comparing system design options for specific applications.

TEWI = Total equivalent warming impact
= (Direct effect) + (Indirect effect)
= (Leakage + Recovery losses) + (Power generation emissions)
$= \{(\text{GWP} \times L \times n) + [\text{GWP} \times m(1-\alpha)]\} + (n \times \text{E} \times \beta]$

Where GWP = Global warming potential
L = Leakage rate per year (kg)
n = Number of years of operation
m = Refrigerant charge (kg)
α = Recycling factor
E = Energy consumption per year (kWh)
β = CO_2 Emission factor, kg CO_2 emitted per kWh of electricity supplied

Comprehensive method details with calculation examples are given in the IOR/BRA TEWI Guidelines.

Example 3.1

Compare the TEWI effect of doubling the charge and alternatively adding 10% energy consumption for a medium temperature 13.5 kW, R134a installation consuming 6 kW electrical power for 5000 h/year. Use the following data:

GWP = 1430 kgCO_2/kg, L = 1 kg/10 kg, n = 15 years, m = 10 kg, $\alpha = 0.75$, $\beta = 0.42$ kg CO_2 /kWh

Taking case (a) as the baseline case, with (b) for additional charge and (c) for additional energy:

	Case (a)	Case (b)	Case (c)
L	1	2	1
m	10	20	10
E	6 × 5000 = 30,000	6 × 5000 = 30,000	6.6 × 5000 = 33,000
Leakage kg $CO_2 \times 10^3$	21.45	42.9	21.45
Recovery kg $CO_2 \times 10^3$	0.025	0.05	0.025
Total Direct kg $CO_2 \times 10^3$	21.475	42.950	21.475
Indirect kg $CO_2 \times 10^3$	189	189	207
TEWI kg $CO_2 \times 10^3$	210.475	231.95	229.375

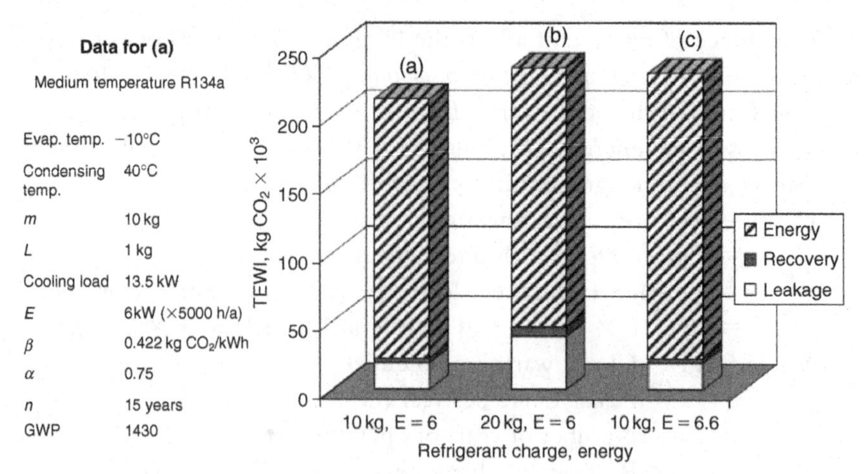

Figure 3.2 ***Comparison of TEWI values, showing the effect of increased refrigerant charge and increased power consumption.***

The largest element of the TEWI systems is energy consumption and this is true for the vast majority of refrigeration and air conditioning systems. Fig. 3.2 shows the dominant effect of the energy consumption element, which if increased by 10% has a similar effect to a doubling of the refrigerant charge and leakage. Column (a) shows the baseline data, with the effect of double charge and 10% energy consumption increase in columns (b) and (c) respectively. The less the amount of energy needed to produce each kW of cooling, the less will be the effect on global warming.

3.5 NOMENCLATURE

Refrigerants are classified by ASHRAE, and their familiar 'R' numbers are assigned according to certain rules. For example, the classification of halogen refrigerants derived from saturated hydrocarbons and consisting of only one substance is illustrated in Fig. 3.3.

Mixtures are designated by their respective refrigerant numbers and mass proportions as in the example Fig. 3.4.

The new class of HFO refrigerants are based on the propylene (propene) molecule R1270, $CH_3CH = CH_2$. Various ways of substituting the atom or group on the carbons can exist and this is denoted by two lower-case letters (see Figure 3.5). The unsaturated carbon bond gives rise to the term olefin. Where a specific molecule can have different geometrical arrangements (stereoisomers) these are designated with an appended upper-case letter.

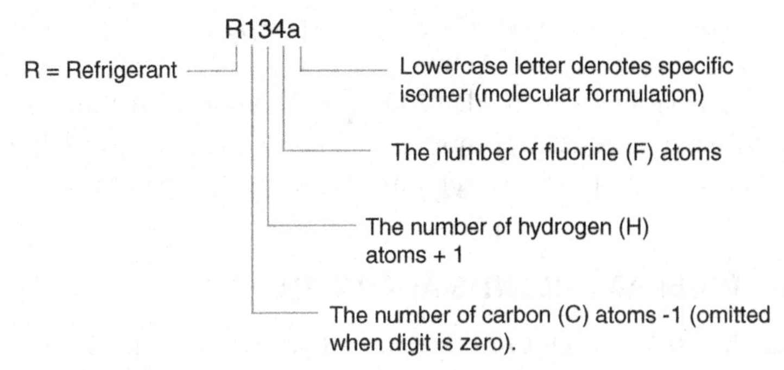

Figure 3.3 ***R134a nomenclature.***

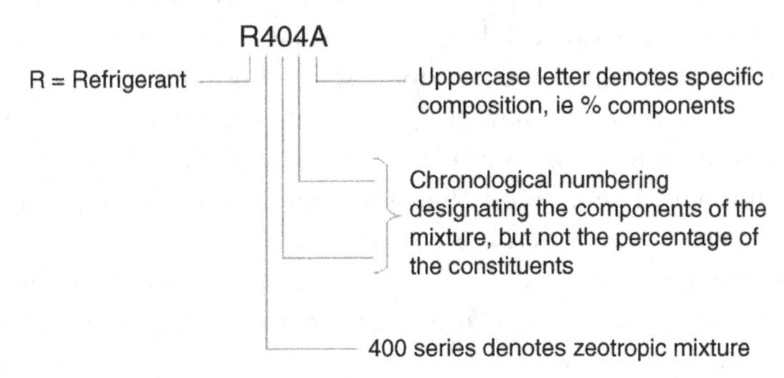

Figure 3.4 ***R404A nomenclature.***

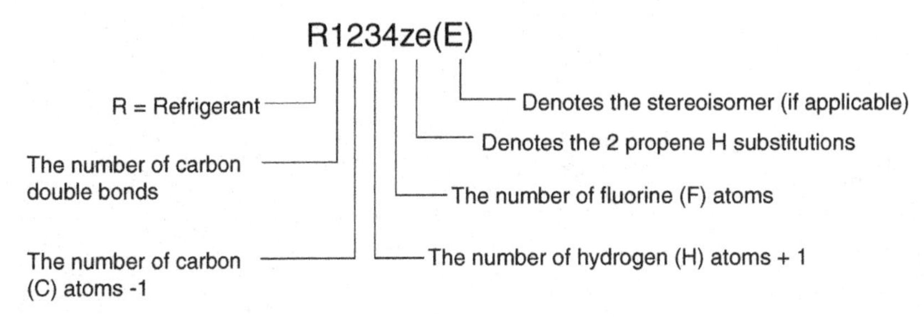

Figure 3.5 ***R1234ze(E) nomenclature.***

Zeotropic mixtures are assigned an identifying number in the 400 series. This number designates which components are in the mixture and the following upper-case letter denotes the proportions. The numbers are in chronological order of the refrigerant's approval by ASHRAE.

For example R407A (R32/R125/R134a (20/40/40)), R407B (R32/R125/R134a (10/70/20)), R407C (R32/R125/R134a (23/25/52)), etc.

Azeotropic mixtures are in the 500 series. For example R507 (R125/R143a (50/50)).

Miscellaneous organic compounds are in the 600 series; numbers are given in numerical order, for example R600a, isobutane

Inorganic compounds are in the 700 series. Identification numbers are formed by adding the relative molecular mass of components to 700. For example R717 corresponds to ammonia which has a molecular mass of 17.

3.6 REFRIGERANT BLENDS AND GLIDE

Many of the HFC and HFO refrigerants are mixtures or blends of two or more individual chemicals. More and more blends are being developed to give the best compromise in performance, GWP, and flammability for various applications. Mixtures can be azeotropes, near azeotropes or zeotropes. Azeotropes exhibit a single boiling point, strictly speaking at one particular pressure, but nevertheless they may be treated as a single substance. The first azeotropic refrigerant was a CFC, R502, so the use of refrigerant blends is not new. A zeotropic refrigerant is one whose boiling point varies throughout the constant pressure boiling process and varying evaporating and condensing temperatures exist in the latent heat of vaporisation phase.

Using zeotrope R407C as an example and referring to Fig. 3.6, the shape of the well-known refrigerant vapour compression cycle is unchanged, as shown. Constant P1, evaporating, and P2 condensing pressure

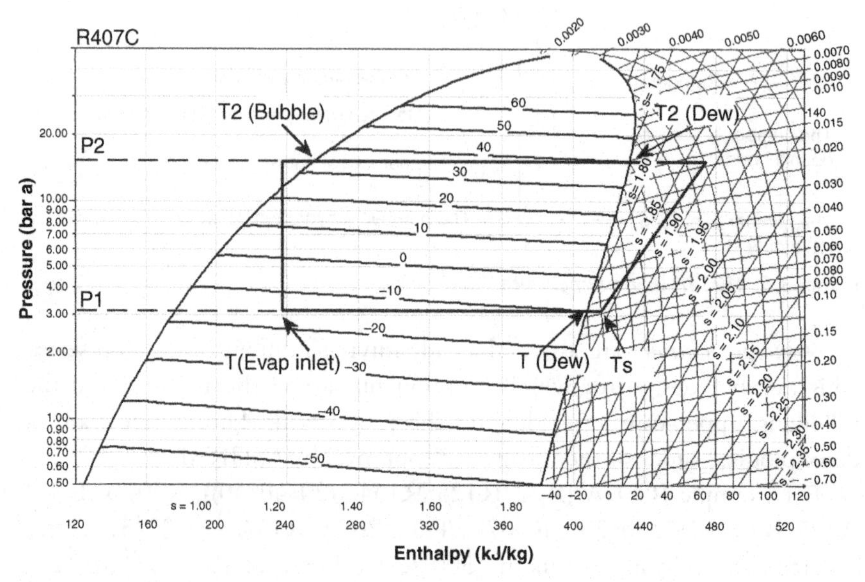

Figure 3.6 *Vapour compression cycle with R407C.*

processes are represented by horizontal lines, but the lines of constant temperature are now sloping. The temperature at which condensation starts is called the dew point, denoted here as T2 (dew). As condensation progresses, the temperature falls to T2 (bubble) so that the dew point temperature T2 of 40°C corresponds to a bubble point temperature of approximately 34°C. The temperature during the evaporation process changes from T1 (evaporator inlet) to T1 (dew), as the lighter components of the mixture, R32 and R125, evaporate preferentially to the R134a. The remaining liquid becomes R134a rich, its boiling point gradually increasing until all the liquid is evaporated. Note that this does not mean that the lighter components boil away leaving liquid R134a at the very end of the process. The composition shift during the process is limited and quite small. Further superheat then occurs after evaporation is complete, raising the temperature to Ts, the suction temperature at the compressor inlet.

The temperature glide can be used to advantage in improving plant performance by correct design of heat exchangers. A problem associated with blends is that refrigerant leakage can result in change in the proportion of components in the blend although this can only happen in system locations where both phases exist. The changes are small and have negligible effect on performance. The following recommendations apply to the use of blends:

- Equipment must always be charged from the liquid phase, or the component concentrations will be incorrect.
- Ingress of air must be avoided.
- Blends which have a large temperature glide, greater than 5 K, should not be used with flooded-type evaporators.

Some mixtures exhibit a glide of less than 2 K, and these are called 'near azeotropes'. For practical purposes they may be treated as single substances. Examples are R404A and R410A.

3.7 LEGISLATION

Some effects of legislation have already been mentioned. This section focuses on the European regulations. Readers should check any updates after 2015. Similar legislation is likely in other countries. Regulations on ozone depleting substances (CFCs and HCFCs) have largely removed these refrigerants from new equipment, and they will not be covered further. Regulations on global warming substances are currently driving changes. The EU F-Gas regulation (N 517/2014) is intended to reduce emissions stemming from F-gases and is part of Europe's climate change agenda. It does this by specifying a phase-down to reduce the amount of HFC, measured in CO_2 equivalent, placed on the market in accordance with the schedule, Fig. 3.7. Additionally there are bans on new HFC equipment, as summarised in Table 3.2. The GWP limit has the effect of excluding certain refrigerants in the various applications. The regulation also covers leakage prevention and detection.

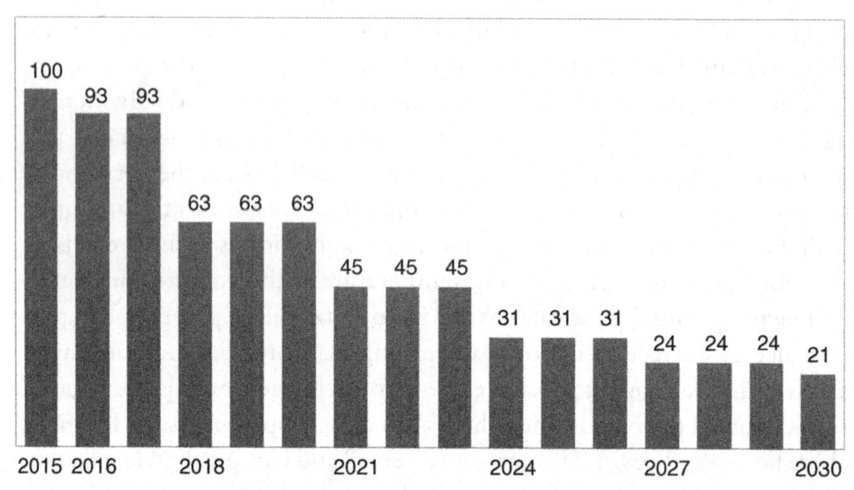

Figure 3.7 ***Percentage of HFC (CO_2 equivalent) that may be placed the market until year 2030 according to EU F-gas Regulation 2014 phase-down requirements.***

Table 3.2 Application bans (excerpt from EU F-Gas Regulations 2014)

Application	**HFC GWP Limit**	**Date of ban**
Domestic refrigerators and freezers	150	1 Jan. 2015
Refrigerators and freezers [...] for commercial use (hermetically sealed systems)	2500	1 Jan. 2020
Refrigerators and freezers [...] for commercial use (hermetically sealed systems)	150	1 Jan. 2022
Stationary refrigeration equipment (Except applications below −50°C)	2500	1 Jan. 2020
Multipack centralised refrigeration systems for commercial use with a capacity of 40 kW or more	150	1 Jan. 2022
Multipack centralised refrigeration systems in the primary refrigerant circuit of cascade systems	1500	1 Jan. 2022
Movable room air-conditioning appliances	150	1 Jan. 2020
Single split air-conditioning systems containing less than 3 kg of fluorinated greenhouse gases	750	1 Jan. 2025

3.8 CONTAINMENT AND TRAINING

Containment to ensure that refrigerant leakage to atmosphere is minimised has received timely attention in recent years. Although the environmental effect of direct emissions is usually much less than the indirect effect (see Section 3.4), it has become unacceptable to introduce new systems containing high GWP refrigerants. This is primarily affecting commercial refrigeration applications, as may be seen from Table 3.2. These systems have been historically using high GWP fluids, and have had high leakage rates. Nevertheless, attention to improved containment is being given by all sectors. The elements of containment improvement are the following:

- *Design*: Components and packaged units incorporate fewer joints with brazed joints replacing compression fittings.
- *Manufacture*: High standards for assembly, with personnel properly trained, and more inspection. Factory leak testing using helium detection methods to locate very small leaks.
- *Installation and site assembly*: Again, properly trained personnel, and more inspection.
- *Logging and maintenance*: Recording refrigerant additions.
- *Leak testing*: Mandatory leak testing requirements, dependent on system size and type are set out in the legislation.
- *Disposal*: Extraction of charge and mandatory use of authorised disposal facilities.

Significant reductions in leakage rates and corresponding TEWI improvements have been achieved. Further detailed information on refrigerant emissions and leakage skills can be found in the REAL Zero material. REAL Zero was a UK-based project which investigated leakage in a variety of sites over a 1-year project. Funded by the Carbon Trust, the IOR-led initiative resulted in a series of guidance notes, monitoring tools, surveys and training booklets to help target leakage reductions. The materials have been updated and presented as e-learning in REAL SKILLS EUROPE co-funded by the Leonardo Lifelong Learning Programme and international partners.

3.9 REFRIGERANT APPLICATIONS

The refrigerants most commonly found today in typical applications are considered in this section. Developments and environmental considerations could further restrict the use of HFCs in the future, develop a share for HFOs and increase the share of the so-called *natural refrigerants.*

These include R717, R744 and hydrocarbons. A move towards virtually zero GWP fluids, where they can be efficiently and safely applied, represents a long-term solution to the environmental issue of refrigerant leakage.

3.9.1 R134a and R407C

These refrigerants are primarily used for air conditioning and heat pumps, and have replaced R22 in many applications. R134a has a relatively low pressure and therefore about 50% larger compressor displacement is required when compared to R22. R134a has been very successfully used in screw chillers where short pipe lengths minimise costs associated with larger tubing. R134a also finds a niche where extra high condensing temperatures are needed and in many transport applications. The HFO low GWP refrigerants R1234yf and R1234ze operate at similar pressures to R134a, and are becoming available as long term alternatives. However they are relatively high cost and are flammable to a limited extent (see section 3.10). R1234yf has been adopted as a replacement for car air conditioning.

R407C is a zeotropic mixture consisting of 23% R32, 25% R125 and 52% R134a. It has properties close to those of R22, and for this reason has been extensively used in Europe due to a rapid R22 phase out. Its glide and heat transfer properties generally penalise the system performance, although counter flow heat exchange can deliver some benefit with plate heat exchangers. To find long-term alternatives, it is necessary to move to R32 HFO blends or R717, all of which require significant system re-design.

3.9.2 R410A

This is a high-pressure fluid with a low critical temperature used mainly in air conditioning and heat pumps. With good system design it has been shown to give equivalent or better performance than R407C. Many air conditioning suppliers switched to R410A from R22, especially for direct-expansion-type systems where an added advantage is that smaller pipe sizes can be used. R32 is seen as a possible longer-term alternative. It is already a 50% component of R410A, but alone it is flammable to a limited extent.

3.9.3 R404A

R404A is an HFC blend designed for commercial refrigeration where has been widely applied. It has superior performance to many other HFCs in low temperature applications, and also exhibits low compressor discharge temperatures which makes it suitable for single-stage compression, avoiding the need for inter-stage cooling. Its high GWP makes it unsuitable for future use, and it is targeted for rapid phase out (Section 3.7). Medium-term replacements are R407A and R407F, together with other HFC blends, and HFO/HFC blends are being proposed. A large proportion of users are investing in R744 technology as a long-term solution.

3.9.4 R717 Ammonia

Ammonia has long been used as a refrigerant for industrial applications. The engineering and servicing requirements to deal with its high toxicity and low flammability are well established. Technical developments are extending the applications for ammonia, for example low-charge packaged liquid chillers for use in air conditioning. Ammonia cannot be used with copper or copper alloys, so refrigerant piping and components have to be steel or aluminium. This may present difficulties for the air conditioning market where copper has been the base material for piping and plant. One property that is unique to ammonia compared to all other refrigerants is that it is less dense than air, so a leakage of ammonia results in it rising upwards and into the atmosphere. If the plant is outside or on the roof of a building, the escaping ammonia can drift away without harming occupants. Ammonia can be detected by its characteristic odour at very low concentrations and this acts as an early warning signal. The safety aspects of ammonia plants are well documented and there is reason to expect a sustained increase in the use of ammonia as a long-term refrigerant.

3.9.5 R290 Propane and Other Hydrocarbons

Hydrocarbons such as propane and butane are being successfully used in new low-charge systems where CFCs and HCFCs have previously been employed. They have obvious flammable characteristics which must to be taken into account. There is a large market for their use in sealed refrigerant systems such as domestic refrigeration and unitary air conditioners.

3.9.6 744 Carbon Dioxide

This refrigerant is gaining more applications, primarily in commercial and industrial refrigeration and also in some air conditioning and heat pump systems. Much work has been done on its use for vehicle air conditioning. High latent heat, and a heat transfer coefficient combined with high pressure and density under operating conditions results in the ability to produce large amounts of cooling with very small displacement compressors and small diameter pipelines. Its main disadvantage is its low critical temperature. The efficiency of a vapour compression cycle becomes very poor if the condensing temperature approaches the critical temperature.

It is being applied in the low stage of cascade systems using the vapour compression cycle, in transcritical cycles and as a booster stage of two-stage cycles, see chapter: The Refrigeration Cycle. The transcritical cycle is particularly effective where a heating requirement calls for water to be heated over a large temperature range, as is the case when hot water for services is being generated from a cold supply. In these cases, the counter flow gas cooler is a good solution.

The R744 transcritical cycle has the ability to revert to a conventional vapour compression cycle in low ambient conditions. Good average seasonal efficiency may be achieved in locations where heat can be rejected at temperatures well below 30°C for most of the year. Operation in the transcritical mode is only necessary under high ambient temperature conditions. High pressures and high pressure differentials allow the use of small tubing. Condensing temperatures can be brought right down under low ambient conditions and there is still adequate pressure difference to drive the refrigerant round the circuit. Use of an economiser can significantly enhance the sub-critical cycle performance.

3.10 REFRIGERANTS AND EFFICIENCY

In chapter: The Refrigeration Cycle, the efficiency shortfall of the vapour compression when compared to an ideal cycle was examined. The effect of the shortfall is dependent on the thermodynamic properties of the refrigerant, and so an ideal COP comparison between an existing refrigerant and its replacement can be informative. The HFCs tend to have slightly poorer thermodynamic properties than the chlorine containing substances they are intended to replace. But a simple ideal COP comparison does not give a true indication of the performance in an actual system. The main additional factors are compressor discharge temperature, governed by the index of compression and the effect of system pressure drops and heat transfer properties. High discharge temperatures tend to increase heat transfer losses and the effect of pressure drop tends to be less with higher pressure refrigerants because, although the loss in absolute terms may be the same, it is less proportionate to the total system pressure difference. System design plays an important role and efficiency can always be improved by increase of heat transfer surface and use of larger pipe diameters, but with cost penalties. These factors have enabled R410A air conditioning and heating systems to attain equal or better performance to previous equivalent cost R22 systems although the ideal COP is slightly lower, and some HFC blends give better performance in a refrigeration system previously charged with R404A. Another factor here is a slight reduction of capacity which tends to proportionately enhance heat exchanger surface. These complexities can result in commercial claims for proprietary refrigerants that need to be closely scrutinised.

In general, ammonia benefits from a very good ideal COP, but high discharge temperature, whereas CO_2 transcritical cycles tend to look poor, but benefit from low pressure drop effects and heat transfer properties.

3.11 HEALTH AND SAFETY

When dealing with any refrigerant, safety of self and that of others are vitally important. People working for service and maintenance need to be familiar with safety procedures and what to do in the event of an emergency. Health and safety information is available from manufacturers of all refrigerants. Local legislation is country dependent.

HFC refrigerants are non-toxic in the traditional sense, but nevertheless great care must be taken to ensure adequate ventilation in areas where heavier than air gases may accumulate. Carbon dioxide is not a simple asphyxiant. Exposure to more than 30% carbon dioxide will rapidly result in death.

Standard EN378 (2008) is the main refrigeration safety standard in Europe and refrigerants are classified by toxicity and flammability hazard categories. Safety codes are available from the IOR for Group A1 (low toxicity, non-flammable), Groups A2/A3 (non-toxic and flammable), ammonia, and carbon dioxide. The next revision of EN378 will cover low flammability refrigerants Group A2L.

In the United Kingdom and most of Europe, it is illegal to dispose of refrigerant in any other way than through an authorised waste disposal company. The UK legislation expects that anyone handling refrigerants is competent to do so and has the correct equipment and containers. Disposal must be through an approved contractor and must be fully documented. Severe penalties may be imposed for failure to implement these laws.

3.12 INNOVATION

Chemical researchers are advising that there are no new molecules that will satisfy the criteria for an ideal refrigerants. We can expect to see innovative use of the existing HFOs in blends targeted at different market sectors as the ability to work with limited flammability (A2L safety category) substances becomes re widespread. In addition, some cost reductions occur as production methods develop. Application developments of ammonia and carbon dioxide with the emphasis on economic efficiency improvement are likely to continue.

Air can be used as a refrigerant as noted in chapter: The Refrigeration Cycle. The use of water (R718) is also gaining attention as an air conditioning refrigerant (see Süß, 2015).

CHAPTER 4

Compressors

4.1 INTRODUCTION

The purpose of the compressor in the vapour compression cycle is to compress the low-pressure dry gas from the evaporator and raise its pressure to that of the condenser.

Compressors may be divided into two types, positive displacement and dynamic, as shown in Fig. 4.1. Positive displacement types compress discrete volumes of low-pressure gas by physically reducing the volumes causing a pressure increase, whereas dynamic types raise the velocity of the low-pressure gas and subsequently reduce it in a way which causes a pressure increase. Fig. 4.2 shows the approximate range of refrigeration capacities covered by various types. The most easily recognisable positive displacement type is the reciprocating or piston compressor, and being easily visualised, it will be used as a reference for descriptions of the compression process, and compressor features before moving on to other types.

The first refrigeration piston compressors were built in the middle of the 19th century, and evolved from the steam engines which provided the prime mover (see Fig. 4.3). Construction at first was double acting, but there was difficulty in maintaining gas-tightness at the piston rod, so the design evolved further into a single-acting machine with the crankcase at suction inlet pressure, leaving only the rotating shaft as a possible source of leakage, and this was sealed with a packed gland. Today, the majority of compressors are completely sealed, with the motor enclosed.

4.2 THE PISTON COMPRESSION PROCESS

The piston type is very widely used, being adaptable in size, number of cylinders, speed and method of drive. It works on the two-stroke cycle (see Fig. 4.4). Automatic pressure-actuated suction and discharge valves are used as shown in Fig. 4.4. As the piston descends on the suction stroke, the suction valve opens to admit gas from the evaporator. At the bottom of the stroke, this valve will close again as the compression stroke begins. When the cylinder pressure becomes higher than that in the discharge pipe, the discharge valve opens and the compressed gas passes to the condenser.

Refrigeration, Air Conditioning and Heat Pumps
http://dx.doi.org/10.1016/B978-0-08-100647-4.00004-8

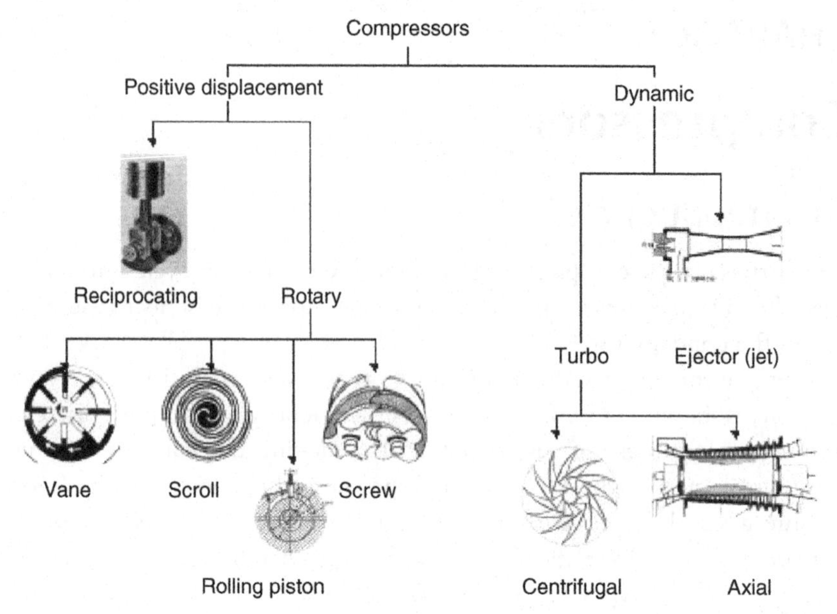

Figure 4.1 ***Chart of compressor types.***

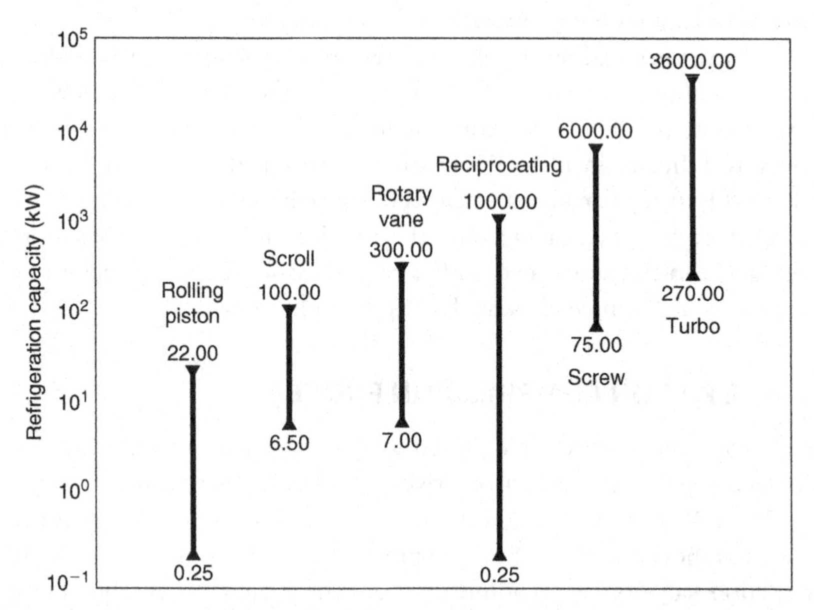

Figure 4.2 ***Approximate range of capacity covered by various compressor types.***

Figure 4.3 *Early Vilter double acting ammonia compressor and steam engine (Emerson Climate Technologies)*.

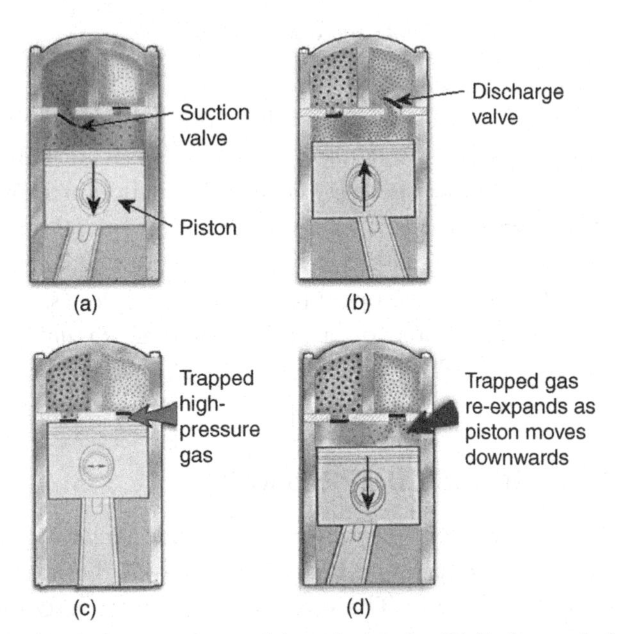

Figure 4.4 *Reciprocating compressor: (a) suction stroke, (b) discharge stroke, (c) piston at top of discharge stroke and (d) re-expansion during first part of suction stroke (Emerson Climate Technologies).*

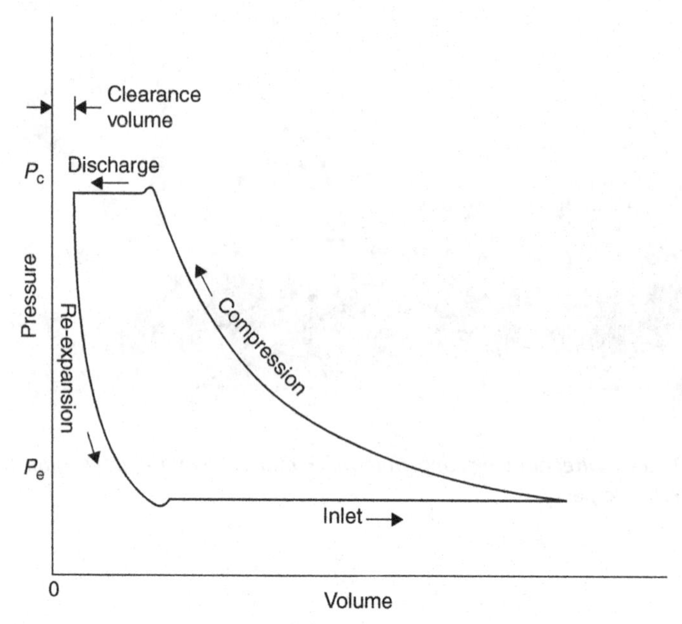

Figure 4.5 ***Reciprocating compressor indicator diagram.***

Gas left in the clearance at the top of the stroke must re-expand before a fresh charge can enter the cylinder, see Fig. 4.4d. The suction valve will not open until the cylinder pressure is lower than the suction pressure. A larger re-expansion or *clearance volume* means the piston must travel further down the cylinder before the pressure falls below suction pressure. The further the piston travel, without the valve opening, the higher the losses.

The process is represented on a pressure–volume or *indicator diagram* (Fig. 4.5). The area of the diagram represents the work required to compress and discharge the gas.

4.3 MULTI-CYLINDER COMPRESSORS

To attain a higher capacity, compressors were made larger during the 1st century of development, having cylinder bores up to 375 mm and running at speeds up to 400 rev/min. The resulting component parts were heavy and cumbersome. To take advantage of larger-scale production methods and provide interchangeability of parts, modern compressors are multi-cylinder, with bores not larger than 175 mm and running at higher shaft speeds. Machines of four, six and eight cylinders are common. Fig. 4.6 shows a large industrial reciprocating compressor built with a welded steel crankcase, the largest of this type being suitable for up to a refrigeration capacity of 900 kW (-10/35°C).

Figure 4.6 *Industrial size multi-cylinder compressor (GEA Refrigeration).*

Cylinders are commonly arranged in banks with two, three or four connecting rods on the same throw of the crankshaft to give a short, rigid machine. This construction gives a large number of common parts – pistons, connecting rods, loose liners and valves – through a range of compressors, and such parts can be replaced if worn or damaged without removing the compressor body from its installation. Compressors for smaller systems are simpler, consisting of two to six cylinders (see Fig. 4.7).

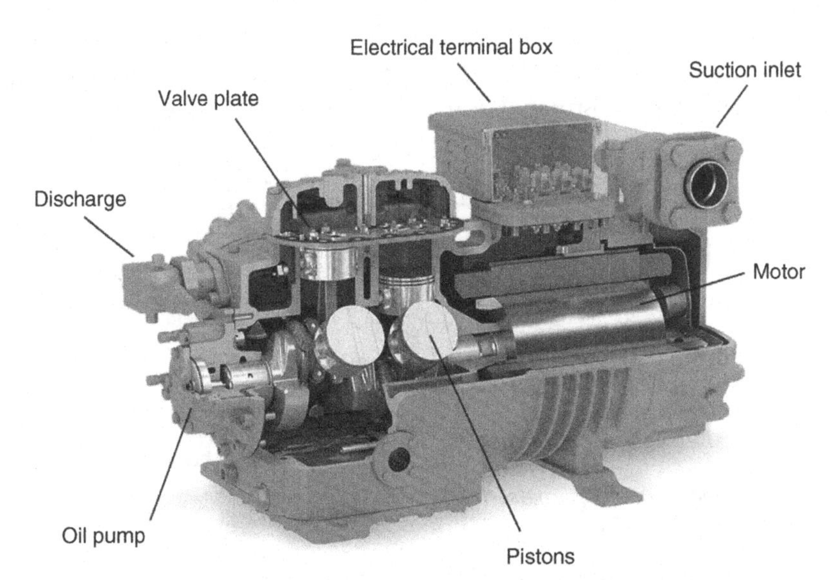

Figure 4.7 *Cut away view of small commercial six-cylinder compressor (Bitzer).*

4.4 VALVES

Piston compressors may be generally classified by the type of valve, and this depends on size, since a small swept volume requires a proportionally smaller inlet and outlet gas port. Small compressors have spring steel reed valves for both inlet and outlet arranged on a valve plate and the differing pressures kept separated by the cylinder head (Fig. 4.8). Above a bore of about 80 mm, the port area available within the head size is insufficient for both inlet and outlet valves, and the inlet is moved to the piston crown or to an annulus surrounding the head. The outlet or discharge valve remains in the central part of the cylinder head. In most makes, both types of valve cover a ring of circular gas ports and so are made in annular form and generally termed ring plate valves (Fig. 4.9). Ring plate valves are made of thin spring steel or titanium, limited in lift and damped by light springs to assist even closure and lessen bouncing.

Although intended to handle only dry gas, droplets of liquid refrigerant or oil may sometimes enter the cylinder and must pass out through the discharge valves. On large compressors with annular valves, these may be arranged on a spring-loaded head, which will lift and relieve excessive pressures.

Valve and cylinder head design is very much influenced by the need to keep the clearance volume to a minimum. A valve design which achieves a small clearance volume uses a conical discharge valve in the centre of the cylinder head, with a ring-shaped suction valve surrounding it (Fig. 4.10). The suction gas enters via passageways within the 'sandwich' valve plate. The piston has a small raised spigot which fits inside the ring-shaped suction valve. When the conical discharge valve lifts, high-pressure gas passes into the cylinder head. This construction is used in compressor bores up to 75 mm.

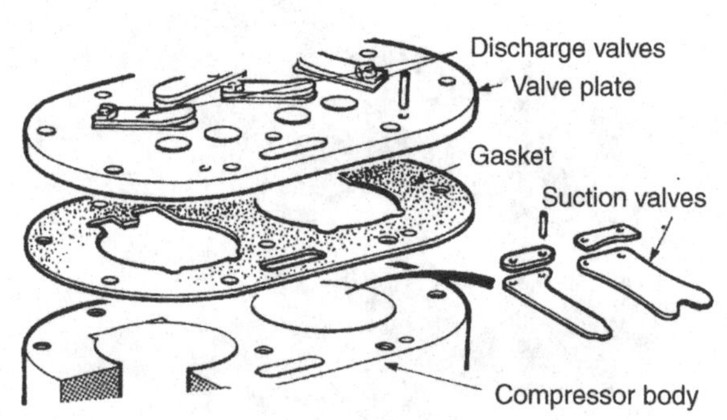

Figure 4.8 *Reed valve plate (Emerson Climate Technologies).*

Figure 4.9 *Ring plate valves (GEA Refrigeration).*

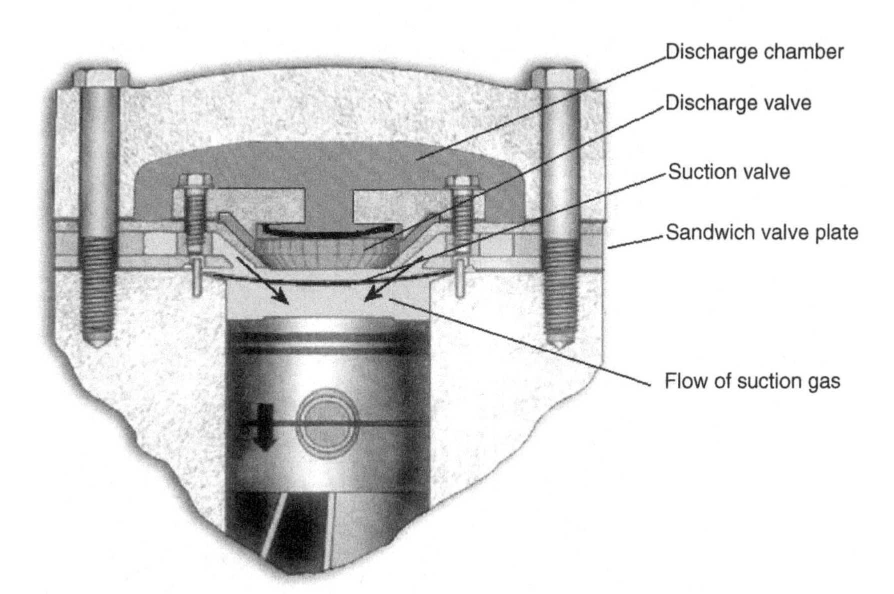

Figure 4.10 *Conical discharge valve and sandwich type valve plate (Emerson Climate Technologies).*

4.5 CAPACITY REDUCTION

A refrigeration system will be designed to have a maximum duty to balance a calculated maximum load, and for much of its life may work at some lower load. Such variations require capacity reduction devices. Speed control is the most obvious method, but this requires an inverter drive (see Section 4.6).

Multi-cylinder machines allow reduction of the working swept volume by taking cylinders out of service with blocked suction or valve-lifting mechanisms. With the blocked suction method, one or more of the cylinder heads incorporate a pressure-actuated valve which closes the supply of suction gas to the cylinder head. In Fig. 4.11a the normal (full capacity) operating position is with the solenoid valve de-energised. The gas pressures across the plunger are equalised, and the plunger is held in the open position by the spring. When the solenoid valve is energised, the needle valve seats on the upper port, and the unloading plunger chamber is exposed to discharge pressure through the discharge pressure port. The differential between discharge and suction pressure forces the plunger down, sealing the suction port in the valve plate, thus preventing the entrance of suction

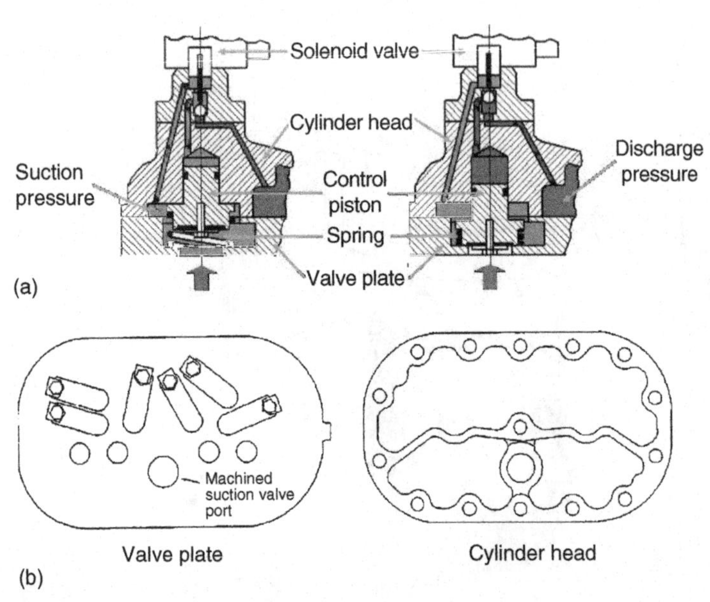

Figure 4.11 ***(a) Blocked suction capacity control mechanism, valve de-energised, suction port open (left); valve energised, suction port closed (right). (b) Valve plate showing machined suction port for blocked suction (Emerson Climate Technologies).***

vapour into the unloaded cylinders. During blocked suction operation, the flow of suction gas is blocked to the cylinders on that bank; therefore there is no gas to compress and the power consumption is reduced. Some designs offer the ability to give between zero and 100% on the unloading bank by varying the duration of the duty cycle.

The valve-lifting method is used with ring plate valves. The ring plate suction valve which is located at the crown of a loose liner can be lifted by various alternative mechanical systems, actuated by pressure of the lubricating oil and controlled by solenoid valves. Typically, an annular piston operates push rods under the valves. Use of these methods allows multi-cylinder machines to have cylinders or banks of cylinders unloaded for capacity reduction. In addition unloaded start can be arranged so that starting current is reduced and build-up of oil pressure occurs before the machine is fully loaded.

Smaller machines may have a valved bypass across the inlet and outlet ports in the cylinder head, or a variable clearance pocket in the head itself. Capacity may be reduced by external bypass piping.

Rotational speed can be reduced by two-speed electric motors or by use of an inverter which provides a variable frequency waveform to vary the motor speed. The lowest speed is usually dictated by the requirements of the in-built lubrication system.

4.6 ENCLOSED MOTORS

The majority of compressors supplied today incorporate an enclosed motor and this avoids any possible slight leakage of refrigerant through an open-drive shaft gland. The wide use of small refrigeration systems led to the evolution of methods of avoiding shaft seals, provided that the working fluid is compatible with the materials of electric motors and has a high dielectric strength.

The *semi-hermetic* or *accessible-hermetic* compressor (Fig. 4.12) has the rotor of its drive motor integral with an extended crankshaft, and the stator is fitted within an extension of the crankcase. Suction gas passes over the motor windings to remove waste heat in all but some of the smallest machines where forced cooling by air or water jackets is used. All starting switches must be outside the crankcase, since any sparking would lead to decomposition of the refrigerant. Electrical leads pass through ceramic or glass seals. Semi-hermetic compressors are built in a very wide range of sizes for the commercial and industrial markets. The motor is specified to suit the

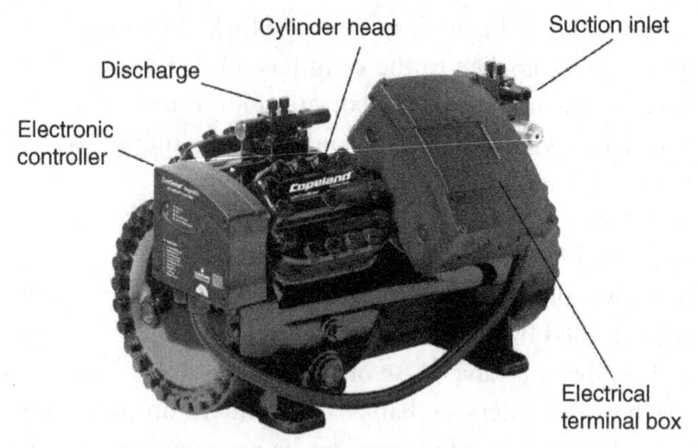

Figure 4.12 ***Exterior view of semi-hermetic piston compressor (Emerson Climate Technologies).***

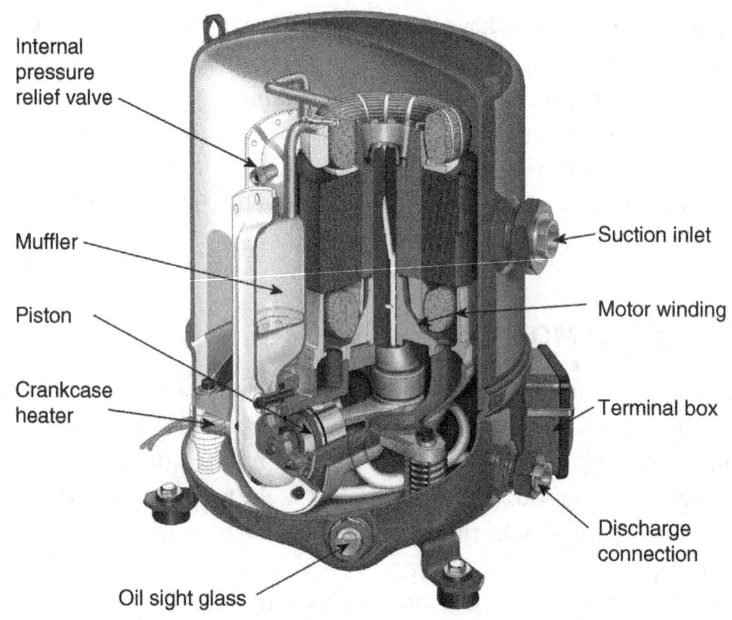

Figure 4.13 ***Hermetic piston compressor (Danfoss).***

compressor and as such can be designed for best efficiency over the application range. Effective refrigerant cooling has the advantage that the motor can be more compact than the corresponding 'standalone' electric motor.

Small compressors can be fully *hermetic*, that is having the motor and all working parts sealed within a steel shell (Fig. 4.13), and so not accessible for repair or maintenance. 'White goods' in the form of domestic refrigerators and freezers accounts for many millions of hermetic compressors, and the concept can be applied in sizes of up to tens of kW. They are generally lighter and more compact than semi-hermetic types and usually operate at two-pole synchronous motor speed (2900 rpm for 50 Hz supply), whereas semi-hermetics most often run at four-pole speed (1450 rpm). Below approximately 5 kW power input single-phase motors are used in locations where a three-phase supply is not available. The upper size limit of the hermetic compressor is determined by the practicality of manufacture – the welded shell design lends itself to volume production methods.

The failure of an in-built motor is likely to lead to products of decomposition and serious contamination of the system, which must then be thoroughly cleaned. Internal and external motor protection devices are fitted with the object of switching off the supply before such damage occurs.

Electronic motor power and speed controls provide the means to vary the speed of synchronous motors by generating electrical waveforms of controllable frequency. These devices are commonly called *inverters* and they can be very useful, particularly in close temperature control applications where stop/start or stepwise capacity changes give rise to unwanted fluctuations. Some over speeding may be possible if the motor is suitable. Care must be taken that both the compressor motor and the compressor itself are suitable for inverter application. There is likely to be a minimum speed limitation due to lubrication requirements. Inverters have internal electrical losses, and because the signal to the motor deviates from the pure waveform, additional motor losses and overheating can occur.

DC motors are now used in some small compressors and a converter is required to convert the supply from the AC source. An advantage of this approach is that only one motor is required for each model regardless of the local power supply voltage. The traditional AC synchronous motor together with its protection system is developed in different versions for the various voltages and frequencies existing in different countries. This results in a single compressor model having a number of motor options, which have to be manufactured and stocked. The DC motor is universal and additionally provides variable speed capability.

There is a need for very small compressors to be driven from low-voltage DC supplies. Typical cases are batteries on small boats and mobile homes, where these do not have a mains voltage alternator. It is also possible to obtain such a supply from a bank of solar cells. This requirement has been met in the past by diaphragm compressors driven by a crank and piston rod from a DC motor, or by vibrating solenoids. Electronic devices now make it possible to obtain the mains voltage AC supply for hermetic compressors from low-voltage DC.

4.7 OPEN COMPRESSORS

Compressors having external drive require a gland, or seal, where the shaft passes out of the crankcase, are termed *open compressors*. The drive comes from an external air-cooled motor or other prime mover. Open-drive compressors are a requirement when using a refrigerant such as ammonia, which is not compatible with copper used in electric motors. Open compressors may be belt driven or directly coupled to the shaft of the electric motor. Belt drives offer the opportunity to match the speed to the capacity requirement, but they must be adequately guarded and since a lateral load is imposed on the compressor bearings, not all compressor models are suitable. This type of drive is widely used in transport applications where compressors are driven from the vehicle engine or from a separate diesel engine.

The shaft seal for open-drive compressors usually comprises a rotating carbon ring in contact with a highly polished metal facing ring, the assembly being well lubricated. To maintain contact under all working crankcase pressures, the carbon ring is spring loaded and to allow for slight movement of the shaft.

4.8 COOLING AND PROTECTION

Heat is generated in the compression process and discharge temperature must be limited to avoid risk of oil or refrigerant decomposition. The temperatures encountered are dependent on the operating conditions and the refrigerant. Under many conditions, cold suction gas combined with heat loss to atmosphere provides sufficient cooling for small machines. For some operating conditions a fan may be specified by the manufacturer. Refrigerants such as ammonia giving high discharge temperatures require the use of water-cooled cylinder heads. Oil coolers may be needed which may be water, air or refrigerant cooled. Discharge temperature will tend to be higher when operating at part load conditions.

Compressors can overheat if the mass flow rate becomes very low, for example at conditions resulting from abnormally low suction pressure. This can be the result of loss of refrigerant charge. Low mass flow rate will also result in loss of oil to the system and the onus is on the system designer to ensure adequate oil return, otherwise lubrication will be impaired. Liquid refrigerant may enter the compressor under fault conditions and this can reduce the lubricant viscosity, and excessive amounts may cause severe damage.

These system issues can have the effect of shortening compressor life, and it is usual to provide compressors with fault protection. Temperature and oil protection are the most effective and usually take the form of 'cut-outs' which stop the compressor in the event of excessive temperature or insufficient oil pressure. Motor protection is normal for enclosed motor types and may take the form of temperature sensors embedded in the windings. For small compressors an internal line break protector which is sensitive to both temperature and electric current is sometimes used (Fig. 4.14). Each leg has a heater and a contact and all three open and close at the same time. It is positioned where the three motor windings meet and all three phases are taken out in the event of overheat and/or overload. Because it is built-in and internal there is no need to bring wires back out of the casing for external connections, and it cannot be accidentally by-passed.

Some applications are known to require the compressor to withstand liquid return. For example, it is not economically feasible to provide preventative system controls on small reversible air conditioners to ensure no liquid return on defrost cycles. Here the compressor manufacturer will ensure that the compressor design has been life tested to give many years of reliable operation under foreseeable conditions.

When first started, a refrigeration system may operate at a higher suction temperature and pressure than at normal operating conditions, and consequently a higher discharge pressure, taking considerably more power. Moreover, during the first seconds of operation, the motor is required to

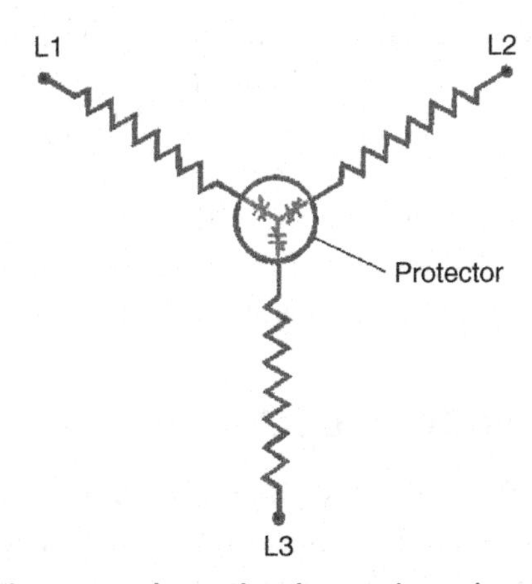

Figure 4.14 ***Line break protector located at the meeting point of the motor windings.***

provide sufficient torque to accelerate the compressor. Assisted start devices are used for most commercial and industrial applications and may take the form of unloaded start bypass, or suction pressure regulation. On the electrical side a frequent requirement is to minimise the electrical surge occurring on start up. Start devices such as star delta or part-winding start motors are used. For open compressors the drive motors must be sized accordingly to provide this pull-down power and an allowance of 25% is usual. As a result, the drive motor will run for the greater part of its life at something under 80% rated output, and so at a lower efficiency, low running current and poor power factor. For semi-hermetic and hermetic compressors the starting characteristics are defined to ensure minimal over-sizing of the motor.

4.9 STRAINERS, LUBRICATION AND CRANKCASE HEATERS

Incoming gas may contain particles of dirt from within the circuit, especially on a new system. Suction strainers or traps are provided to catch such dirt and will be readily accessible for cleaning on the larger machines.

All but the smallest compressors will have a strainer or a filter in the lubricating oil circuit. Strainers within the sump are commonly of the self-cleaning slot disc type. Larger machines may also have a filter of the fabric throwaway type, as in automobile practice. Reciprocating compressors have splash lubrication in the small sizes and forced oil feed with gear or crescent pumps on all others. A sight glass is normally fitted to semi-hermetic and open compressors to indicate oil level.

When the compressor is idle, the lubricating oil will contain a certain amount of dissolved refrigerant, depending on the pressure, temperature and the refrigerant itself. At the moment of starting, the oil will be diluted by this refrigerant and, as the suction pressure falls, gas will boil out of the oil, causing it to foam.

To restrict the refrigerant quantity in the oil to an acceptable amount, heating devices are commonly fitted to crankcases and connected so as to remain in operation whenever the compressor is idle.

4.10 COMPRESSOR EFFICIENCY

The amount of gas pumped by the compressor will always be less than the physical displacement of the pistons in the cylinders. The re-expansion loss has already been mentioned and the other losses are illustrated in Fig. 4.15.

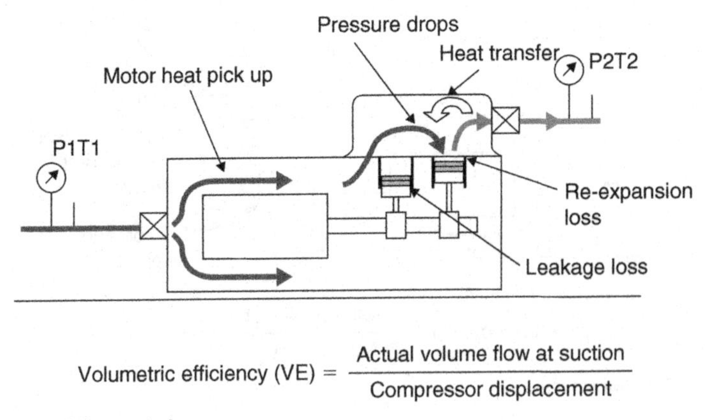

Figure 4.15 ***Volumetric losses.***

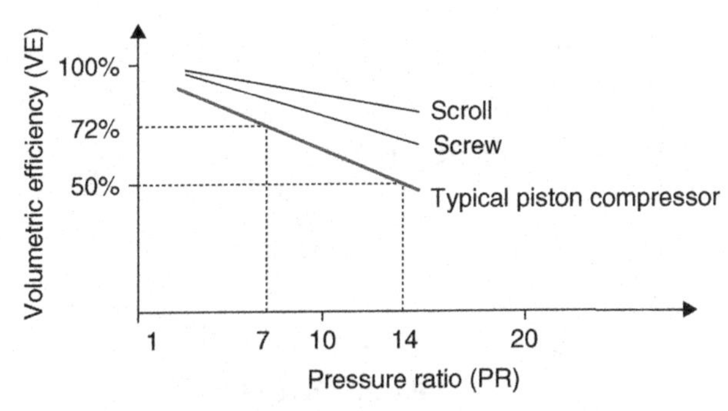

Figure 4.16 ***Typical VE characteristics for various compressor types.***

Volumetric efficiency (VE) accounts for all the losses affecting the flow rate of the compressor. The reference point is the condition of the gas, pressure P1 and temperature T1 at the inlet or suction. Heat pick up due to motor losses is included in the case of enclosed types. In practice VE is close to linear with pressure ratio as illustrated in Fig. 4.16. The major element of the volumetric loss in a piston machine is due to re-expansion. Its effect is not too serious because work is recovered in the re-expansion process.

The energy efficiency of compression is defined with reference to the ideal adiabatic compression process. The isentropic power input is the minimum amount of power required to compress the gas, mass flow rate, m, from P1, T1, to P2. The actual power will always exceed the isentropic power because of the losses shown in Fig. 4.17. The magnitude of the losses will depend on the compressor type, and an approximate order of magnitude is 10% motor loss (motor efficiency 90%), 10% friction losses and 10% flow and heat transfer losses. In practice values of isentropic efficiency (IE) above

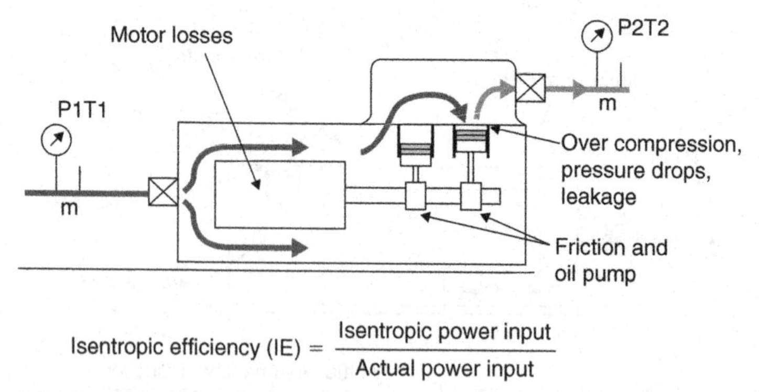

Figure 4.17 ***Power losses.***

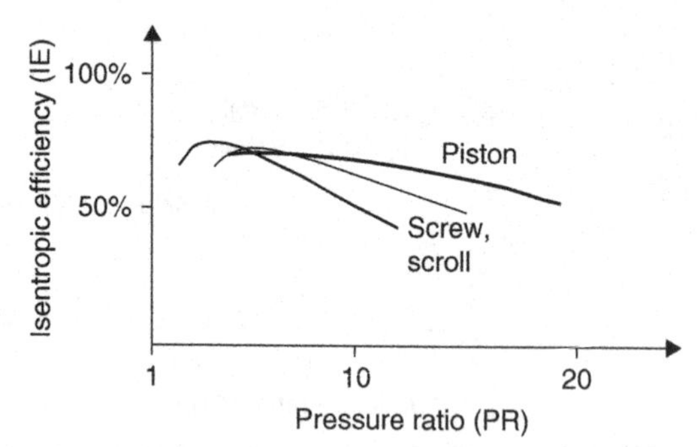

Figure 4.18 ***Typical IE characteristics for various compressor types.***

70% are very good compressor efficiencies. It is difficult for designers to get much more out of the compressor. General IE trends are illustrated in Fig. 4.18. These do not relate to specific products, and manufacturer's data must be consulted to establish specific values. Published data quote performance in terms of refrigeration capacity, power input and COP, and these values reflect the underlying efficiency characteristics.

4.11 SCREW COMPRESSORS

The *screw compressor* can be visualised as a development of the gear pump. For gas pumping the rotor profiles are designed to give maximum swept volume and no clearance volume where the rotors mesh together. The pitch of the helix is such that the inlet and the outlet ports can be arranged at the ends instead of at the side. The solid portions of the screws slide over the gas ports to separate one stroke from the next so that no inlet or outlet valves are needed.

The more usual form has twin-meshing rotors on parallel shafts (see Figs 4.19 and 4.24). As these turn inside the closely fitting casing, the space between two grooves comes opposite the inlet port, and gas enters. On further rotation, this pocket of gas becomes sealed from the inlet port and moved down the barrels. A meshing lobe of the male rotor, seen on the left in Fig 4.19, then reduces the pocket volume compressing the gas, which is

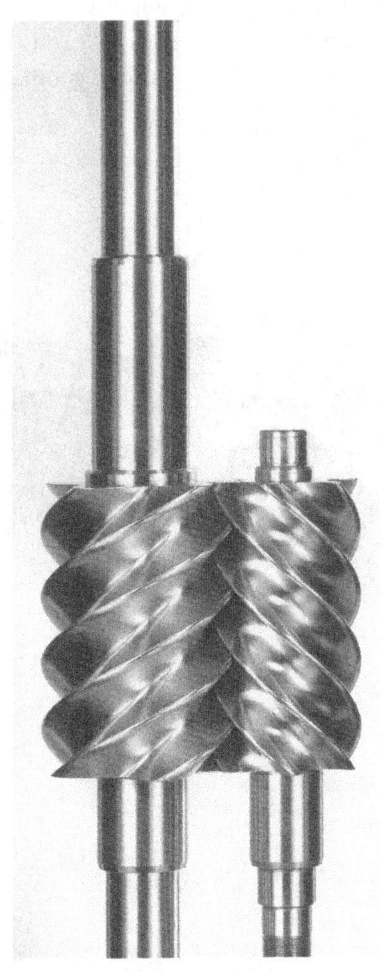

Figure 4.19 ***Twin screw compressor rotors (Bitzer).***

finally released at the opposite end, where the exhaust port is uncovered by the movement of the rotors. Various combinations of rotor sizes and number of lobes have been successfully employed. In most designs the female rotor is driven by the male rotor. Studies on optimising rotor design for screw machines carried out at City University, London have been documented in the proceedings of the International Conferences on Compressors and their systems. Maintenance of adequate lubrication is essential. Lubrication, cooling and sealing between the working parts is usually assisted by the injection of oil along the length of the barrels. This oil must be separated from the discharge gas and is then cooled and filtered before returning to the lubrication circuit (see chapter: Oil in Refrigerant Circuits).

The single screw compressor has a single grooved rotor, with rotating star tooth seal vanes to confine the pockets of gas as they move along the rotor flutes (see Fig. 4.20). Once again, various geometries are possible, but

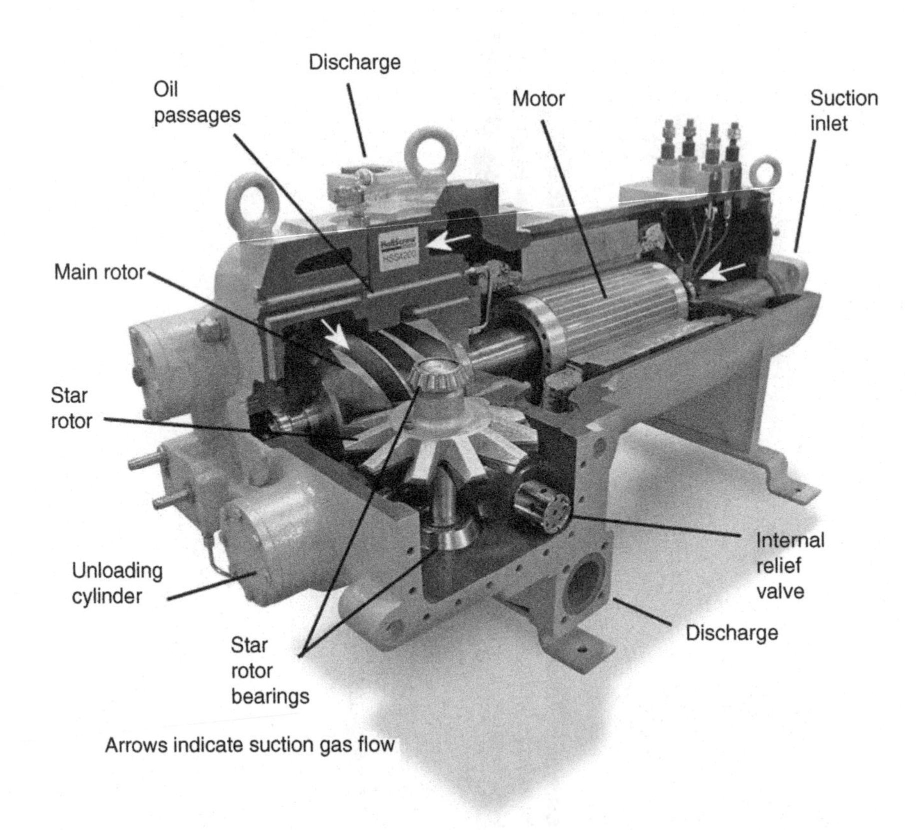

Figure 4.20 ***Single screw compressor (J&E Hall).***

compressors currently being manufactured have a rotor with six flutes and stars with eleven teeth. The normal arrangement is two stars, one on either side of the rotor. Each rotor flute is thus used twice in each revolution of the main rotor, and the gas pressure loading on the rotor is balanced out, resulting in much lighter bearing loads than for the corresponding twin screw design (see Fig. 4.21). The stars are driven by the rotor, and because no torque is transmitted the lubrication requirements in the mesh are lighter also. Oil cooling and sealing is usual and the oil circuit is similar to that of the twin screw.

Screw compressors have no clearance volume, and there is no loss ofVE due to re-expansion as in a piston machine. Volumetric losses result mainly from leakage of refrigerant back to the suction via in-built clearances. Oil is used for sealing, but leakage of oil containing dissolved refrigerant to the intake plenum reduces VE both by release of refrigerant and by heating the incoming gas. The VE decreases with increasing pressure ratio, but less than with some piston types (Fig. 4.16). Leakage losses are a function of tip speed, so that smaller machines need to operate a higher speed to maintain efficiency. With synchronous motor drives, this sets a lower practical limit on the size (Fig. 4.2).

In all screw compressors, the gas volume will have been reduced to a preset proportion of the inlet volume by the time the outlet port is uncovered, and this is termed the *built-in volume ratio*. At this point the gas within

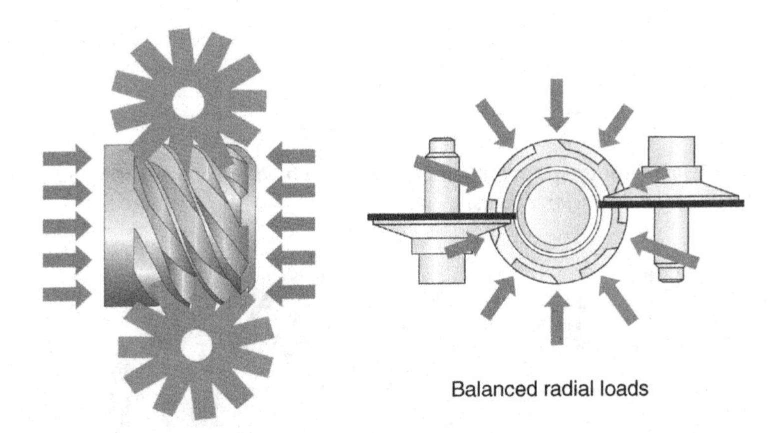

Figure 4.21 ***Balanced rotor pressure loading with single screw compressor (Emerson Climate Technologies).***

the screws is opened to condenser pressure, and gas will flow inwards or outwards through the discharge port if the pressures are not equal.

The absorbed power of the screw compressor will be at its optimum only when the working pressure ratio corresponds to the built-in volume ratio. The over and under compression losses can be visualised as additional areas on an indicator diagram as in Fig. 4.22. This results in an IE characteristic having a strongly defined peak, as shown in Fig. 4.18. To the left of the peak, over-compression of the gas results in loss of efficiency, whereas to the right, there is under-compression with back flow of gas into the compression pocket when the discharge port is uncovered. Changing the size of the discharge port changes the position of the peak, and this is illustrated by the two curves in Fig. 4.18. A screw compressor should be chosen to have a volume ratio suitable for the application. Leakage also contributes towards efficiency loss, but friction effects are quite small.

Capacity reduction of the screw compressor is effected by a sliding block covering part of the barrel wall, which permits gas to pass back to the suction, so varying the working stroke (Fig. 4.23). The valve is moved by an oil pressure acting on a piston, shown on the right. It is usual for the

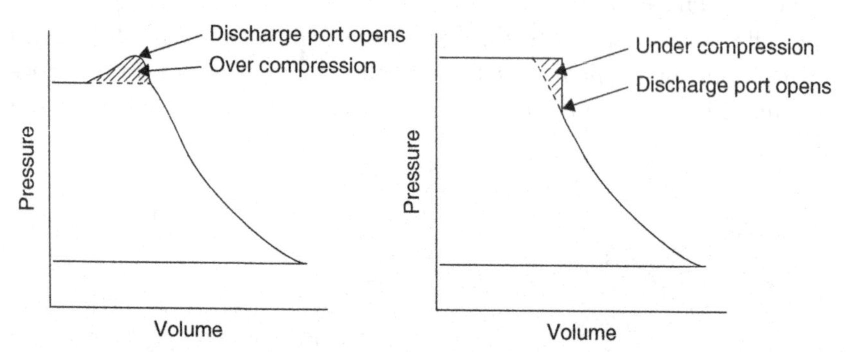

Figure 4.22 ***Indicator diagram for compressors with built-in volume ratio, to illustrate over and under compression effects.***

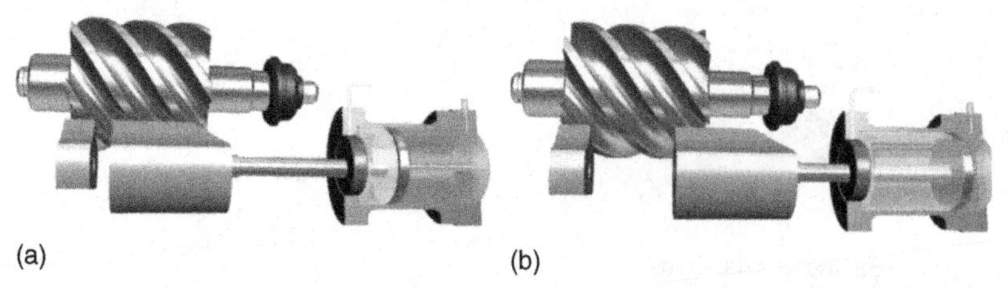

Figure 4.23 ***Capacity reduction slide for twin screw compressor (a) just starting to open, (b) at minimum load (Howden).***

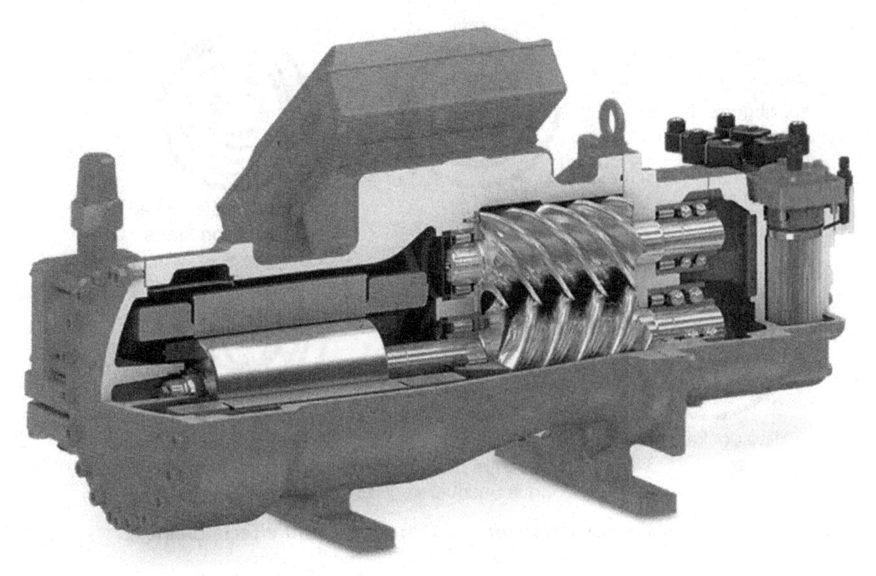

Figure 4.24 ***Semi-hermetic screw compressor (Bitzer).***

sliding part of the barrel to adjust the size of the discharge port at the same time, so that the volume ratio is at least approximately maintained at part load. Many design variations and control methods exist. The single screw type will generally have two sliding valves; lifting valves are sometimes used instead of slides. Reduction down to 10% of maximum capacity is usual.

The oil separation, cooling and filtering for a screw compressor add to the complexity of an otherwise simple machine. Liquid injection is sometimes used instead of an external oil cooler. Some commercial screw compressors have the oil-handling circuit built into the assembly. In Fig. 4.24 the suction gas enters at the suction connection on the left, passes over the motor, through the compressor, to the discharge connection on the right.

4.12 SCROLL COMPRESSORS

Although the scroll mechanism has been known for many years, having been patented in France in 1905, it was not until the latter part of the last century that it first appeared in commercially available compressors. Manufacturing technology had by this time developed sufficiently to enable the precision spiral forms to be made economically.

Scroll compressors are positive displacement machines that compress refrigerants with two inter-fitting spiral-shaped scroll members as shown

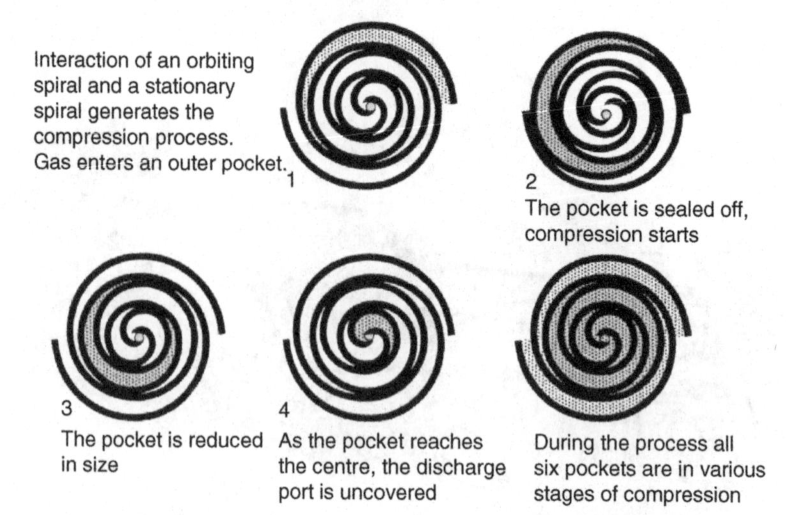

Interaction of an orbiting spiral and a stationary spiral generates the compression process. Gas enters an outer pocket.
1
2
The pocket is sealed off, compression starts
3
The pocket is reduced in size
4
As the pocket reaches the centre, the discharge port is uncovered
During the process all six pockets are in various stages of compression

Figure 4.25 ***Scroll gas compression process (Emerson Climate Technologies).***

in Fig. 4.25. One scroll remains fixed whilst the second scroll moves in orbit inside it. Note that the moving scroll does not rotate but orbits with a circular motion. Typically two to three orbits, or crankshaft revolutions, are required to complete the compression cycle.

The scroll has certain common features with the screw. Both types have a built in volume ratio and therefore the scroll exhibits IE curves similar in shape to those of the screw (Fig. 4.18). There is no clearance volume and hence no re-expansion loss. However, there is a very important difference in the sealing of the compression pockets. The screw relies on clearance between rotors and casing whereas the scroll can be built with contacting seals, that is the scrolls touch each other at the pocket boundaries. This is possible because the orbiting motion gives rise to much lower velocities than rotating motion, and also the load on the flanks and tips of the scrolls can be controlled. In addition, there is no direct path between the discharge port area at the centre of the scrolls and the suction. The result of this is very low leakage and heat transfer losses, giving better VE characteristic than most other types (Fig. 4.16). This enables the scroll to function efficiently in much smaller displacements than the screw (Fig. 4.2), with the upper size limit being effectively determined by the economics of manufacture.

Almost all production scrolls are of the hermetic type and a typical configuration is shown in Fig. 4.26. These compressors have advantages

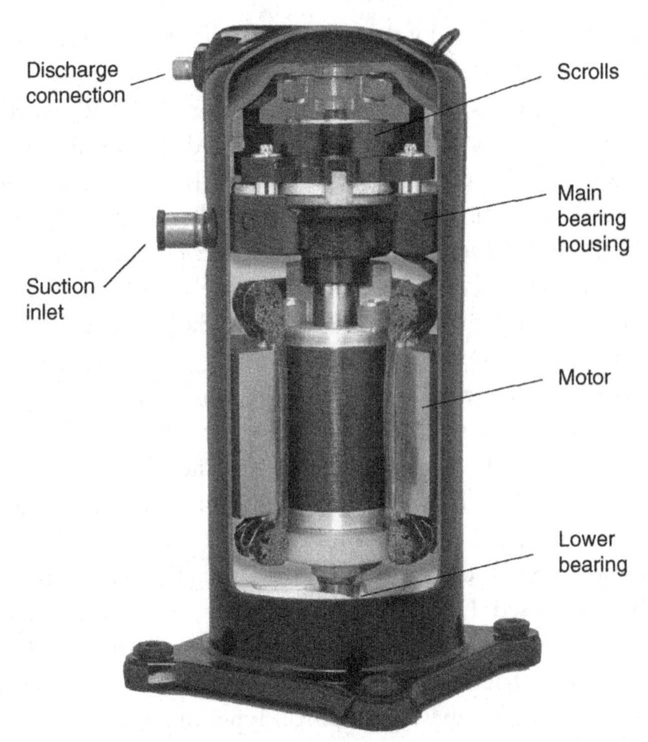

Figure 4.26 ***Cut away view of scroll compressor (Emerson Climate Technologies).***

over similar sized piston hermetics in air-conditioning applications and this has encouraged investment in production facilities, building millions worldwide.

The flat volumetric curve enables the scroll to deliver more cooling and heating capacity at extreme conditions, the compression process is smoother and quieter, and there are many fewer moving parts, ensuring very high reliability. In addition, the scroll has excellent resistance to fault conditions such as liquid floodback, and compliance mechanisms can deliver unloaded starting and extreme pressure protection (Elson et al., 1991).

Although no oil injection into the compression process is needed, bearing and thrust surface lubrication is vital. Oil can be fed to the upper drive bearings and other surfaces using the centrifugal forces generated by an offset drilling along the length of the shaft. Capacity control using variable speed inverter drive is possible for many scrolls. A method using intermittent and frequent scroll separation has been introduced (Hundy, 2002). When the scrolls are separated axially the capacity is zero. The motor continues to run at normal speed but with very low power and back flow of gas from the high-pressure side is prevented by a discharge valve. When the scrolls are brought together normal pumping is resumed. The axial movement of the fixed scroll is powered by a hydraulically actuated piston, in response to a pulse width modulated signal from a controller. The total cycle time is typically 20 s, and the duration of the loaded period within that cycle time is infinitely variable. Because the cycle time is relatively short, the thermal inertia of the system has the effect of damping the fluctuations so that the effect is very similar to continuous operation at reduced capacity.

The 'take-off' of air-conditioning scrolls prompted the introduction of many variants, the most important of which is the refrigeration or low-temperature version. As with the screw, use of a smaller discharge port enables the compressor to be optimised for the higher-pressure ratios applicable to lower temperature applications. By introducing a discharge valve, similar to those employed in reciprocating compressors, the effects of under compression can be minimised. Liquid injection is used for cooling where necessary, and the economiser cycle can be used to boost capacity and efficiency (see chapter: The Refrigerant Cycle). These developments have enabled the refrigeration scroll to compete with piston types in a wide variety of commercial applications.

4.13 SLIDING AND ROTARY VANE COMPRESSORS

The volumes between an eccentric rotor and sliding vanes vary with angular position, to form a positive displacement compressor, larger types, generally referred to as *Rotary Vane* (Fig. 4.27) can have eight or more blades and do not require inlet or outlet valves. The blades are held in close contact with the outer shell by centrifugal force, and sealing is improved by the injection of lubricating oil along the length of the blades. Rotary vane machines have no clearance volume, but they are limited in application by the stresses set up by the thrust on the tips of the blades. Although they have been used at low discharge pressures such as the first stage of a compound cycle, they are no longer widely applied in refrigerant compression.

Smaller machines, termed *Rotary Compressors* or *Rotaries* (Fig. 4.28) have fewer vanes and with virtually no inbuilt compression, utilise a discharge valve so that at the pocket reduces in size, the pressure rise until it exceeds that necessary to open the valve, similar to a piston compressor discharge valve. Again, they are best suited to low compression ratios.

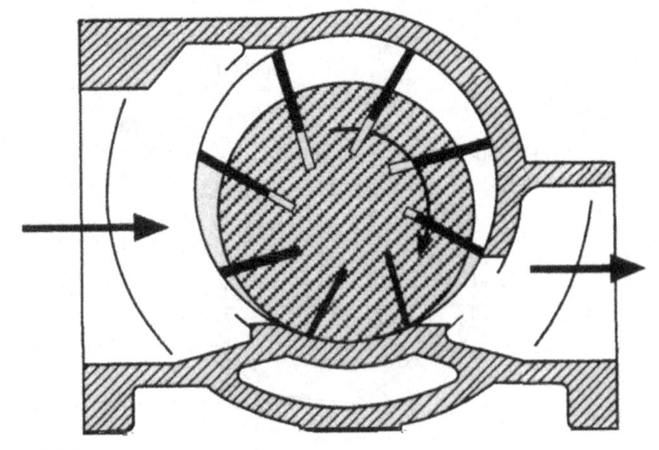

Figure 4.27 *Sliding vane compression.*

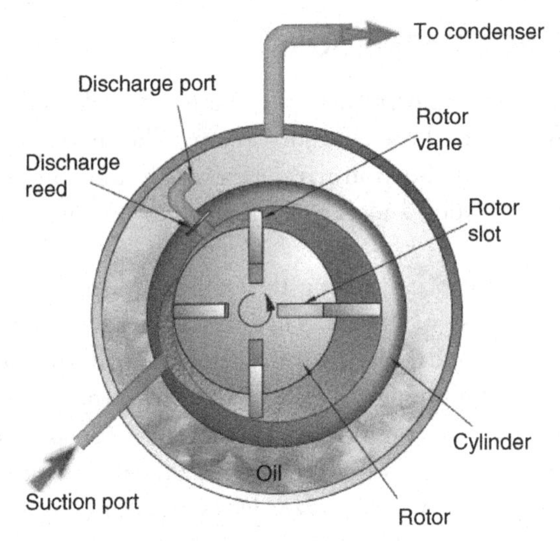

Figure 4.28 *Rotary compressor (Business Edge).*

Sliding vane or *rolling piston* compressors have one or two blades, which do not rotate, but are held by springs against an eccentric, rotating roller. These compressors require discharge valves. This type has been developed extensively for domestic appliances, packaged air-conditioners and similar applications, up to a cooling duty of 15 kW (see Fig. 4.29).

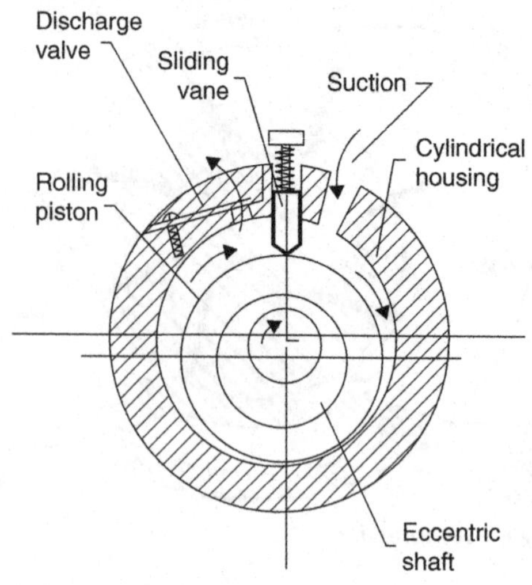

Figure 4.29 ***Rolling piston compression.***

4.14 DYNAMIC COMPRESSORS

Dynamic compressors impart energy to the gas by velocity or centrifugal force and then convert this to pressure energy. The most common type is the *centrifugal compressor*. Suction gas enters axially into the eye of a rotor which has curved blades, and is thrown out tangentially from the blade circumference.

The energy given to gas passing through such a machine depends on the velocity and density of the gas. Since the density is already fixed by the working conditions, the design performance of a centrifugal compressor will be decided by the rotor tip speed. Owing to the low density of gases used, tip speeds up to 300 m/s are common. At an electric motor speed of 2900 rev/min, a single-stage machine would require an impeller 2 m in diameter. To reduce this to a more manageable size, drives are geared up from standard-speed motors or the supply frequency is changed to get higher motor speeds. The drive motor is integral with the compressor assembly and may be of the open or hermetic type. On single-stage centrifugal compressors for air-conditioning duty, rotor speeds are usually about 10,000 rev/min.

Gas may be compressed in two or more stages. The impellers are on the same shaft, giving a compact tandem arrangement with the gas from one stage passing directly to the next. The steps of compression are not very great and, if two-stage is used, the gas may pass from the first to the second without any inter-cooling.

Centrifugal machines can be built for industrial use with ammonia and other refrigerants, and these may have up to seven compression stages. With the high tip speeds in use, it is not practical to build a small machine, and the smallest available centrifugal compressor for refrigeration duty has a capacity of some 260 kW. Semi-hermetic compressors are made up to 7000 kW and open drive machines up to 21,000 kW capacity. There are no components which require lubrication, with the exception of the main bearings. As a result, the machine can run almost oil free.

Systems of this size require large-diameter refrigerant suction and discharge pipes to connect the components of the complete system. As a result, and apart from large-scale industrial plants, they are almost invariably built up as liquid-cooling, water-cooled packages with the condenser and evaporator complete as part of a factory-built package.

The pumping characteristic of the centrifugal machine differs from the positive displacement compressor since, at excessively high discharge pressure, gas can slip backwards past the rotor. This characteristic makes the centrifugal compressor sensitive to the condensing condition, giving higher duty and a better coefficient of performance if the head pressure drops, whilst heavily penalising performance if the head pressure rises. This will vary also with the angle of the capacity reduction blades. Excessive pressure will result in a reverse flow condition, which is followed a fraction of a second later by a boosted flow as the head pressure falls. The vapour surges, with alternate forward and reverse gas flow, throwing extra stress on the impeller and drive motor. Such running conditions are to be avoided as far as possible, by designing with an adequately low head pressure and by good maintenance of the condenser system. Rating curves indicate the stall or surge limit.

Since centrifugal machines are too big to control by frequent stopping and restarting, some form of capacity reduction must be inbuilt. The general method is to throttle or deflect the flow of suction gas into the impeller. With most models it is possible to reduce the pumping capacity down to 10–15% of full flow. The availability of low-cost inverters has led to the use of variable speed drive which offers increased centrifugal compressor efficiency. However this cannot totally replace the need for variable inlet guide vanes because of early (low head) surge arising from low flow system head requirements (Brasz, 2007). An example of a centrifugal compressor with variable geometry is shown in Fig. 4.30. A recent entry to the field of compressors is a variable speed centrifugal compressor with a DC drive motor and magnetic bearings Fig. 4.31. This opens up the possibility of oil free systems.

Figure 4.30 *Centrifugal compressor with variable geometry, showing inlet guide vanes (labelled 3) and moveable wall diffuser (labelled 4) (Carrier).*

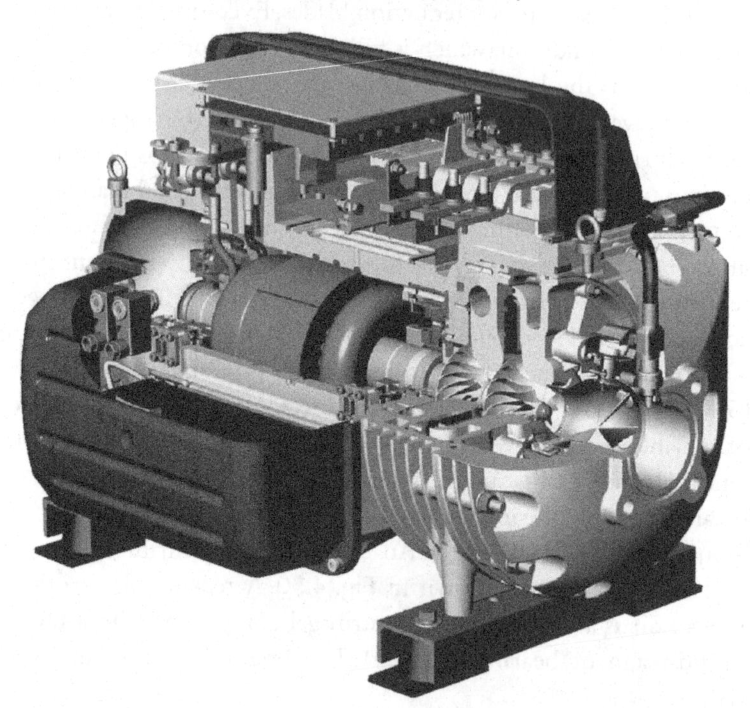

Figure 4.31 *Centrifugal compressor with variable high-speed DC drive and magnetic bearings (Danfoss).*

The jet compressor is a dynamic compressor at the other end of the size scale. At present it is a subject of research work and commercial introduction has not occurred. Its use as a way of enhancing the absorption cycle has been successfully tested (Eames, 2005).

4.15 REFRIGERANT SUITABILITY

An indicative summary is given in Table 4.1. The reciprocating type is the most versatile, can be designed for all refrigerants, and many models can be applied with several refrigerants of similar pressures. R410A operates at higher pressures and in the main, is used by scrolls in air conditioning and heating. It is less suitable for screws because the higher pressure differentials. R744 transcritical operation requires a much higher pressure design, with a very small displacement for a given motor size for which reciprocating technology is most suited. Dedicated R744 booster designs operate at similar pressures to R410A bringing in the scroll option.

Ammonia, incompatible with copper, requires open compressors and this exclude scrolls with hermetic motors. There has been some investigation of aluminium for motor windings but size and efficiency are issues. Ammonia also needs extra cooling in low temperature applications. Piston machines commonly use water-cooled heads and screws machines are can be oil cooled.

As may be seen, the main criterion affecting suitability is the pressure range and centrifugal compressors are limited in the pressure difference that can be generated. This makes them most suitable for R134a and the equivalent HFOs.

None of the aforementioned precludes innovative development of any type in any application.

Table 4.1 Indicative summary of refrigerant suitability

Refrigerant type	Recip.	Screw	Scroll	Centrifugal
Medium pressure HFCs and blends (R407 series)	√	√	√	×
High pressure HFCs (R410A, R32)	√	★	√	×
Low pressure HFCs, HFOs, and blends (R134a, R1234 series)	√	√	√	√
R744 Transcritical	√	×	×	×
R744 Booster	√	★	√	×
R717	√	√	★	×

√, Well suited; ★, less suitable (see text); ×, poorly suited.

CHAPTER 5

Oil in Refrigerant Circuits

5.1 INTRODUCTION

The primary purpose of the oil is for compressor lubrication; also sealing and cooling for oil-injected types. The oil specified by the compressor manufacturer should be used whenever possible. Various mineral and synthetic types of lubricant are available and Fig. 5.1 gives some indication as to their suitability with refrigerant type. A blank indicates that the oil type is generally unsuited. It is necessary to emphasise that the oil types shown each represent a family of products, which can be blended to give the required viscosity, and it is necessary to use an appropriate grade or product within the family. For example, polyolester (POE) oils are shown as suitable for hydrocarbons, but a higher viscosity grade will generally be required for hydrocarbons when compared with that used for HFCs. Moreover, specific additive packages to enhance lubricity or to act as inhibitors may be present in branded products, and this is why the compressor manufacturer should be consulted with regards to changing the specified oil for a particular compressor.

5.2 REQUIREMENTS AND CHARACTERISTICS

The properties of the oil must take proper account of its behaviour in the system, namely oil return from the system to the compressor, and the effect of oil on the heat transfer process in the evaporator and condenser. Part of the compressor designer's task is to ensure lubricant fitness for both the compressor and the system.

In the working environment the lubricant is always a mixture of oil and refrigerant and therefore its composition and properties are governed by solubility characteristics that are pressure and temperature dependent. The CFC, HCFC and ammonia refrigerant/mineral oil combinations are backed by many years of experience; their properties are well known. Compressor designers have utilised the combination of viscosity and the excellent boundary lubrication (lubricity) properties of the chlorine-containing refrigerants to good effect in the design of the moving parts. With the advent of HFC refrigerants came the need to move from mineral oils to synthetic

Refrigeration, Air Conditioning and Heat Pumps
http://dx.doi.org/10.1016/B978-0-08-100647-4.00005-X

Refrigerant type	Traditional mineral oil (MO)	Alkyl-benzine (AB)	MO + AB	Polyolester (POE)	Polyalpha-olefin	Poly-alkalene-glycol (PAG)
CFCs and HCFCs	✓	✓	✓	(✓)	(✓)	
HCFC blends	(✓)	✓	✓	(✓)		
HFCs and HFC blends		(✓)		✓		(✓)
HFCs and HFO blends		(✓)		✓		(✓)
Hydrocarbons	✓	(✓)	(✓)	✓	✓	(✓)
Ammonia	✓	(✓)	(✓)		✓	(✓)
CO_2 Carbon dioxide				(✓)		(✓)

✓ Good suitability; (✓) applicable with limitations.

Figure 5.1 ***Lubricant types.***

oils in order to ensure miscibility with the refrigerant and hence adequate oil return from the system. POE oils were chosen for most applications, on the basis of their properties, cost and availability. POEs are made from organic acids and alcohols, which combine to produce esters and water. The formulation of the ester is determined by the original acid structure. As its name implies, a POE is a mixture of esters derived from a mixture of acids.

The behaviour of lubricating oil in a refrigerant circuit and its physical interaction with the refrigerant itself are dominant factors in the design of circuits in general and evaporators in particular. It should be noted that the solubility of ammonia in most lubricants is very low.

A degree of solubility of refrigerant in oil is desirable because viscosity of the mixture in the evaporator is reduced, allowing it to become more mobile, which aids transport back to the compressor. The most important property for compressor operation is the viscosity of the solution for bearing lubrication. As the low-side pressure changes with evaporator temperature the refrigerant concentration changes also and this in turn affects viscosity. A typical behaviour for halocarbon refrigerants is shown in Fig. 5.2.

At low pressure of 1 bar corresponding to evaporation at say 40°C a small amount is refrigerant is dissolved and this has negligible effect on viscosity. With a higher evaporation pressure at, say 6 bar, corresponding to 10°C the oil absorbs 10% refrigerant which effectively reduces the viscosity to half that of the base oil and the bearing load carrying capacity is

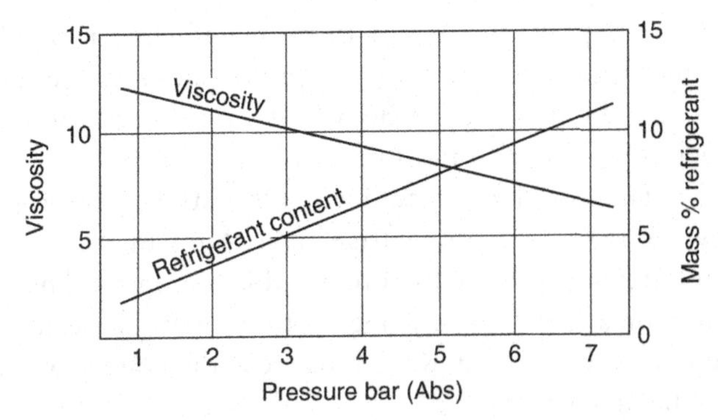

Figure 5.2 ***Typical low-side sump refrigerant content and mixture viscosity.***

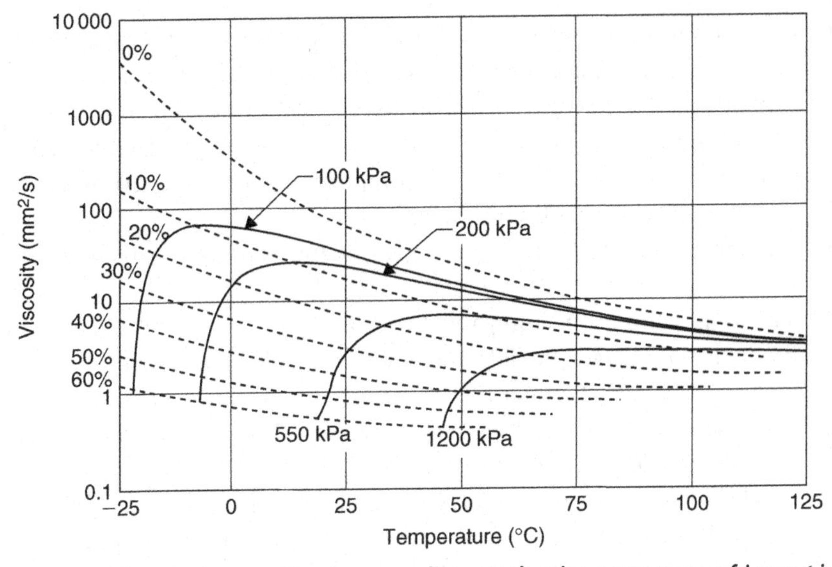

Figure 5.3 ***Viscosity/temperature/pressure diagram showing percentage refrigerant in solution and corresponding viscosity.***

reduced. These effects are studied with the aid of viscosity/temperature/pressure diagrams, the format of which is shown in Fig. 5.3. This diagram is just intended to illustrate the general form of the characteristic and is based on refrigerant R134a and POE oil. Specific data may be found in the *ASHRAE Refrigeration Handbook*.

When liquid lubricants and liquid refrigerants are mixed together and allowed to settle out, a homogeneous mixture may be formed. In this case the pair are said to be *miscible* at the prevailing temperature and pressure.

Alternatively, two separate phases may form; one of which is an oil-rich solution, the other being a refrigerant-rich solution. In most cases the heavier refrigerant-rich solution is at the bottom. This can cause problems with systems where the compressor is in a cold location and refrigerant condenses in the crankcase during shutdown. On start-up, the oil pump will tend to draw a very low-viscosity refrigerant-rich mixture. Crankcase heaters and pump down cycles are used to avoid this problem. This is not the case with ammonia as it generally does not mix with lubricants except in small amounts and the oil tends to accumulate at the bottom of evaporators where it can be drained.

Desired properties for lubricant may be summarised as follows. Many of the characteristics are influenced by the refrigerant and so the oil property cannot be considered in isolation.

1. Adequate lubrication viscosity at the temperatures and pressures in the bearings, and adequate lubricity for sliding contacts.
2. Stability, so that chemical reactions or decomposition do not occur at the conditions to be encountered. Normally the highest temperature and pressure is at the discharge of compression. Resistance to oxidation is measured with a flash point test.
3. The lubricant must be moisture and contamination free, as far as is possible.
4. The lubricant must be compatible with the materials used in the system. Particular points are flexible non-metallic rubber and plastic components such as seals. Copper cannot be used with ammonia.
5. The viscosity of the solution on the low-temperature side should be low enough for adequate oil return.
6. Solids should not be precipitated. Mineral oils can sometimes precipitate waxes at low temperature; this is identified with a floc point test.
7. High electrical resistance is necessary for enclosed motors.
8. Foaming characteristics must be considered.
9. Availability at an acceptable cost is essential.

By way of comment on some of the aforementioned points, it should be noted that the lubrication characteristics in the actual working environment can only be proven by actual experience and/or life tests. It is a tribute to the engineering efforts of compressor builders and system installers that the changeover to HFC refrigerants and POE oils had been a very smooth and trouble-free process. Chemical stability has to be adequate in the presence of moisture and air although the aim is always to exclude these contaminants from the system. Traces are nevertheless present in practice and this is dealt with further. Excessive foaming is undesirable when it is caused by rapid refrigerant release when the compressor is started and the crankcase pressure reduces because it tends to give rise to loss of oil to the system. Some foaming during normal running can assist oil distribution in the compressor and reduce sound level.

5.3 MOISTURE AND AIR CONTAMINATION

In the past the main problem with moisture was ice formation in critical areas such as expansion valves, but it also caused corrosion and damage to motor windings. Lubricants play an important role in determining the effect of contaminants on a refrigeration system. The oil should remain as clean as it is when it entered the compressor (unlike that of the automobile engine which is quickly contaminated by fuel, water, carbon and atmospheric dust). The condition of the compressor oil is therefore a direct indication of the physical and chemical cleanliness of the system. Lubricating oil should be kept in tightly sealed containers to exclude atmospheric moisture. Oil drained from ammonia systems is not used again unless it can be properly filtered and kept dry.

When mineral oils become overheated inorganic halogen acids can be formed and cause damage. Overheating or an electrical fault in the winding of a hermetic or semi-hermetic compressor motor can be the cause. Eye goggles and rubber gloves should be worn when handling such suspect oil. If shown to be acid, the oil must be removed and carefully disposed of and the system thoroughly cleaned out.

Moisture reacts with POE lubricants giving rise to organic acids. This hydrolysis of the ester lubricant can be visualised as the reverse process used in ester manufacture:

$$\text{Acid} + \text{Alcohol} \Leftrightarrow \text{Ester} + \text{Water}$$

The reaction can go either way and the equilibrium composition of the mixture depends on the component proportions and temperature. Addition of water and heat tends to drive the equilibrium point towards the left. The organic acids are much weaker than the halogen acids and this needs to be taken into account when the acidity level is measured using an acid test kit. Nevertheless any breakdown of the POE is undesirable and it is recommended that the moisture level be kept below 50 ppm in POE systems. Because POEs have a high affinity for moisture it is essential to keep exposure of the oil to atmospheric air to an absolute minimum. The moisture cannot

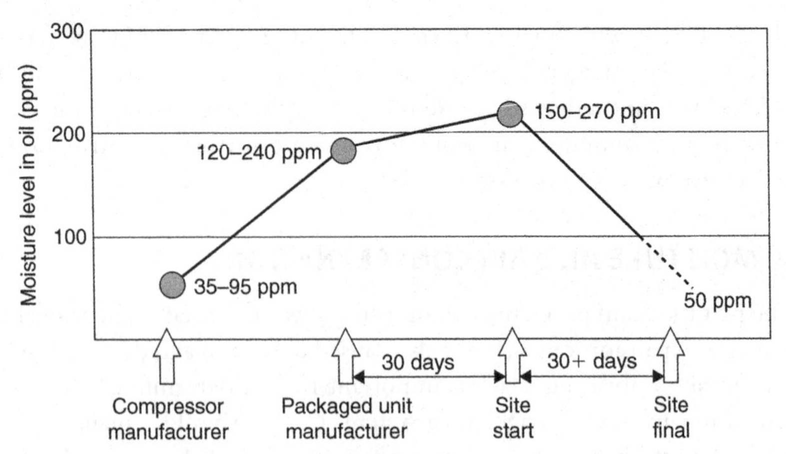

Figure 5.4 ***Typical moisture ingress and subsequent removal by the action of the drier in a POE system.***

be fully removed by vacuum procedures and hence appropriate filter–driers are always recommended. Fig. 5.4 illustrates a typical expected moisture content variance during the process of installing and commissioning. A compressor which is pre-charged with POE oil is connected first to a factory assembled unit, such as a compressor pack, installed on site observing correct evacuation and sealing procedures, and finally with subsequent running the moisture content is reduced towards 50 ppm by the drier.

Proper evacuation will remove the air content to an acceptable minimum. Oxidation and other high-temperature chemical reactions are more likely to cause damage in the presence of moisture.

When systems are changed over from CFCs or HCFCs to HFC refrigerants, the oil also generally needs to be changed to POE. The original equipment manufacturers recommendation should be sought prior to attempting such a conversion.

5.4 OIL SEPARATORS

During the compression stroke of a reciprocating machine, the gas becomes hotter and some of the oil on the cylinder wall will pass out with the discharge gas. Some oil carry-over will occur with all lubricated compressor types, and in small self-contained systems it quickly finds its way back to the compressor. Start-up after a long idle period can result in a large amount of oil carry-over for a short period due to foaming. To reduce the amount of oil

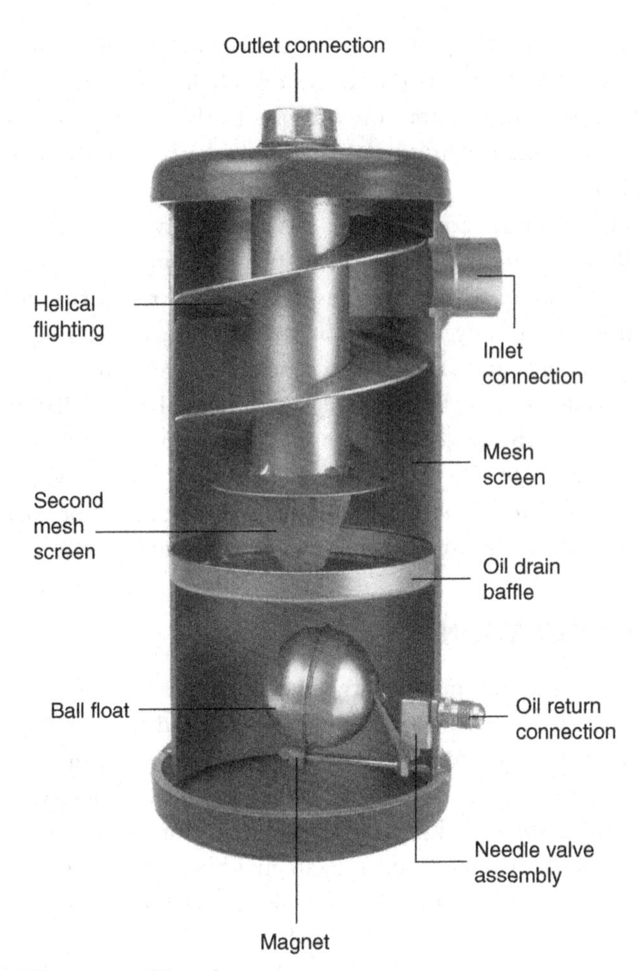

Figure 5.5 ***Oil separator (Henry).***

carried over to the system, particularly where remote low temperature evaporators are in use, an oil separator (Fig. 5.5) can be used. It is located in the discharge line between the compressor and condenser.

The hot entering gas is made to impinge on a spiral to lose much of the oil on the surface by centrifugal force. Some 95–98% of the entrained oil may be separated from the hot gas and fall to the bottom and can be returned to the crankcase. The oil return line is controlled by the float valve, or it may have a bleed orifice. In either case, this metering device must be backed up by a solenoid valve to give tight shut-off when the compressor stops, since the separator is at discharge pressure and the compressor oil sump at suction pressure.

On shutdown, high-pressure gas in the separator will cool and some will condense into liquid, to dilute the oil left in the bottom. When the compressor restarts, this diluted oil will pass to the sump. In order to limit this dilution, a heater is commonly fitted into the base of the separator on large installations.

Oil-injected screw compressors invariably have oil separators and these handle continuous oil carry-over from the injection process. They are frequently built-in to the compressor assembly, particularly with semi-hermetic air-conditioning types. Recirculation back to the injection ports and bearings is continuous. For low-temperature screw compressors the oil is normally cooled during the recirculation process. For installations, which might be very sensitive to accumulations of oil, a two-stage oil separator can be fitted and up to 99.7% of the entrained oil can be removed. However efficient the separation process, a small quantity of oil will be carried over, and the system design must accommodate oil circulation.

5.5 OIL CIRCULATION

Leaving the compressor, the discharge line should be led downwards where possible. This will avoid accumulation of oil or liquid refrigerant in the discharge head of the compressor during idle periods. Where the condenser is above the compressor, it is necessary to have adequate velocity in the line from the compressor, or oil separator, to ensure that oil is carried forward. A non-return valve may be fitted as a safeguard to prevent oil flowing back. Horizontal lines should sloped downwards in the flow direction. Traces of oil, which enter the condenser, will settle on the cooling surfaces and fall to the bottom as a liquid or become dissolved in the condensed refrigerant. Either way, the two liquids will then pass to the expansion valve and into the evaporator. Here, the refrigerant will change to a vapour but most of the oil will remain as a liquid, containing dissolved refrigerant. Slight traces of oil pass out as a low-pressure vapour with the suction gas. It is necessary to limit the build-up of liquid oil in the evaporator, since it would quickly accumulate, reducing heat transfer and causing malfunction.

With ammonia, oil sinks to the bottom and does not go into solution with the refrigerant. Ammonia condensers, receivers and flooded evaporators can be distinguished by the provision of oil drainage pots and connections at the lowest point. Automatic drainage and return of the oil from these would have to depend on small different densities and is very rarely fitted. The removal of oil from collection pots and low-point drains is a periodic manual function and is carried out as part of the routine maintenance. Some ammonia systems are designed to return the oil by maintaining gas velocities, for example direct expansion and liquid overfeed evaporators. The halocarbons are all sufficiently miscible with oil to preclude the possibility of separate drainage and they are in any case rarely used with flooded evaporators.

The most common method of returning oil from the evaporator to the compressor is to keep it moving, by ensuring a minimum continuous fluid velocity in all parts of the circuit by using direct expansion evaporators. This dynamic circulation method is the decisive factor in the design of nearly all halocarbon evaporators.

The critical section of the circuit is where there is no liquid refrigerant left to help move the oil, that is the evaporator outlet and the suction line back to the compressor. Suction lines should be sloped downwards towards the compressor but not as to allow liquid to flood the compressor. A minimum of about 3.5 m/s of gas velocities are required. Where the evaporator is below the compressor vertical sections which provide sufficient entrainment velocities, typically at least 7 m/s are required to ensure that oil droplets will be carried back by the dry refrigerant gas to the compressor. Fig. 11.1 illustrates piping arrangements, which can be used to convey oil upwards.

In some situations a suction accumulator (see chapter: Controls and Other Circuit Components) is used to prevent large quantities of oil and/or refrigerant from suddenly entering the compressor on start-up, or immediately after defrost for example. This system behaviour is termed *liquid slugging* and compressors are invariably designed to handle this to some extent.

With some small cooling circuits which can work at reduced gas flow when capacity controlled, it may not be possible to maintain the minimum velocity to carry oil back to the compressor, and it will settle in the circuit. This is particularly true with speed-controlled compressors where much reduced speed is allowed. Reversing refrigerant flow type circuits (ie cooling/heat pump) are another example of this. Arrangements must be made to increase or reverse the gas flow periodically to move this oil.

5.6 OIL PROPERTIES SUMMARY

The main properties of refrigeration oils which may be quoted by commercial products or test laboratories are given in Table 5.1. These are useful benchmarks, but actual behaviour in a system will be influenced by factors described in this chapter.

Table 5.1 Oil properties summary

Property	Units	Definition
Viscosity	ISO grade	Viscosity (centistokes) at 40°C
Acidity	Total acid number (TAN)	mg potassium hydroxide per g of oil required to neutralise the acidity in the oil.
Moisture	ppm	Proportion of water in the oil
Floc point	°C	Temperature at which wax starts to precipitate from mineral oils
Pour point	°C	Temperature at which the oil ceases to flow
Dielectric strength	kV	The resistance to electric current, that is electrical insulation property, determined by the voltage necessary to cause an arc between poles 0.1" apart.

CHAPTER 6

Condensers and Cooling Towers

6.1 INTRODUCTION

The purpose of the condenser in a vapour compression cycle is to accept the hot, high-pressure gas from the compressor and cool it to remove first the superheat and then the latent heat, so that the refrigerant will condense back to a liquid. In addition, the liquid is usually slightly subcooled. In nearly all cases, the cooling medium will be air or water. For transcritical R744 heat rejection conditions the term *Gas Cooler* is used since R744 does not condense under these conditions.

The topics of Section 6.2–6.11 refer to the main application of condensers, which is to reject unwanted heat from cooling systems. Additional comments on heat pump condensers are in Section 6.13.

Cascade system condensers include a heat exchanger that is the evaporator of the high stage and the condenser for the low stage (see chapter: The Refrigeration Cycle). Construction of this heat exchanger will be a combination of the design factors for evaporators and condensers.

The small condensing surface required by a domestic appliance such as a 'fridge/freezer' may allow the use of the outside metal skin of the appliance itself as a surface condenser. In such a construction, the condenser tube is held in close mechanical contact with the skin, so that heat is conducted through to the outside air, where it is lost by natural convection. This system is restricted to a few hundred watts.

The term *coil* is sometimes referred to a heat exchanger, particularly a condenser, even when the physical shape is not coiled. The context will make the meaning clear.

6.2 HEAT TO BE REMOVED

The heat to be removed in the condenser is shown in the *p–h* diagram (Fig. 6.1). Part of this heat is de-superheating. Apart from comparatively small heat losses and gains through the circuit, the total heat rejection will be

$$\text{Heat taken in by evaporator} + \text{heat of compression.}$$

Refrigeration, Air Conditioning and Heat Pumps
http://dx.doi.org/10.1016/B978-0-08-100647-4.00006-1

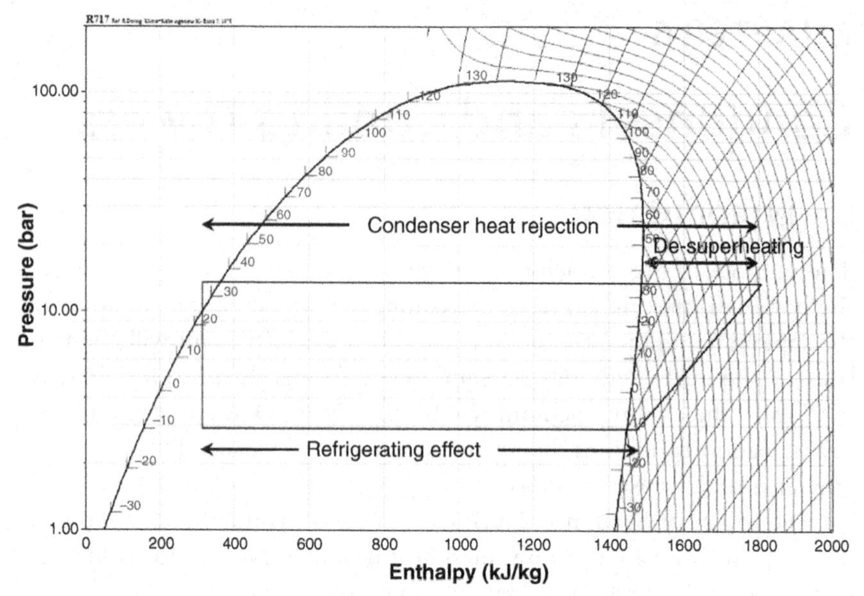

Figure 6.1 ***Condenser load*** **p–h** ***diagram.***

This latter, again ignoring small heat gains and losses, will be the power input to the compressor, giving

$$\text{Evaporator load} + \text{compressor input power} = \text{condenser load}$$

Condenser load is stated as the rate of heat rejection. Some manufacturers give ratings in terms of the evaporator load, together with a 'de-rating' factor, which depends on the evaporating and condensing temperatures.

$$\text{Evaporator load} \times \text{factor} = \text{condenser load}$$

Separate de-superheaters may be applied to recover higher-grade sensible heat prior to the condensation proper. In the R744 transcritical cycle the gas is simply cooled without condensation, but a separate heat recovery unit can reclaim useful heat in both transcritical and subcritical modes. The intercooler of a two-stage or compound system desuperheats the discharge gas from the first stage so that it will not be too hot on entering the high stage. The provision of a separate oil cooler will reduce condenser load by the amount of heat lost to the oil and removed in the oil cooler. This is of special note with oil-injected screw compressors, where a high proportion of the compressor energy is taken away in the oil. The proportion varies with the method of oil cooling, and figures should be obtained from the compressor manufacturer for a particular application.

6.3 AIR-COOLED CONDENSERS

The simplest air-cooled condenser consists of a plain tube containing the refrigerant, placed in still air and relying on natural air circulation. An example is the condenser of the domestic refrigerator, which may also have some secondary surface in the form of supporting and spacer wires.

Above this size, the flow of air over the condenser surface will be by forced convection, that is fans. The high thermal resistance of the boundary layer on the air side of the heat exchanger leads to the use, in all but the very smallest condensers, of an extended surface.

In conventional condensers this takes the form of plate fins, usually aluminium, mechanically bonded onto the refrigerant tubes, usually copper. The ratio of outside to inside surface is normally between 5:1 and 10:1. Flow of the liquefied refrigerant is assisted by gravity, so the inlet will be at the top of the condenser and the outlet at the bottom. Rising pipes should be avoided in the design, and care is needed in installation to get the pipes level. The flow of air may be vertically upwards or horizontal, and the configuration of the condenser will follow from this (see Fig. 6.2). Small cylindrical matrices are also used, the air flowing radially inwards and out through a fan at the top.

Forced convection of the large volumes of air at low resistance leads to the general use of propeller or single-stage axial flow fans. Where a single fan would be too big, multiple smaller fans give the advantages of lower tip speed and noise, and flexibility of operation in winter (see Section 6.11). In residential areas slower-speed fans may be specified to reduce noise levels. A smaller air flow will de-rate the condenser, and manufacturers may give ratings for 'standard' and 'quiet' products.

The low specific heat capacity and high specific volume of air implies a large volume to remove the condenser heat. If the mass flow is reduced, the temperature rise must increase, raising the condensing temperature and pressure to give lower plant efficiency. In practice, the temperature rise of the air is kept between 9 and 12 K. The mass flow of air per kW heat rejection, assuming a rise of 10.5 K, is then

$$\frac{1}{10.5 \times 1.02} = 0.093\,\text{kg/s}$$

where 1.02 is the specific heat capacity of ambient air.

Figure 6.2 ***Air-cooled condenser (Searle).***

As an example of these large air flows required, the condenser for an air-conditioning plant for an office block, having a cooling capacity of 350 kW and rejecting 430 kW, would need 40.85 kg/s or about 36 m^3/s of air. This cooling air should be as cold as possible, so the condenser needs to be mounted where such a flow of fresh ambient air is available without recirculation.

The large air flows needed, the power to move them, and the resulting noise levels are the factors limiting the use of air-cooled condensers.

As the condenser load increases the temperature difference between the air inlet (ambient) temperature and the condensing temperature will increase in order to reject heat at a faster rate with the same surface. This is with a constant air flow. A condenser rating, kW/K, where the condenser load is in kW and the K is the temperature difference, can be considered to be constant, as a first approximation.

Example 6.1

A condenser is sized to reject 12 kW heat at a condensing temperature of 50°C when the maximum outdoor temperature is 35°C, what is the rating and what will be the approximate condensing temperature when the outdoor temperature is at 15°C and the load is reduced to 8 kW?

$$\text{Condenser rating} = \frac{\text{Load}}{\text{Temperature difference}} = \frac{12}{15} = 0.8\,\text{kW/K}$$

$$\text{Temperature difference at } 15°\text{C} = \frac{8}{0.8} = 10$$

$$\text{Condensing temperature at } 15°\text{C} = 15 + 10 = 25°\text{C}$$

The condenser must be sized to meet the design load at the maximum ambient condition, but during typical running conditions with the air temperature at 15°C, the load will fall because the cooling load will tend to be less, and the compressor power will certainly be less. A condenser, which may appear to be small and to require a high condensing temperature at the design condition, balances out to give an acceptable condensing temperature most of the time.

Materials of construction for finned tube condensers can be aluminium fins on stainless steel tube for ammonia, or aluminium or copper fins on copper or aluminium tube for other refrigerants.

In recent years, coil technology has increasingly focused on microchannel technology. Microchannel condenser coils use all aluminium brazed fin construction. The coil is composed of three components: the flat microchannel tube, the fins located between the microchannel tubes, and two refrigerant headers. Please see Fig. 6.3. This approach originated in the car industry in the 1980s and development of the manufacturing techniques has allowed larger units suitable for commercial systems to be economically constructed. The components are joined together into a single coil using a nitrogen-charged brazing furnace. Product quality and integrity are maximised since only one braze is required compared to 200 or 300 manually brazed connections with traditional copper/aluminium coils. Circuiting is accomplished by placing baffles in the distribution headers to feed the refrigerant through the flat tubes. Microchannel condensers can offer the following advantages:

- *Thermal performance*: It is significantly better than a standard, Al/Cu coil and is obtained by the flat tubes, which maximise airside heat transfer, and the microchannels within the tubes maximise refrigerant side heat

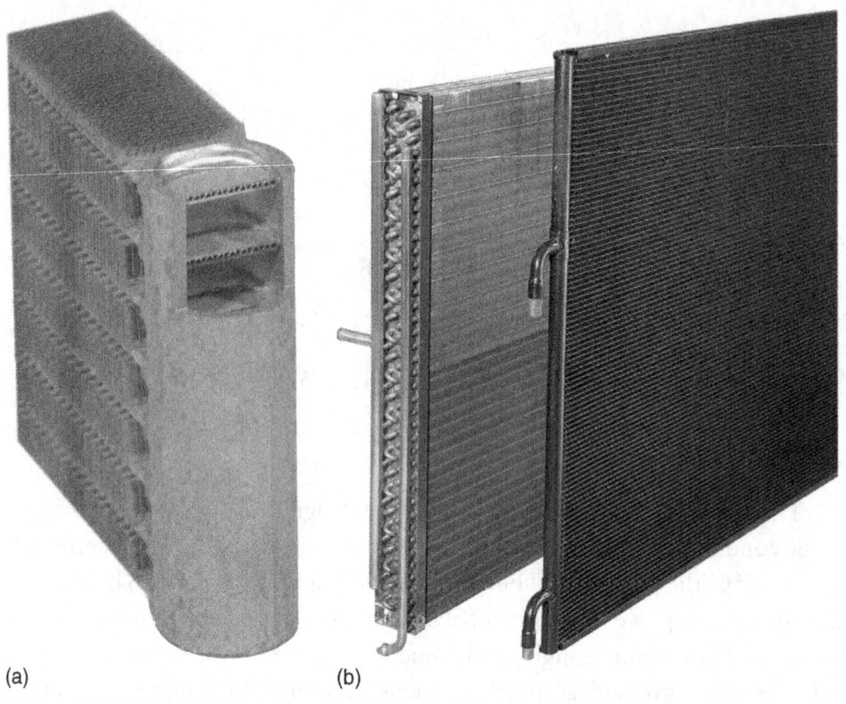

(a) (b)

Figure 6.3 (a) Microchannel condenser cut away view, (b) microchannel condenser (right) alongside conventional finned tube condenser (Airedale).

transfer. The tiny refrigerant channels provide increased primary surface area. In addition, the metallurgical fin-tube bond resulting from the braze operation maximises surface contact and increases the heat transfer surface area, further improving the heat transfer performance.

- *Corrosion protection*: There is less potential for corrosion as there are no dissimilar metals to initiate galvanic processes.
- *Reduced refrigerant charge*: The condenser charge can be less than 50% of a conventional unit, resulting in cost savings and lower system TEWI values.
- *Durability*: Microchannel coil construction is more durable and provides significant weight reduction. The likelihood for leaks is significantly reduced. They can be repaired using a two-part epoxy process. Coils with thickness less than 1 in. allow easy removal of any debris. They can be washed with a high-pressure sprayer without bending the fins.

The reason for the slow take up of microchannel is cost. Because a microchannel unit is 'cast' in a defined block it lends itself to a few small, standard sizes and high-volume production. This is why the microchannel became commercially viable in the car industry. The commercial RACHP markets demand far fewer units per year with different configurations. This affects the volume equation, and subsequent cost per unit in a significant way.

In view of the high material cost for air-cooled condensers, a higher temperature difference than for water cooled is usually accepted, and condensing temperatures may be 5–8 K higher for a given cooling medium temperature. Microchannel technology can give several degrees reduction compared to conventional coils.

Air-cooled condensers are very widely used in sizes ranging from a few kW to several hundred kW. They can be seen as wall mounted fan-coil units on air conditioners and on large roof mounted systems (Fig. 6.4). They must, of course, be used on land transport systems and in areas where the supply of cooling water is unreliable.

It is frequently necessary to vary the air flow, for example to ensure that condensing pressure does not fall too low for proper control of the low side system, to reduce sound levels at night time, or to reduce the fan power required under low ambient conditions. The control parameter is usually condensing pressure, and an intelligent controller will reduce fan power when this reduces the total power consumption, including that of the compressors. The air flow reduction can be achieved by switching off fans on multiple fan units or by varying fan speed. Single-phase fan motors can often be speed controlled with a simple pressure-sensing controller that varies the voltage to the motor. A hot coil will induce an air flow even with the fan idle. Large condensers may be arranged in two or more sections to

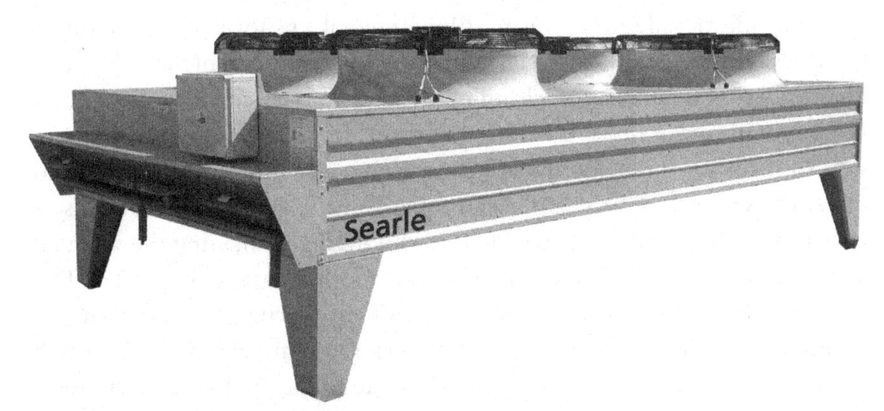

Figure 6.4 ***Multiple section air cooled condenser (Searle).***

overcome over-capacity situations. The effective size of the condenser is reduced by shutting off the appropriate section.

Arranging the coil in sections allows the condenser to serve more than one refrigeration system. They can have different operating conditions or refrigerants. Most manufacturers offer units with two rows of fans: a two-section coil can be incorporated for this purpose.

6.4 WATER-COOLED CONDENSERS

The higher heat capacity and density of water make it an ideal medium for condenser cooling and, by comparison with the 350 kW plant cited previously, the flow is only 9.8 L/s. Small water-cooled condensers may comprise two concentric pipes (double pipe), the refrigerant being in either the inner tube or the annulus. Configurations may be straight, with return bends or headers, or coiled. The double-pipe condenser is circuited in counterflow (media flowing in opposite directions) to get the most sub-cooling, since the coldest water will meet the outgoing liquid refrigerant.

Larger sizes of water-cooled condenser require closer packing of the tubes to minimise the overall size, and the general form is shell-and-tube, having the water in the tubes (Fig. 6.5). This construction is a very adaptable mechanical design and is found in all sizes from 100 mm to 1.5 m diameter and in lengths from 600 mm to 6 m, the latter being the length of commercially available tubing. Materials can be selected for the application and refrigerant, but all mild steel is common for fresh water, with cupronickel or aluminium brass tubes for salt water.

Some economy in size can be effected by extended surfaces on the refrigerant side, usually in the form of low integral fins formed on the tubes. On the water side, swirl strips can be fitted to promote turbulence, but these interfere with maintenance cleaning and are not much in favour. Water velocity within the tubes is of the order of 1 m/s, depending on the bore size. To maintain this velocity, baffles are arranged within the end covers to direct the water flow to a number of tubes in each 'pass'. Some condensers have two separate water circuits (double bundle, Fig. 6.6), using the warmed water from one circuit as reclaimed heat in another part of the system. The main bundle rejects the unwanted heat. Where the mass flow of water is unlimited (sea, lake, river or cooling tower), the temperature rise through the condenser may be kept as low as 5 K, since this will reduce the head pressure at the cost only of larger water pumps and pipes.

Figure 6.5 *Shell-and-tube condenser (Titan).*

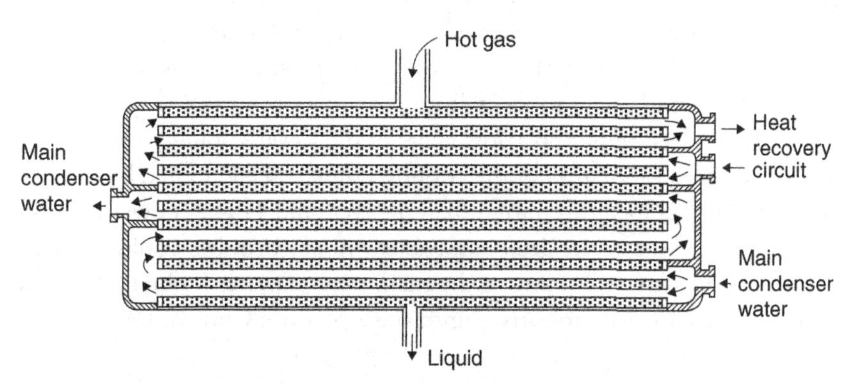

Figure 6.6 *Double-bundle shell-and-tube condenser.*

Example 6.2

A condenser uses water from a river with a temperature rise of 5.2 K. Total duty at the condenser is 930 kW. How much water flow is required?

$$\frac{930}{5.2 \times 4.187} = 43\,\text{kg/s}$$

Brazed plate heat exchangers are used as condensers and can be a lower cost alternative to shell and tube. The construction and characteristics are discussed in chapter: Evaporators, and illustrated in Fig. 7.9. Because the refrigerant volume is small, they can work with a lower charge. To be fully effective a BPHX needs to be kept fully drained into a liquid receiver.

The supply of water is usually limited and requires the use of a cooling tower. Other possibilities are worth investigation; for example, in the food industries, large quantities of water are used for processing the product, and this could be passed first through the condensers if precautions are taken to avoid contamination. Also, where ground water is present, it could be taken from a borehole and afterwards returned to the ground at some distance from the suction. In both these cases, water would be available at a steady temperature and some 8–10 K colder than summer water from a cooling tower.

6.5 COOLING TOWERS AND DRY COOLERS

A cooling tower is the final link in the process of heat rejection to the atmosphere for many water cooled systems. An alternative that avoids the need for water treatment is a *dry cooler* that uses a sealed water system. They are of similar construction to air cooled condensers, illustrated in Fig. 6.2. The fluid is circulated through the tubes of the heat exchanger, whilst blowing air over the outside. The tubes will usually have an extended surface. The use of dry coolers cannot take advantage of the lower cooling temperatures available by the evaporation of the cooling water, and are limited by the ambient dry bulb temperature, rather than the wet bulb. Higher power is therefore required, and a given size compressor will perform less cooling duty.

In a cooling tower, cooling of the main mass of water is obtained by the evaporation of a small proportion into the airstream. Cooled water leaving the tower will be 3–8 K warmer than the incoming air *wet bulb* temperature (see also chapters: Air Treatment Fundamentals; Practical Air Treatment).

The quantity of water evaporated will take up its latent heat equal to the condenser duty, at the rate of about 2430 kJ/kg evaporated, and, per kW heat rejection, will be approximately

$$\frac{1}{2430} = 0.41 \times 10^{-3} \text{ kg/s}$$

For a condenser load of 400 kW, evaporation would be at a rate of 0.16 kg/s.

Cooled water from the drain tank is taken by the pump and passed through the condenser. The warmed water then passes back to sprays or distribution troughs at the top of the tower and falls in the up-going airstream, passing over packings which present a large surface to the air. Evaporation takes place, the vapour obtaining its latent heat from the body of the water, which is therefore cooled (see Fig. 6.7).

The power consumption of the tower can be reduced by fan cycling or fan speed control under light load conditions. An induced draught tower in which the fan is in the outlet air stream can achieve 10–15% performance

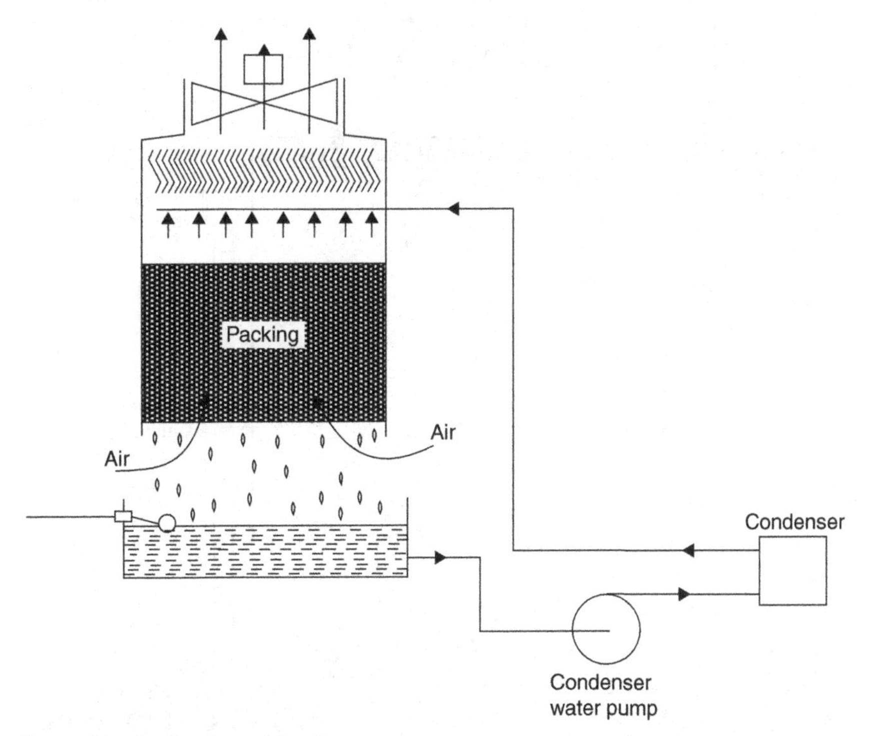

Figure 6.7 ***Cooling tower circuit.***

with the fan idle. This is not the case with a forced draught tower where the fan is located in the inlet air stream.

6.6 EVAPORATIVE CONDENSERS

The evaporative condenser can be an efficient method of heat rejection. The heat is rejected at a lower temperature than with simple air-cooling. However this advantage is only realised above a certain size and loading conditions. Clark and Gillies (2014) have shown that operating costs can be higher than for air cooled, taking account of cost of water supply, disposal and treatment.

The cooling effect of the evaporation of water is applied directly to the condenser refrigerant pipes in the evaporative condenser (Figs 6.8 and 6.9). The mass flow of water over the condenser tubes must be enough to ensure wetting of the tube surface, and is of the order of 80–160 times the quantity evaporated. The mass flow of air must be sufficient to carry away the water vapour formed, and a compromise must be reached with expected variations in ambient conditions. An average air mass flow per kW heat rejection is 0.06 kg/s.

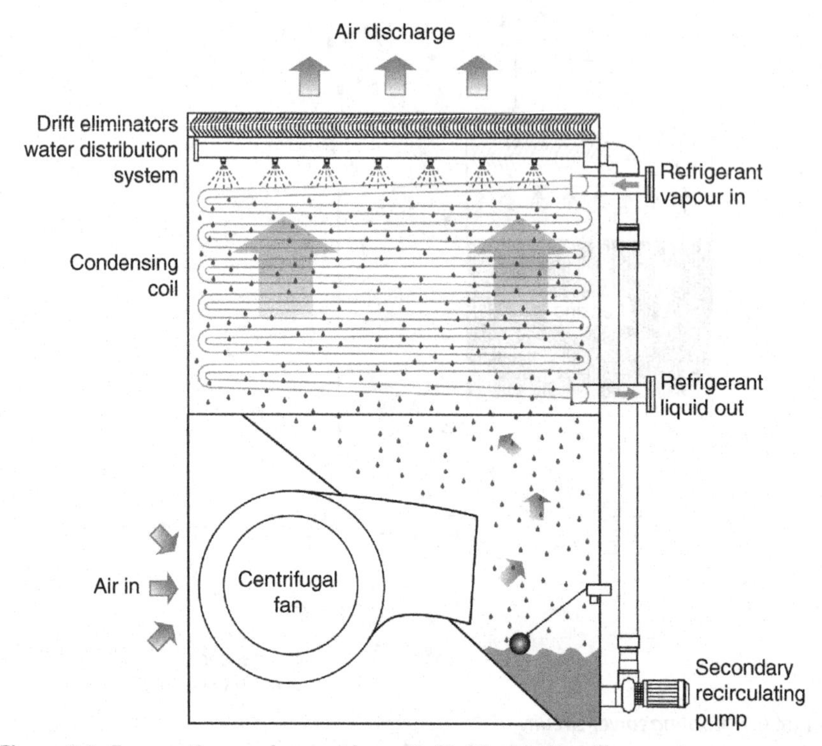

Figure 6.8 ***Evaporative condenser schematic (Baltimore Aircoil).***

Figure 6.9 ***Evaporative condenser installation (Baltimore Aircoil).***

Example 6.3

An evaporative condenser rated at 880 kW and the water-circulating pump takes another 15 kW. What will be the evaporation rate, the approximate circulation rate, and the air mass flow?

$$\text{Total tower duty} = 880 + 15$$
$$= 895\,\text{kW}$$
$$\text{Evaporation rate} = 895 \times 0.41 \times 10^{-3} = 0.37\,\text{kg/s}$$
$$\text{Circulation rate, 80 times} = 30\,\text{kg/s}\ (\Delta T = 7.1\,\text{K})$$
$$160\,\text{times} = 60\,\text{kg/s}\ (\Delta T = 3.6\,\text{K})$$
$$\text{Air flow} = 895 \times 0.06 = 54\,\text{kg/s}$$

It will be seen that the water and air mass flow rates over an evaporative condenser are roughly equal.

Evaporative condensers have a higher resistance to air flow than cooling towers and centrifugal fans are often used, ganged together to obtain the required mass flow without undue size. This arrangement is also quieter in operation than axial flow fans. Most types use forced draught fans.

Cooling towers and evaporative condensers may freeze in winter if left operating on a light load. A common arrangement is to switch off the fan(s) with a thermostat, to prevent the formation of ice. The water-collection tank will have an immersion heater to reduce the risk of freezing when the equipment is not in use or the tank may be located inside the building under the tower structure, if such space is conveniently available.

Materials of construction must be corrosion resistant. Steel should be hot galvanised, although some resin coatings may suffice. GRP casings are used by some manufacturers. The water-dispersal packing of a cooling tower is made of treated timber or corrugated plastic sheet.

6.7 WATER TREATMENT

All water supplies contain a proportion of dissolved salts. These will tend to be deposited at the hottest part of the system, for example the furring of a kettle or hot water pipes. Also, these impurities do not evaporate into an airstream, so where water is being evaporated as part of the cooling process, the salts will remain in the circuit and increase in concentration, thus hastening the furring process.

It is possible to remove all solids from the make-up water, but it is much cheaper to check the concentration by other means. Two general methods are employed. The first relies on physical or chemical effects to delay deposition of scale on the hot surfaces; the second restricts the concentration to a level at which precipitation will not occur. In both cases, the accumulation of solids is removed by bleeding off water from the circuit to drain, in addition to that which is evaporated (see Fig. 6.10).

The concentration of solids in the circulating water will increase until the amount carried away by the bleed water compensates for that not carried away in the water vapour. So, if

c_m = concentration of solids in make-up water (kg/kg)
c_b = concentration of solids in bleed-off (kg/kg)
w_e = mass flow of water evaporated (kg/s)
w_m = mass flow of make-up water (kg/s)

Mass of solids entering = mass of solids leaving

$$c_m \times w_m = c_b \times (w_m - w_e)$$

$$w_m = w_e \left(\frac{c_b}{c_b - c_m} \right)$$

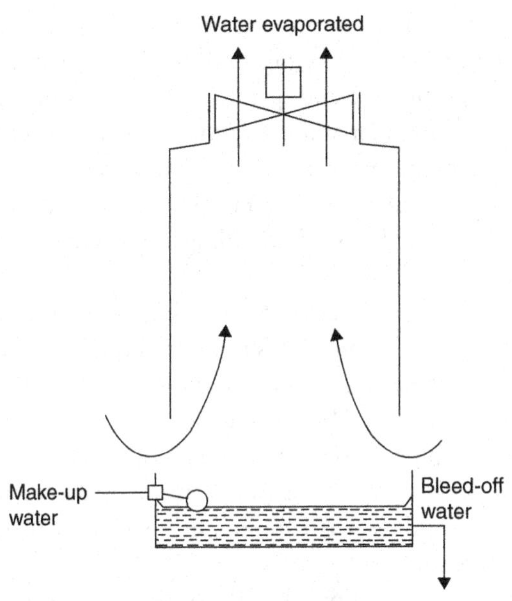

Figure 6.10 ***Limitation of solids concentration by bleed-off.***

The concentration of mains make-up water, c_m, is obtained from the water supply authority. The permissible concentration, c_b, will be decided by the method of water treatment or the assumed concentration of untreated water which will prevent precipitation.

Example 6.4

If water hardness is 560 ppm (parts per million), the water treatment can permit a concentration of solids to 1200 ppm, the cooling capacity is 700 kW and the compressor power 170 kW, how much water should be bled to waste and what is the total make-up required?

$$\text{Cooling tower capacity} = 700 + 170 = 870\,\text{kW}$$

$$\text{Latent heat of water vapour} = 2420\,\text{kJ/kg}$$

$$\text{Rate of evaporation} = \frac{870}{2420} = 0.36\,\text{kg/s}$$

$$\text{Rate of make-up} = 0.36\left(\frac{0.0012}{0.0012 - 0.00056}\right) = 0.68\,\text{kg/s}$$

$$\text{Rate of bleed-off} = 0.68 - 0.36 = 0.32\,\text{kg/s}$$

In all cases where water is used for cooling, but more especially where it is being evaporated, the hardness figure should be obtained from the local water supply authority. Enquiries should also be made as to possible variations in the supply, since many cities draw their water from two or more catchment areas, and the type and quantity of hardness may change.

Many suppliers now offer water treatment for use in refrigeration condenser circuits, and the merits of different methods need to be assessed before making a choice. The reader is referred to specialist works on the subject.

There are several methods of providing a percentage 'bleed-off' from the water circuit:

1. The make-up ball valve can be set a little high so that some water always goes down the overflow pipe. This is rather difficult to set initially, but is reliable and cannot easily be tampered with. It will work at all times, and so will waste water if the plant is not running.
2. A small bleed-off pipe is taken from the pump discharge, with an adjusting valve, and led to waste. This can be more easily adjusted and works only when the condenser is running, but is subject to interference by unauthorised persons.
3. A tundish, having an area possibly 1% of the cross-sectional area of the tower, is located just above the water level and is led to the drain, forming part of the overflow fitting. This will bleed off 1% of the water falling through the tower.

All these methods provide the maximum required rate of bleed-off at all times of the year, and so will waste water at light load conditions. The user should be aware of the essential nature of bleed-off, since cases often occur in dry weather of misguided persons closing off the bleed to 'save water'.

In some locations, it is necessary to drain the tank frequently to clear other contaminants. With careful control, this can be used as the necessary bleed-off.

6.8 CONDENSER MAINTENANCE

As with any mechanical equipment, condensers should never be located where they are difficult of access, since this encourages omission of routine maintenance. Periodic maintenance of a condenser is limited to attention to the moving parts – fans, motors, belts, pumps – and cleaning of water filters, if fitted.

The overall performance will be monitored from the plant running log (see chapter: Commissioning and Maintenance) and the heat exchange surfaces must be kept clean for maximum efficiency – meaning the lowest head pressure and lowest power.

Air-cooled surfaces may be cleaned by brushing off the accumulation of dust and fluff where the air enters the coil, by the combination of a high-pressure air hose and a vacuum cleaner, or, with the obvious precautions, by a water hose. Foaming detergents are also used.

Advance warning should be had from the plant running log of any build-up of scale on water-cooled surfaces. Scale within the tubes of a straight double-pipe or shell-and-tube condenser can be mechanically removed with suitable wire brushes or high-pressure water lances, once the end covers have been removed. Tubes which cannot be dealt with in this way must be chemically cleaned (see also chapter: Commissioning and Maintenance).

It will be appreciated that, where air and water are present, as in a water cooling tower or evaporative condenser, the apparatus will act as an air washer, removing much of the dust from the air passing through it. Such dirt may be caught in a fine water filter, but is more commonly allowed to settle into the bottom of the tank and must be flushed out once or twice a year, depending on the severity of local contamination. Where heavy contamination is expected, it is good practice to provide a deeper tank than usual, the pump suction coming out well clear of the bottom, and tanks 3 m deep are in use. Where plant security is vital, the tank is divided into two parts, which may be cleaned alternately.

Algae and other organisms will tend to grow on wet surfaces, in particular those in daylight. Control of these can be effected by various proprietary chemicals.

6.9 LEGIONELLA

Cooling towers and evaporative condensers tend to be looked on unfavourably because they are inevitably linked with legionella. Legionnaires' disease is caused by inhalation small droplets of water, such as may be dissipated from these heat exchangers when they are contaminated by legionella bacterium. The bacterium is very common and may be found in mains water so there is no chance of preventing their access. It is only when conditions allow rapid multiplication that there is any risk. Because of the need for mandatory inspection and fear of forced shut down on suspicion that the unit may have caused an incidence of Legionnaires' disease, there has been a move towards air-cooled units. Controls to reduce the risk of legionella are effective in United Kingdom, but nevertheless there remains a need for vigilance (Oughton, 1987). A type of evaporative condenser which does not require a water reservoir, known as a 'once-through' type, can eliminate the risk. Developments are reported by Pearson (2006).

6.10 DESIGN CONSIDERATIONS

The inlet pipe bringing high-pressure gas from the compressor must enter at the top of the condenser, and adjacent piping should slope in the direction of flow so that oil droplets and any liquid refrigerant which may form will continue in the right direction and not back to the compressor.

The outlet pipe must always be from the lowest point, but may have a short internal upstand so that any dirt such as pipe scale or metal swarf will be trapped and not taken around the circuit.

Condensers for ammonia systems may have an oil trap, usually in the form of a drain pot, and the liquid outlet will be above this.

Water connections to a shell-and-tube condenser must always be arranged so that the end covers can easily be removed for inspection, cleaning, and repair of the tubes. Heavy end covers require the use of lifting tackle, and supports above the lifting points should be provided on installation to facilitate this work.

Condensers contain pressurised refrigerant and where they exceed certain volumes they will be subject to the requirements of the Pressure Vessel Directive (PED) and EN378. Manufacturers will be aware of these requirements, and proprietary products will be correctly equipped.

6.11 LOW AMBIENT OPERATION

Condensers are sized so that they can reject the system heat load under maximum conditions of air or water temperature. In colder weather, the condensing temperature will fall with that of the cooling medium and this may cause system control difficulties. In particular, the pressure across the expansion valve (see also chapter: Expansion Valves) may be too low to circulate the required mass flow of the refrigerant. Under such circumstances, artificial means must be used to keep the head pressure up, always remembering that the condensing pressure should be kept as low as practical for power economy.

Various systems are used which are given as follows:

1. Air-cooled condensers having two or more fans (Fig. 6.4) may have a pressure switch or thermostatic control to stop the fans one by

one as mentioned in Section 6.3. This method is simple, cheap and effective.

2. The fans on such condensers may be fitted with two-speed motors or other speed control. It should be borne in mind that, if one fan of a pair stops, the noise level will fall by 3 dB, but if both fans drop to half speed, the noise drops by 15 dB. This method is of special use in residential areas where the greater noise level will be tolerated in the daytime when condensing air is warmest, but a lower fan speed can be used at night.
3. Evaporative condensers and water cooling towers with two or more fans on separate drive may be controlled in the same way. If a single motor drives several fans on one shaft, speed control or dampers will be required. Evaporative condensers and cooling towers should be fitted with antifreeze thermostats which will stop all fans before the water reaches freezing point.
4. Cooling air flow can be restricted by blanking flaps, baffles or winter enclosures, providing that the air flow is restored when required.
5. Water flow may be restricted by throttling valves. One such device is operated directly by head pressure, but electric or pneumatic throttling or flow diversion valves can be applied for the purpose (see chapter: Controls and Other Circuit Components).
6. A set pressure bypass valve can be fitted across the condenser, so that hot gas will pass directly to the receiver in cooler weather. This will cause the condenser to partially fill with liquid refrigerant, thus decreasing the heat transfer surface available for condensation. Sufficient refrigerant must be available for this, without starving the rest of the circuit (see chapter: Controls and Other Circuit Components).
7. Where a complex system is served by two or more condensers, a complete condenser can be taken off line by a pressure switch.
8. Liquid pressure may be increased by using a liquid pump downstream of the receiver. This is sometimes termed liquid pressure amplification.

Apart from such requirements for head pressure control, winter precautions are needed to prevent freezing of the water whilst the plant is not rejecting heat to it. These commonly take the form of an electric immersion heater in the water tank, together with lagging and possible trace heating of exposed pipes. In some systems, the evaporative condenser itself may be within the building, with air ducts to the outside. In severe climates, external tanks need to be lagged to conserve the heat provided by the immersion heater.

6.12 RECEIVERS

A liquid receiver will be required if it is necessary to temporarily store refrigerant charge within the system, or to accommodate the excess refrigerant arising from changing operating conditions. The total refrigerant charge required in a circuit will vary with different operating loads and ambient temperatures, and must be sufficient at all times so that only liquid enters the expansion valve.

A receiver requires a minimum operating charge which adds to overall charge and cost, and also increases system complexity. Hence receivers are avoided on many smaller systems.

A typical receiver suitable for a large system is shown in Fig. 6.11.

Receivers also act as pump-down reservoirs, and should be capable of holding enough of the total refrigerant charge to permit evacuation of any one vessel for maintenance, inspection or repair. They should never be more than 85% full, to allow for expansion and safety.

Receivers are commonly made of steel tube with welded dished ends, and are located horizontally. Small receivers may be vertical, for convenience of location. The liquid drain pipe from the condenser to the receiver should be amply sized, and any horizontal runs sloped to promote easy drainage. Shut-off valves in this line should not be in a horizontal outlet from the condenser, since their slight frictional resistance will cause liquid back-up in the condenser. Outlet pipes from the receiver may be from the bottom or, by means of an internal standpipe, may leave at the top. A valve is invariably fitted at this point.

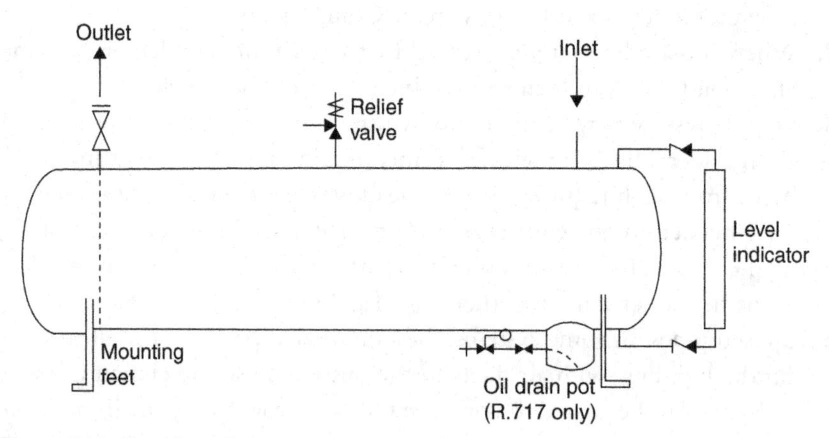

Figure 6.11 ***Liquid receiver.***

Ammonia receivers may have an oil drum pot, and the receiver will slope slightly down towards this.

Receivers are pressure vessels covered by the provisions of EN378 and require appropriate safety pressure relief devices. In cases where there is no shut-off valve between the condenser and receiver, such protection may be fitted to one or the other, providing the total volume is considered.

In practice, receivers will operate about one-sixth full during normal running. Some means are usually provided to indicate the liquid level inside. A few examples are as follows:

1. An external, vertical sight glass, of suitable pattern, having self-closing shut-off valves.
2. A number of sight glasses arranged at different heights in the shell.
3. A pair of sight glasses, arranged on the same cross-section and some 45° up from the horizontal diameter. A light is shone through one and the observer looks through the other.

A receiver should ensure that the condenser is fully drained and the entire surface is available for condensation, that is no back up of liquid in the condenser. Holding of liquid in a condenser can give rise to useful sub-cooling, but with a receiver the liquid at the outlet will be saturated. This is because both liquid and gaseous phase are present in the vessel.

6.13 HEAT-PUMP CONDENSERS

Heat pump condensers are designed for delivery of air or water for heating purposes. Direct air heating takes the form of indoor units, and in most cases they are part of reversible or VRF systems covered elsewhere. Many UK requirements are for water heating, both for space heating and domestic hot water (DHW). These are sometimes referred to as hydronic systems. The condensers for these systems are commonly plate heat exchangers sized to provide water at the desired flow temperature.

It is desirable for the condensing temperature to be as low as possible in order maximise the heating COP and also, in the case of DHW to maximise the water delivery temperature within the maximum refrigerant pressure and compressor discharge temperature limits. The following example is used to illustrate possible restrictions.

Condenser temperature profiles with R410A and R407C, a glide refrigerant, are shown in Fig. 6.12. The condenser is to heat water from 30 to 35°C. The water temperature follows the lower line from left to right as it passes through the exchanger whilst the refrigerant follows the upper line

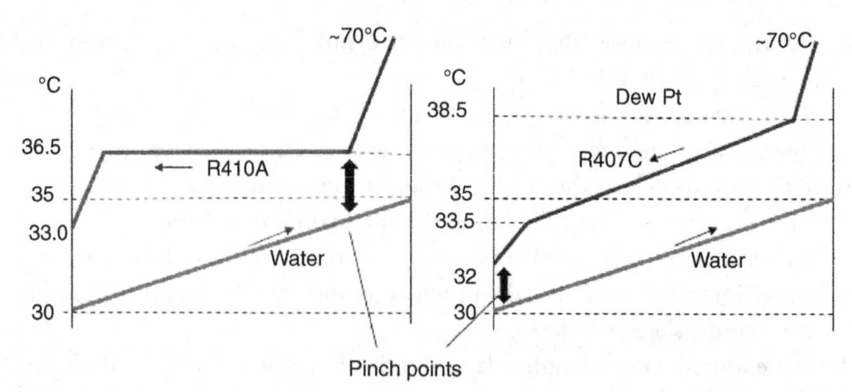

Figure 6.12 ***Temperature profile example for R410A and R407C (not to scale).***

in a counter flow manner from right to left. The lines show refrigerant entering at relatively high temperature, de-superheated, condensing and subcooled prior to leaving. With R410A minimum condensing takes place at 36.5°C. Anything below that could compromise the temperature difference where condensation starts. This is sometimes referred to as a *pinch point*. The pinch point is at the water outlet. With R407C minimum condensing dew point is 38.5°C. Anything below that could compromise the temperature difference where the refrigerant leaves the condenser, allowing for a small amount of sub-cooling. The pinch point is at the water inlet. This type of analysis can be useful in setting the controls for maximum efficiency, and in the decision to use a receiver.

CHAPTER 7

Evaporators

7.1 INTRODUCTION

The purpose of the evaporator is to receive low-pressure, low-temperature fluid from the expansion valve and to bring it in close thermal contact with the load. The refrigerant takes up its latent heat from the load and leaves the evaporator as a dry gas. Evaporators are classified according to their refrigerant flow pattern and their function.

The flow pattern can be one of two types. Either the refrigerant flows continuously through the heat exchanger whilst it evaporates and becomes superheated, or alternatively it resides in a vessel at low pressure whilst it evaporates or from which it is taken to individual coolers, returning as liquid/vapour mixture. The most common type by far is the continuous flow type, referred to as a direct expansion evaporator.

In cooling applications the function of the evaporator is to cool air or liquid in almost all cases. In heating applications it is to extract heat from a low-grade heat source. The cooled air or liquid can then cool the load. For example in a refrigerated display cabinet, the air is cooled and circulated to keep the contents at the required temperature; in a water chiller system, the water is circulated to individual fan-coil units to provide air conditioning.

7.2 AIR COOLING EVAPORATORS

Air cooling evaporators for display cases, coldrooms, blast freezers, air conditioning and heat pumps have finned pipe coils (see Fig. 7.1). In all but very small coolers such as domestic and small retail units, fans blow the air over the coil.

Construction materials are the same as for air-cooled condensers. Aluminium fins on copper tube are the most common for the halocarbons, with stainless steel or aluminium tube for ammonia. Frost or condensed water will form on the fin surface and must be drained away. To permit this, fins are vertical and the air flow horizontal, with a drain tray provided under.

The size of the tube will be such that the velocity of the boiling fluid within it will cause turbulence to promote heat transfer. Tube diameters will vary from 9 mm to 32 mm, according to the size of coil.

Refrigeration, Air Conditioning and Heat Pumps
http://dx.doi.org/10.1016/B978-0-08-100647-4.00007-3

Figure 7.1 ***Air cooling evaporators. (a) Floor mounted, (b) ceiling mounted (Searle).***

Fin spacing will be a compromise between compactness (and cost) and the tendency for the interfin spaces to block with condensed moisture or frost. Spacings will vary from 2 mm on a compact air-conditioner to 12 mm on a low-temperature coldroom coil.

Microchannel evaporators are starting to become available. They are similar to those used for condensers (Section 6.3) and can offer similar advantages.

All direct expansion types are susceptible to mal-distribution of either refrigerant or air flow which leads to reduced heat transfer effectiveness and thus lower system efficiency. It is a normal practice for the refrigerant, after leaving the expansion valve, to enter a distributor consisting of a number of individual feeds to various sections of the coil. Good air flow is maintained by cleaning and defrosting. Microchannel evaporators with many parallel channels fed by a common header are especially susceptible to refrigerant mal-distribution. A study of this topic is given by Bowers et al. (2012).

7.3 LIQUID COOLING EVAPORATORS

Liquid cooling evaporators may be direct expansion or flooded type. Ground source heat pumps usually use brine in a ground collector. The evaporator cools the brine, when it sent back to the ground again to gather heat. This is dealt with more fully in chapter: Heat Pumps and Integrated Systems. Flooded evaporators (Fig. 7.2) have a body of refrigerant boiling in a random manner, the vapour leaving at the top. In the case of ammonia,

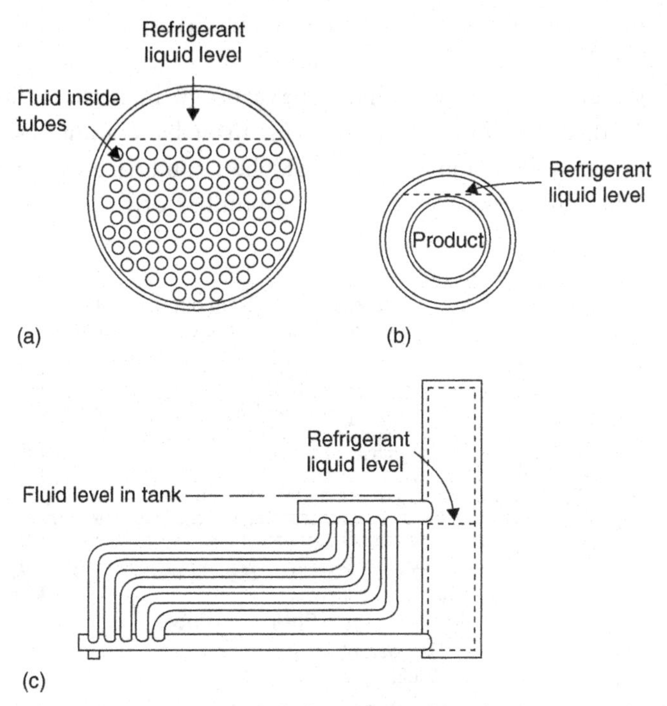

Figure 7.2 ***Flooded evaporators. (a) Shell-and-tube, (b) jacketed, (c) raceway.***

any oil present will fall to the bottom and be drawn off from the drain pot or oil drain connection.

In the shell-and-tube type, the liquid to be cooled is usually in the pipes and the shell is some three-quarters full of boiling refrigerant, (Fig. 7.2a). A number of tubes is omitted at the top of the shell to give space for the suction gas to escape clear of the surface without entraining liquid. Further features such as multiple outlet headers, suction trap domes and baffles will help to avoid liquid droplets entering the main suction pipe. Gas velocities should not exceed 3 m/s and lower figures are used by some designers.

A sectional arrangement of a flooded shell and tube type is shown in Fig. 7.3. The speed of the liquid within the tubes should be about 1 m/s or more, to promote internal turbulence for good heat transfer. End cover baffles will constrain the flow to a number of passes, as with the shell-and-tube condenser.

Liquid cooling evaporators may comprise a pipe coil in an open tank, and can have flooded or direct expansion circuitry. Flooded coils will be connected to a combined liquid accumulator and suction separator (usually termed the surge drum), in the form of a horizontal or vertical drum (see Figs 7.2c and 7.4). The expansion valve maintains a liquid level in this drum and a natural circulation is set up by the bubbles escaping from the liquid refrigerant at the heat exchanger surface.

Shell and tube evaporators with direct expansion circuits have the refrigerant within the tubes, in order to maintain a suitable continuous velocity for oil transport, and the liquid in the shell. These can be made as shell-and-tube, with the refrigerant constrained to a number of passes (Fig. 7.5), or

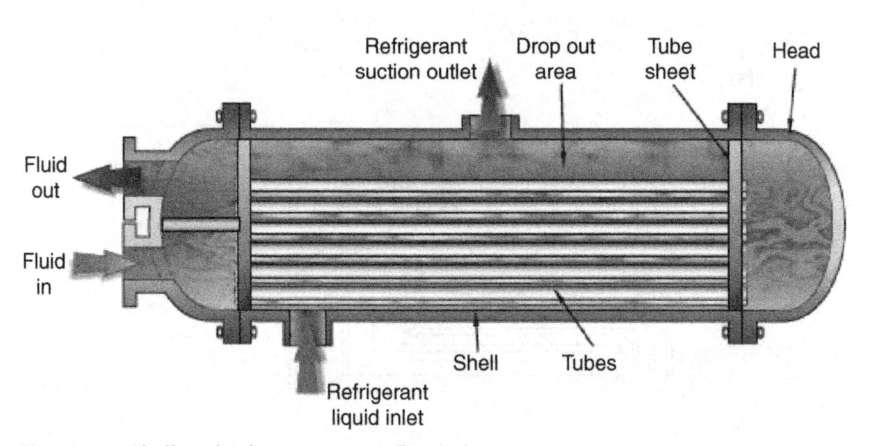

Figure 7.3 ***Shell-and-tube evaporator, flooded.***

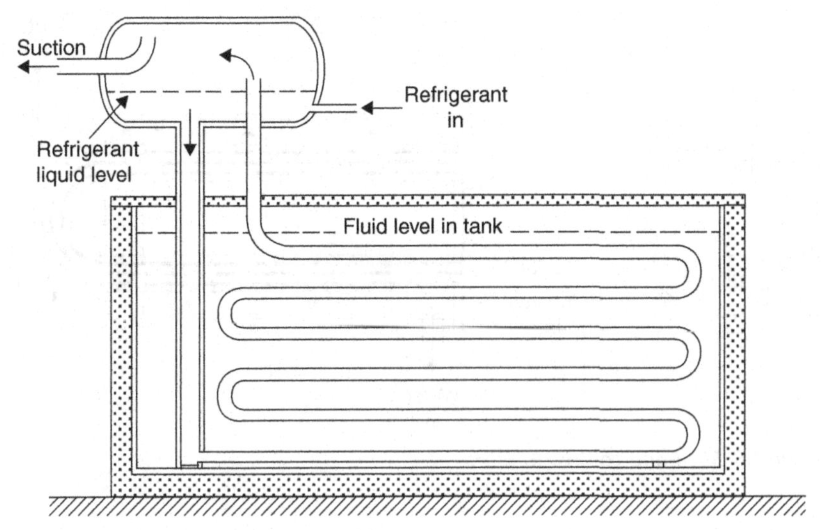

Figure 7.4 *Flooded tank evaporator.*

Figure 7.5 *Onda shell and tube direct expansion evaporators (Titan).*

may be shell-and-coil (see Fig. 7.6). In both these configurations, baffles are needed on the water side to improve the turbulence, and the tubes may be finned on the outside. Internal swirl strips or wires will help to keep liquid refrigerant in contact with the tube wall.

The spray chiller operates with a much lower refrigerant charge than a conventional flooded evaporator. The liquid refrigerant level in a surge drum shell is kept below the tubes and liquid is pumped to spray nozzles

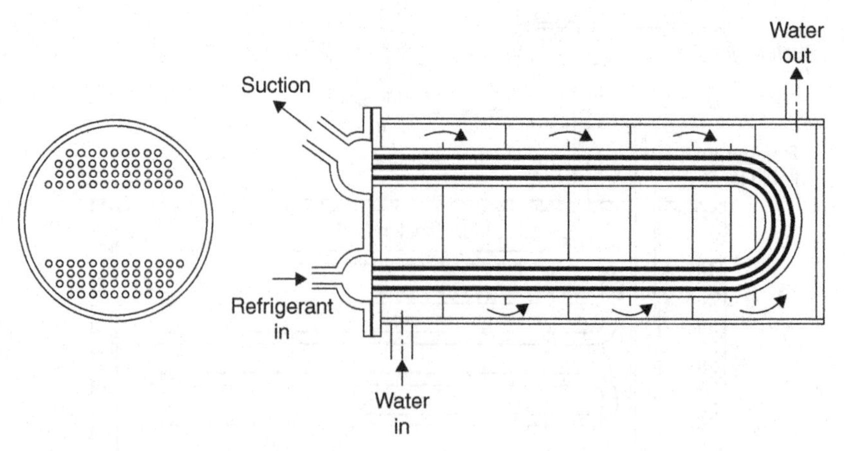

Figure 7.6 ***Shell-and-coil evaporator.***

which ensure that the tube surfaces are covered with an evaporating liquid film. Water or brine passes through the tubes. The gas outlet to the compressor suction is in the upper part of the shell and a baffle arrangement prevents entrainment of liquid droplets. Due to the distribution of refrigerant very close control of the evaporation can be obtained. Evaporation ceases immediately when the liquid spray is stopped. For these reasons the brine can be cooled to a temperature close to its freezing point. Water can be chilled to a temperature below 1°C with an evaporating temperature close to −2°C.

Direct expansion coils for immersion in an open tank will be in a continuous circuit or a number of parallel circuits (see Fig. 7.7). Liquid velocity over such coils can be increased by tank baffles and there may be special purpose agitators, as in an ice-making tank. Coils within an open tank can be allowed to collect a layer of ice during off-load periods, thus providing thermal storage and giving a reserve of cooling capacity at peak load times (see also chapter: Distributed Cooling and Heating).

Plate heat exchanger evaporators are also widely used. A heat exchanger of this type consists of a number of herringbone-corrugated plates assembled to form a pack (Fig. 7.8). The herringbone indentations are set in opposite directions to each other in relation to each facing plate. Brazed plate heat exchangers (BPHX) (Fig. 7.9) have plates made from stainless steel with a copper coating on one side. During manufacture they are assembled and held together by the end plates, and heated under vacuum conditions. The copper melts and coagulates at the contact points and seals the edge joints. When cooled a structure of

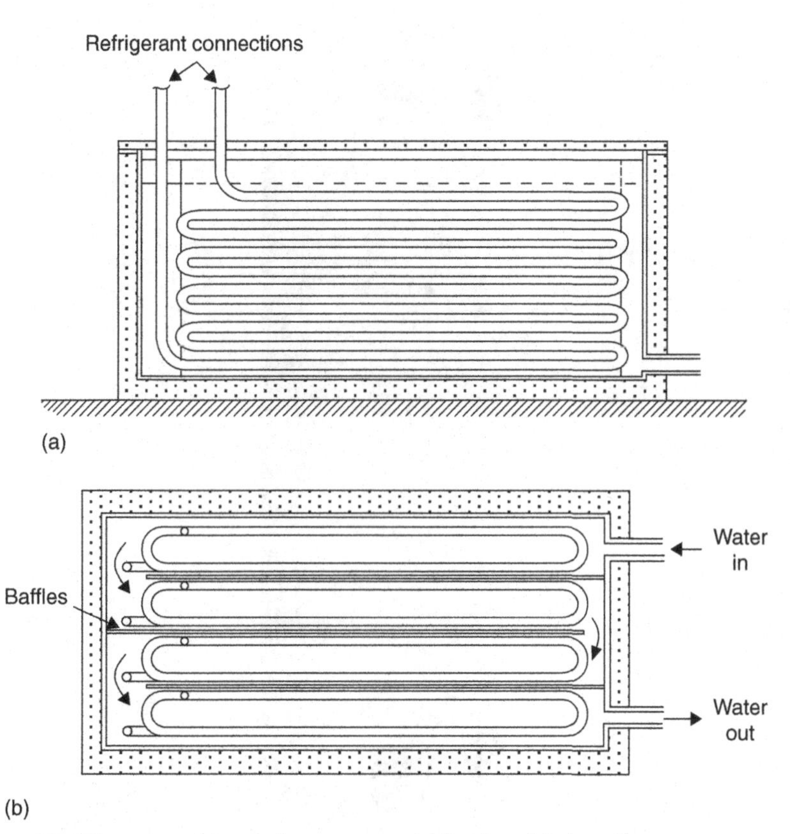

Figure 7.7 ***Direct expansion tank evaporator.*** (a) Section, (b) elevation.

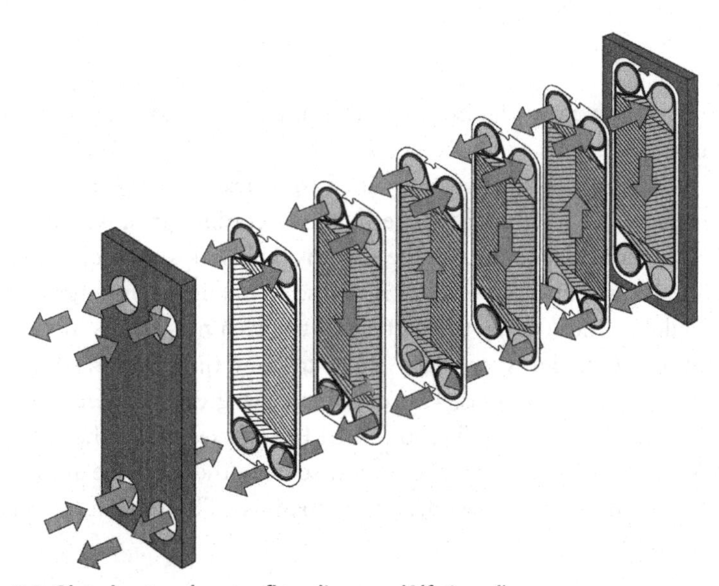

Figure 7.8 ***Plate heat exchanger flow diagram (Alfa Laval).***

Figure 7.9 ***Brazed plate heat exchanger assembly (Alfa Laval).***

alternate counter flow channels is formed, separated only by a thin layer of stainless steel.

The volume of refrigerant contained in a heat exchanger of this type is approximately 2 l for each square metre of cooling area, which is up to 10 times lower than for multi-tube designs. This helps to keep refrigerant charge level low and offers a rapid response to changes in energy demand. The turbulence induced by the pattern of the channels results in very high heat transfer coefficients, typically three to four times greater than with conventional tubular designs. The counter flow gives temperature differences close to the ideal. A refrigeration BPHX always has all the refrigerant channels surrounded by water channels so that there is one more water channel than the number of refrigerant channels. The outermost ones are water channels.

Many configurations are possible. If the channel height is decreased or the corrugation angle is increased the pressure drop and heat transfer rises. Increasing the length of the plates has a similar effect.

When used as a direct expansion evaporator the refrigerant velocity should be high enough to entrain oil that remains after evaporation is complete. Where conditions give rise to non-miscibility, the formation of oil film on the wetted surface can impair heat transfer. On the superheating section there is less effect because this region is sensible heat transfer and velocity to carry the oil droplets upwards is the requirement.

It is important to ensure good distribution so that the refrigerant enters all the channels evenly. The BPHX should be mounted vertically with the refrigerant entry at the bottom. The pipe between the expansion valve and the entry point should be short and of small diameter so that the liquid velocity carries the droplets through to the far plate. Some designs incorporate distributors to aid this process. Mal-distribution can cause erratic expansion valve behaviour and depress the evaporation pressure. Electronic expansion valves, which provide continuous flow, such as variable orifice types, are suitable, but due to the small internal volume of a BPHX, a pulse-modulated valve may give rise to unacceptable pressure fluctuations.

Larger installations can use plate and shell heat exchangers (Fig. 7.10), which work on a similar principle. In this case a welded construction is used making this type of evaporator suitable for ammonia and carbon dioxide refrigerants. They can be kept flooded with refrigerant, working in conjunction with a surge drum into which the liquid is metered by a float expansion valve. An example of this assembly is shown in Fig. 7.11.

Where water is to be cooled close to its freezing point without risk of damage to the evaporator, the latter is commonly arranged above the

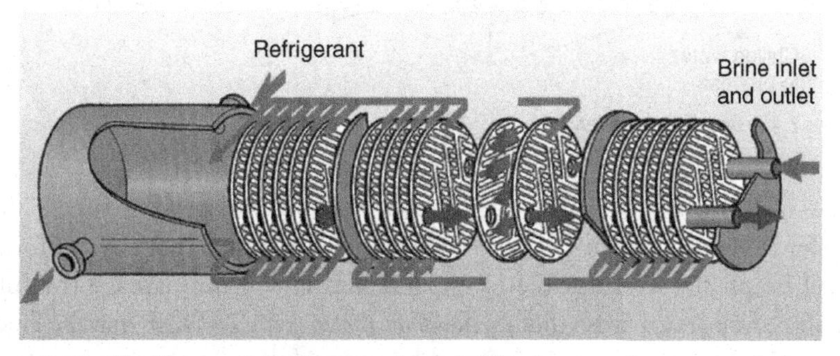

Figure 7.10 ***Witt Plate and shell heat exchanger (Titan).***

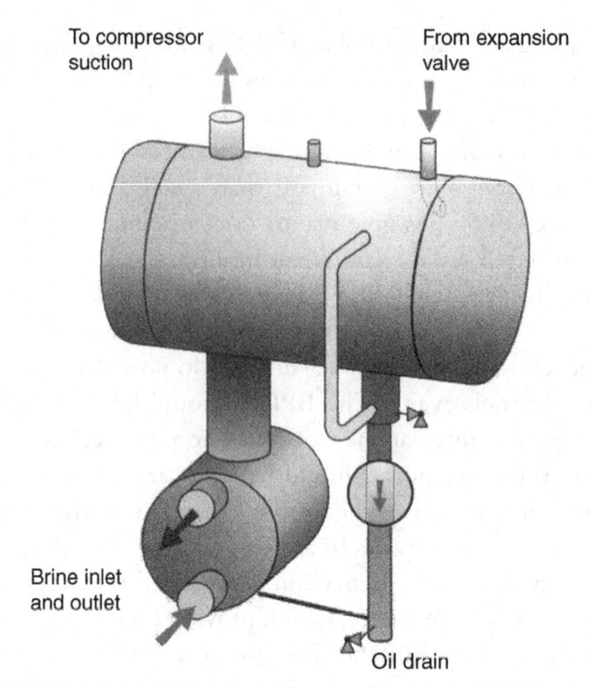

Figure 7.11 ***Witt Plate heat exchanger and surge drum assembly (Titan).***

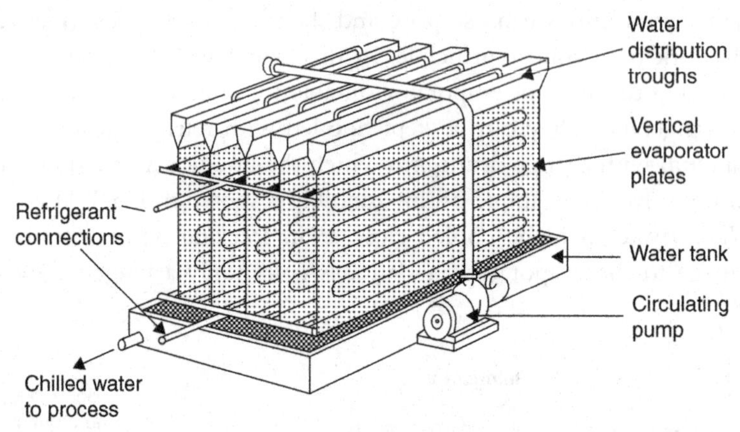

Figure 7.12 ***Baudelot cooler.***

water-collection tank and a thin film of water runs over the tubes. Heat transfer is very high with a thin moving film of liquid and, if any ice forms, it will be on the outside, free to expand, and it will not damage the tube. Such an evaporator is termed a *Baudelot cooler* (Fig. 7.12). It may be open, enclosed in dust-tight shields to avoid contamination of the product (as in surface milk and cream coolers), or may be enclosed in a pressure vessel as in the Mojonniér cooler for soft drinks, which pressurises with carbon dioxide at the same time.

Some liquids, such as vegetable fats and ice-cream mixes, increase considerably in viscosity as they are cooled, sticking to the heat exchanger surface. Evaporators for this duty are arranged in the form of a hollow drum (see Fig. 7.2b) surrounded by the refrigerant and having internal rotating blades which scrape the product off as it thickens, presenting a clean surface to the flow of product and impelling the cold paste towards the outlet.

7.4 PLATE EVAPORATORS FOR FREEZING

Plate evaporators are formed by cladding a tubular coil with sheet metal, welding together two embossed plates, or from aluminium extrusions.

The extended flat face may be used for air cooling, for liquid cooling if immersed in a tank, or as a Baudelot cooler, but the major use for flat plate evaporators is to cool a solid product by conduction, the product being formed in rectangular packages and held close between a pair of adjacent plates.

In the horizontal plate freezer (Fig. 7.13a), the plates are arranged in a stack on slides, so that the intermediate spaces can be opened and closed. Trays, boxes or cartons of the product are loaded between the plates and the stack is closed to give good contact on both sides. When the necessary cooling is complete, the plates are opened and the product removed.

The vertical plate freezer (Fig. 7.13b) is used to form solid blocks of a wet product, typically fish. When frozen solid, the surfaces are thawed and the blocks pushed up and out of the bank.

To ensure good heat transfer on the inner surface of the plates and achieve a high rate of usage, liquid refrigerant is circulated by a pump at a rate 5–12 times the rate of evaporation.

Figure 7.13 ***Plate freezers.*** (a) Horizontal, (b) vertical (Jackstone).

7.5 DEFROSTING

Air cooling evaporators working below 0°C will accumulate frost which must be removed periodically, since it will obstruct heat transfer.

Where the surrounding air is always at 4°C or higher, it can be sufficient to stop the refrigerant flow for a period and allow the frost to melt off (as in the auto-defrost domestic refrigerator). This method can be used for coldrooms, packaged air-conditioners etc., where the service period can be interrupted.

For lower temperatures, heat must be applied to melt the frost within a reasonable time and ensure that it drains away. Methods used are as follows:

1. *Electric resistance heaters*: Elements are within the coil or directly under it.
2. *Hot gas*: A branch pipe from the compressor discharge feeds superheated gas to the coil. The compressor must still be working on another evaporator to make hot gas available. Heat storage capsules can be built into the circuit to provide a limited reserve of heat for a small installation.
3. *Reverse cycle*: The direction of flow of the refrigerant is reversed to make the evaporator act as a condenser. Heat storage or another evaporator are needed as a heat source.

In each of these cases, arrangements must be made to remove cold refrigerant from the coil whilst defrosting is in progress. Drip trays and drain pipes may require supplementary heating.

An innovative energy saving defrost method using stored heat in a phase change material has been reported by Davies et al. (2014).

7.6 CONDENSATE PUMPS

Condensed water will run down the evaporator fins to a collection tray below the coil. From there, drain pipes will take this water to a drain. If plastic pipe is used, it should be black to exclude daylight, or slime will grow inside the tube. Drain pipes passing through rooms below freezing point need to be fitted with trace heaters.

Where the outlet drain is higher than the coil, the water needs to be pumped away for disposal. Condensate pumps are fitted to lift this water to drain by gravity. Such pumps are usually of the peristaltic type.

CHAPTER 8

Expansion Valves

8.1 INTRODUCTION

The purpose of the expansion valve is to control the flow of refrigerant from the high-pressure condensing side of the system into the low-pressure evaporator. In most cases, the pressure reduction is achieved through a variable flow orifice, either modulating or two-position. Expansion valves may be classified according to the method of control.

8.2 THERMOSTATIC EXPANSION VALVES

Direct expansion circuits must be designed and installed so that there is no risk of liquid refrigerant returning to the compressor. To ensure this state, heat exchange surface in the evaporator is used to heat the dry saturated gas so that it becomes superheated. The amount of superheat is usually of the order of 5 K.

Thermostatic expansion valves (TEVs) for such circuits embody a mechanism which will detect the superheat of the gas leaving the evaporator (Fig. 8.1). Refrigerant boils in the evaporator at T_e and p_e, until it is all vapour, point *A* and then superheats to a condition T_s, p_e, at which it passes to the suction line, point *B*. A separate container of the same refrigerant at temperature T_s would have pressure p_s, and the difference $p_s - p_e$ represented by *C–B* in Fig. 8.1 is a signal directly related to the amount of superheat.

The basic thermostatic expansion valve (Fig. 8.2) has a detector and power element, charged with the same refrigerant as in the circuit. The pressure p_s generated in the phial by the superheated gas equalizes through the capillary tube to the top of the diaphragm. An adjustable spring provides the balance of $p_s - p_e$ at the diaphragm, and the valve stem is attached at the centre. Should the superheat fall for any reason, there will be a risk of liquid reaching the compressor. T_s will decrease with a corresponding drop in p_s. The forces on the diaphragm are now out of balance and the spring will start to close the valve.

Conversely if the load on the evaporator increases, refrigerant will evaporate earlier and there will be more superheat at the phial position. Then p_s will increase and open the valve wider to meet the new demand.

The phial must be larger in capacity than the rest of the power element or the charge within it may all pass into the valve capsule and tube, if these

Refrigeration, Air Conditioning and Heat Pumps
http://dx.doi.org/10.1016/B978-0-08-100647-4.00008-5

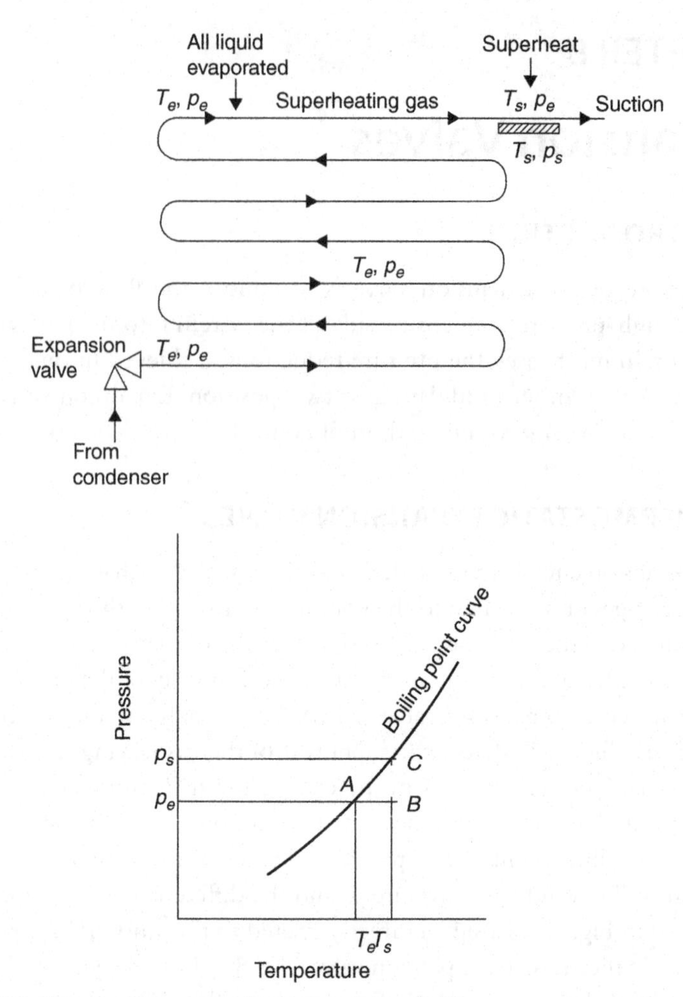

Figure 8.1 ***Superheat sensor on direct expansion circuit.***

are colder. If this happened, the phial at T_s would contain only vapour and would not respond to a position T_s, p_s on the $T - p$ curve.

Use can be made of this latter effect. The power element can be limit charged so that all the refrigerant within it has vaporised by a predetermined temperature (commonly 0°C). Above this point, the pressure within it will cease of follow the boiling point curve but will follow the gas laws as shown in Fig. 8.3:

$$\frac{p_1}{p_2} = \frac{T_1}{T_2}$$

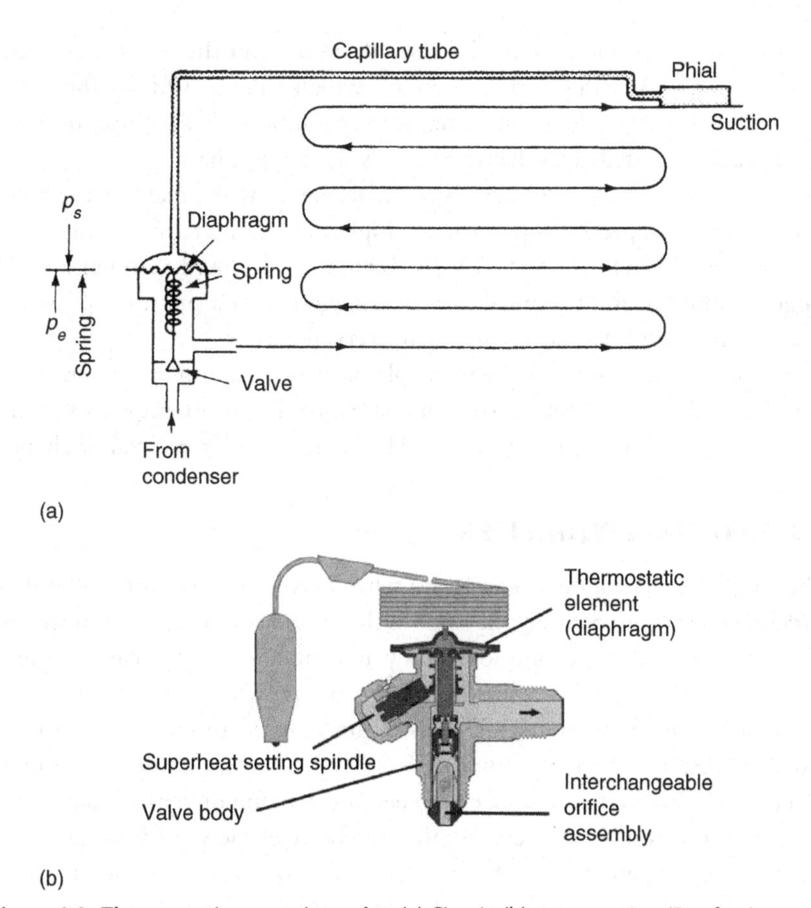

Figure 8.2 ***Thermostatic expansion valve.*** (a) Circuit, (b) cross-section (Danfoss).

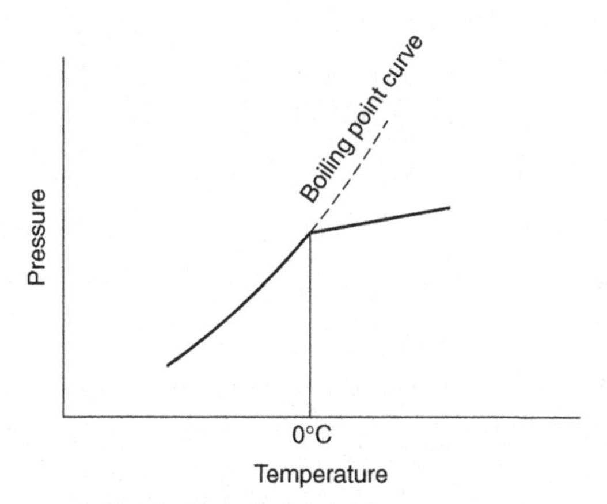

Figure 8.3 ***Detector pressure for limit charged valve.***

and the valve will remain closed. This is done to limit the evaporator pressure when first starting a warm system, which might overload the drive motor. This is termed limit charging or maximum operating pressure. Such valves must be installed so that the phial is the coldest part.

The slope of the $T - p$ curve is not constant, so that a fixed spring pressure will result in greater superheat at a higher operating temperature range. To allow for this and provide a valve which can be used through a wide range of applications, the phial may be charged with a mixture of two or more volatile fluids to modify the characteristic curve.

Some manufacturers use the principle of the adsorption of a gas by a porous material such as silica gel or charcoal. Since the adsorbent is a solid and cannot migrate from the phial, these valves cannot suffer reversal of charge.

8.3 EXTERNAL EQUALISER

The simple *thermostatic expansion valve* relies on the pressure under the diaphragm being approximately the same as that at the coil outlet, and small coil pressure drops can be accommodated by adjustments to the spring setting.

Where an evaporator coil is divided into a number of parallel passes, a distribution device with a small pressure loss is used to ensure equal flow through each pass. Pressure drops of 1–2 bar are common. There will now be a much larger finite difference between the pressure under the diaphragm and that at the coil inlet. To correct this, the body of the valve is modified to accommodate a middle chamber and an *equalising connection* which is taken to the coil outlet, close to the phial position. Most thermostatic expansion valves have provision for an external equaliser connection (see Fig. 8.4).

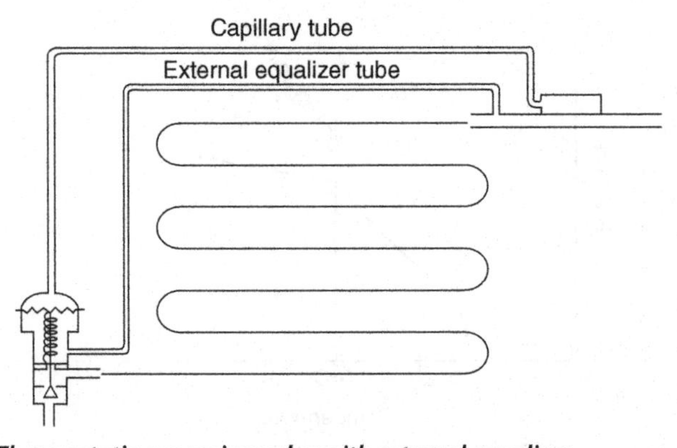

Figure 8.4 ***Thermostatic expansion valve with external equaliser.***

The thermostatic expansion valve is substantially an undamped proportional control and hunts continuously, although the amplitude of this swing can be limited by correct selection and installation, and if the valve always works within its design range of mass flow. Difficulties arise when compressors are run at reduced load and the refrigerant mass flow falls below the valve design range. It is helpful to keep the condensing pressure steady, although it does not have to be constant and can usually be allowed to fall in colder weather to save compressor power. Valves on small systems may be seen to fully close and fully open at times. Excessive hunting of the thermostatic expansion valve means that the evaporator surface has an irregular refrigerant feed with a resulting slight loss of heat transfer effectiveness. If the hunting is caused by a time lag between the change of valve position and the effect at the evaporator outlet, a solution can be to increase the mass of the sensor phial which will increase damping. Over-sized valves and incorrect phial position can also give rise to hunting. The phial should always he located on the horizontal outlet, as close to the evaporator as possible and not on the underside of the pipe.

8.4 ELECTRONIC EXPANSION VALVES

The electronic expansion valve offers a finer degree of control and system protection. The benefits can be summarised as follows:

1. Precise flow control over a wide range of capacities.
2. Rapid response to load changes.
3. Better control at low superheats so that less evaporator surface is required for superheat. More surface for evaporation results in higher evaporating temperature and better efficiency.
4. Electrical connection between components offers greater flexibility in system layout, which is important for compact systems.
5. The valve can close when the system shuts down, which eliminates the need for an additional shut off solenoid valve.

Types of electronic valve in use include a continuous flow type in which the orifice size is varied by a stepper motor, and a pulse width modulated (PWM) type. In each case a controller is used in conjunction with the valve. The controller is pre-configured for the refrigerant and valve type and it receives the information from sensors, for example, pressure and temperature at the evaporator outlet. This enables the superheat to be determined. The output signal to the valve initiates the orifice adjustment. In the case of the PWM valve it is the relationship between the opening and closing which determines the capacity of the valve. The valve is either open or closed and each time interval of a few seconds will include an opening period depending on the signal.

There is a third type of valve that combines both features. A modulating voltage is sent to the actuator, and as the voltage increases the pressure in the actuator's container increases, resulting in an increased valve opening during an 'on cycle' of fixed duration.

In each case the control can be configured so that the valve remains closed in the event of power loss. Under partial load condition or floating condensing pressure, which happens at low ambient temperature, the condensing pressure decreases. Thermostatic expansion valves tend to hunt, but systems with electronic components operate at partial load in exactly the same and stable manner as at full load.

A continuous flow type valve is shown in Fig. 8.5. The valve seat and slider are made out of solid ceramic. The form of the valve slide provides for a highly linear capacity characteristic between 10 and 100%. Depending on the controller and its configuration, a single control valve can be used for different control tasks. Possible uses include: expansion valve for superheat control, suction pressure control for capacity control, liquid injection for de-superheating of compressor, condensing pressure control and hot gas bypass control to compensate excess compressor capacity and to ensure evaporating pressure does not go below a set point.

8.5 CAPILLARY TUBES AND RESTRICTORS

The variable orifice of the expansion valve can be replaced, in small systems, by a long thin tube. This is a non-modulating device and has certain limitations, but will give reasonably effective control over a wide range of conditions if correctly selected and applied. Mass flow is a function of pressure difference and the degree of liquid subcooling on entry. The capillary tube is used almost exclusively in small air conditioning systems and is self-regulating within certain parameters. Increasing ambient temperature results in increasing load on the conditioned space and the condensing pressure will rise, forcing more refrigerant flow.

Tube bores of 0.8–2 mm with lengths of 1–4 m are common. The capillary tube is only fitted on factory-built and tested equipment, with exact refrigerant charges. It is not applicable to field-installed systems.

The restrictor expansion device overcomes some of the limitations of the capillary tube. The orifice can be precision-drilled whereas capillary

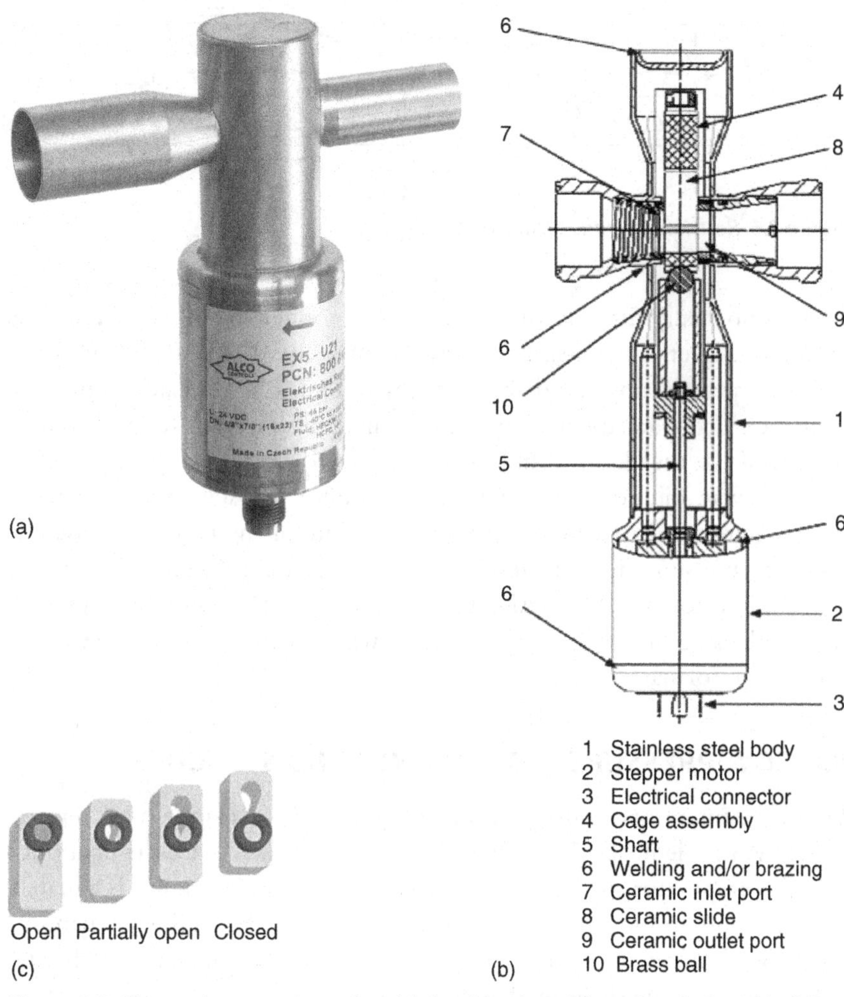

Figure 8.5 ***Electronic expansion valve.*** (a) Outside view, (b) sectional view, (c) sliding orifice (Emerson Climate Technologies).

tubes can suffer from variations in internal diameter over their length which results in changes to predicted performance. Fig. 8.6 shows how the device is applied in a reversible air conditioner. In Fig. 8.6a the device is shown in normal cooling mode. A bullet which is free to move horizontally by a small amount is pressed against a seat-forcing the refrigerant through the central restriction which acts as an expansion device. When the flow reverses, Fig. 8.6b, the bullet moves back to the other seat, but grooving allows flow around the outside as well as through it, so that the restriction is very small.

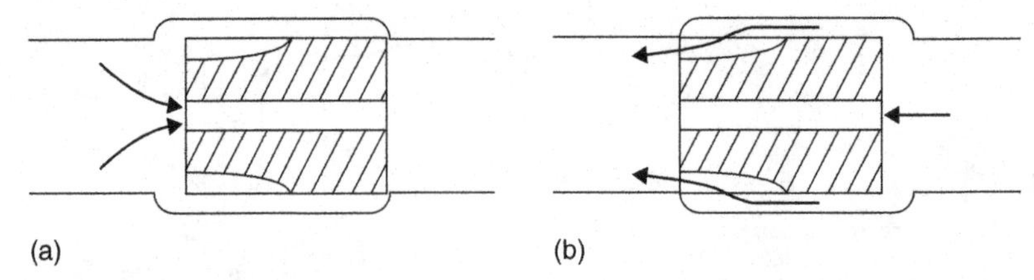

Figure 8.6 ***Restrictor expansion device.***

It is normally fitted at the outlet of the condenser rather than at the evaporator inlet. This means that instead of a liquid line to the evaporator, the pipe contains liquid and flash gas and must be insulated. Although heat pick up is detrimental to performance, the pressure drop, which is used to drive the fluid, would normally have occurred in the expansion valve anyway. Liquid lines to remote evaporators on split systems can be quite lengthy and in a high-pressure liquid line of the type more usually used, the pressure drop can result in an increased condenser pressure and tendency to form bubbles. Also the restrictor can be delivered as part of the condensing unit and is removable, allowing changes to be made to give optimal performance.

8.6 LOW-PRESSURE FLOAT VALVES AND SWITCHES

Flooded evaporators require a constant liquid level, so that the tubes remain wetted. A simple float valve suffices, but must be located with the float outside the evaporator shell, since the surface of the boiling liquid is agitated and the constant movement would cause excessive wear in the mechanism. The float is therefore contained within a separate chamber, coupled with balance lines to the shell (see Fig. 8.7).

Such a valve is a metering device and may not provide positive shut-off when the compressor is stopped. Under these circumstances, refrigerant will continue to leak into the evaporator until pressures have equalised, and the liquid level might rise too close to the suction outlet. To provide this shut-off, a solenoid valve is needed in the liquid line.

Since the low-pressure float needs a solenoid valve for tight closure, this valve can be used as an on–off control in conjunction with a pre-set orifice and controlled by a float switch (Fig. 8.8).

The commonest form of level detector is a metallic float carrying an iron core which rises and falls within a sealing sleeve. An induction coil surrounds the sleeve and is used to detect the position of the core. The resulting signal

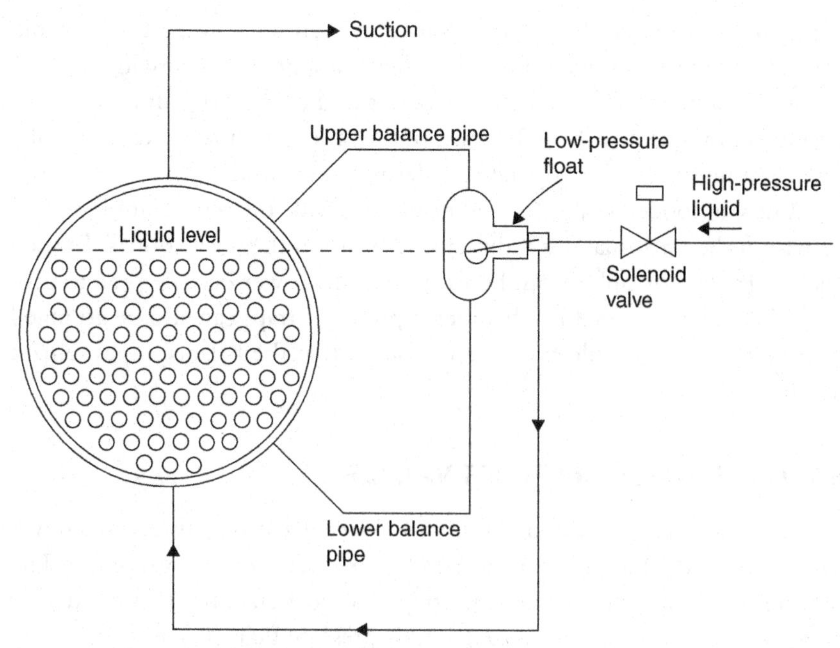

Figure 8.7 ***Low-pressure float valve on flooded cooler.***

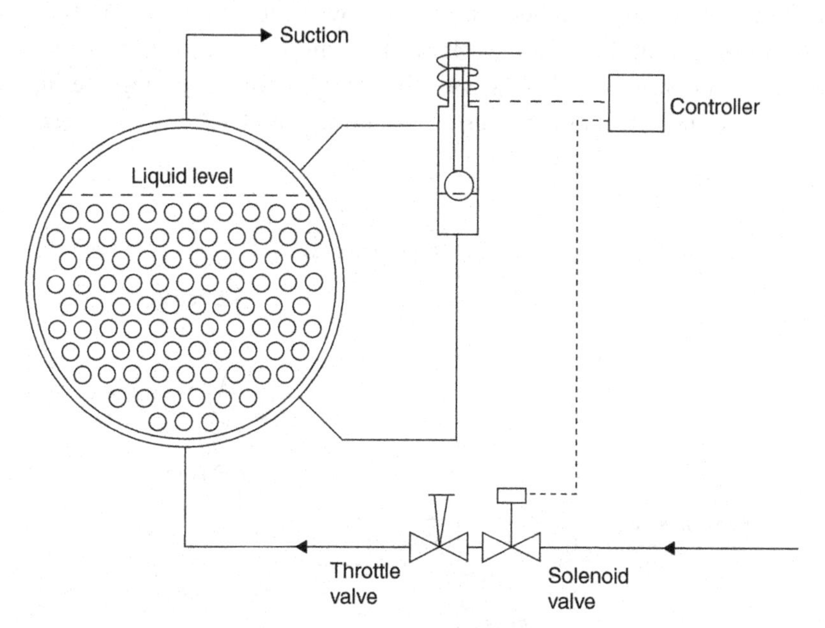

Figure 8.8 ***Low-pressure float switch.***

is amplified to switch the solenoid valve, and can be adjusted for level and sensitivity. A throttle valve is fitted to provide the pressure-reducing device.

Should a float control fail, the level in the shell may rise and liquid pass into the compressor suction. To warn of this, a second float switch is usually fitted at a higher level, to operate an alarm and cut-out.

Where a flooded coil is located in a liquid tank, the refrigerant level will be within the tank, making it difficult to position the level control. In such cases, a gas trap or siphon can be formed in the lower balance pipe to give an indirect level in the float chamber. Siphons or traps can also be arranged to contain a non-volatile fluid such as oil, so that the balance pipes remain free from frost.

8.7 HIGH-PRESSURE FLOAT VALVES

On a single-evaporator flooded system, a float valve can be fitted which will pass any drained liquid from the condenser direct to the evaporator. The action is the same as that of a steam trap. The float chamber is at condenser pressure and the control is termed a high-pressure float (Fig. 8.9).

The high-pressure float switch keeps the condenser drained without the need for a high-pressure receiver. The level in the evaporator is fixed by the system charge. Low charge systems using shell and plate heat exchangers and spray chillers are possible with this method. The type of float valve in Fig. 8.10 can work with ammonia or carbon dioxide refrigerants. Economiser circuits with the float switch expanding the liquid to an intermediate flash expansion vessel are used for low-temperature

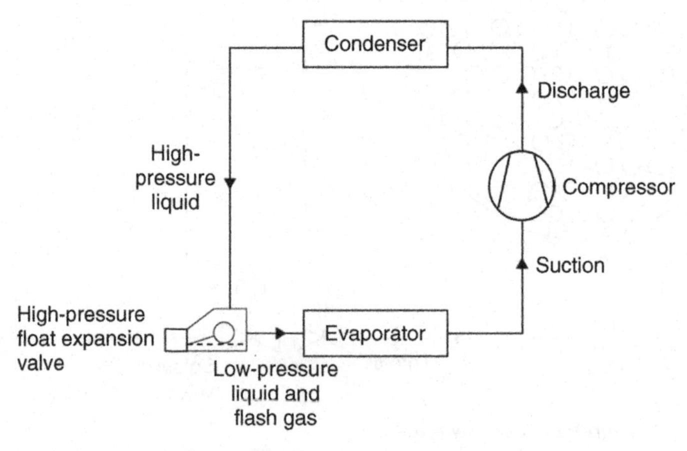

Figure 8.9 ***High-pressure float valve circuit.***

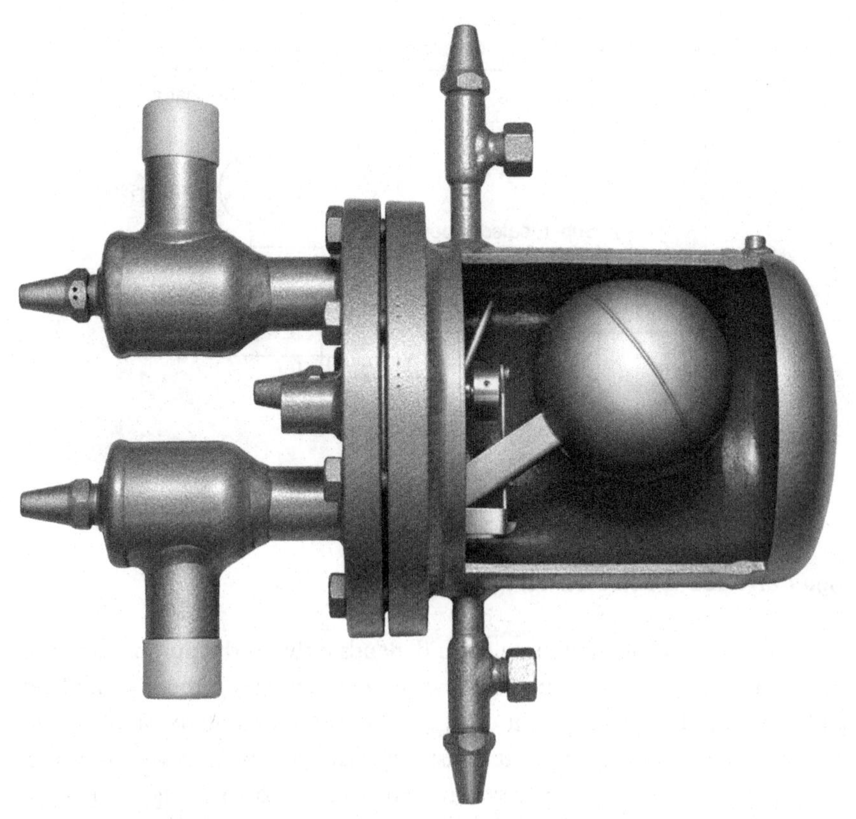

Figure 8.10 ***Sectioned view of Witt high-pressure float valve (Titan).***

applications. This control cannot feed more than one evaporator, since it cannot detect the needs of either.

The difficulty of the critical charge can be overcome by allowing any surplus liquid refrigerant leaving the evaporator to spill over into a receiver or accumulator in the suction line, and boiling this off with the warm liquid leaving the condenser. In this system, the *low-pressure receiver circuit,* liquid is drained from the condenser through the high-pressure float, but the final step of pressure drop takes place in a secondary expansion valve after the warm liquid has passed through coils within the receiver. In this way, heat is available to boil off surplus liquid leaving the evaporator (see Fig. 8.11). Two heat exchangers carry the warm liquid from the condenser within this vessel. The first coil is in the upper part of the receiver, and provides enough superheat to ensure that gas enters the compressor in a dry condition. The lower coil boils off surplus liquid, leaving the evaporator itself. With this method of refrigerant feed, the evaporator has a better internal wetted surface, with an improvement in heat transfer.

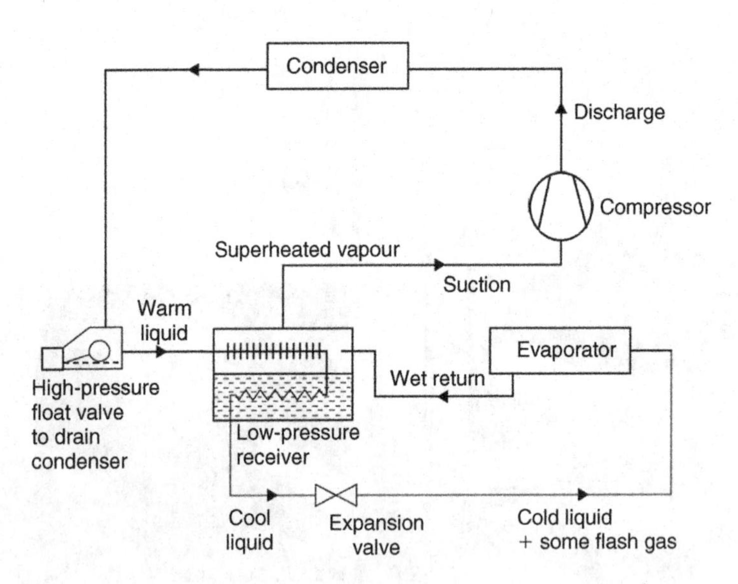

Figure 8.11 ***Low-pressure receiver circuit.***

The rate of refrigerant circulation depends only on the pumping rate of the compressor. The quantity evaporating will depend on the cooling load and provided there is sufficient charge to flood the evaporator at all conditions, and there is sufficient heat transfer surface, the system is self-adjusting.

The low-pressure receiver system can be adapted to compound compression and can be fitted with hot gas defrost by reverse gas flow. In both circuits the low-pressure receiver provides the safety vessel to prevent liquid entering the compressor. Providing the high-pressure float is correctly sized, this system can operate at low condenser pressures, saving compressor energy in cool weather. Where the halocarbon refrigerants are used in this system, an oil-distillation device is fitted.

8.8 OTHER LEVEL CONTROLS

If a small heater element is placed at the required liquid level of a flooded evaporator, together with a heat-sensing element, then the latter will detect a greater temperature if liquid refrigerant is not present. This signal can be used to operate a solenoid valve.

The thermostatic or electronic expansion valve can also be used to maintain a liquid level. The phial and a heater element are both clamped to a bulb at the required liquid level. If liquid is not present, the heater warms the phial to a superheat condition and the valve opens to admit more liquid.

CHAPTER 9

Controls and Other Circuit Components

9.1 INTRODUCTION

A refrigeration system can be built with only the following four essential components:

1. Evaporator
2. Compressor
3. Condenser
4. Expansion valve

Other system controls and components are, however, necessary for safety, maintenance and control purposes. Fig. 9.1 illustrates the functionalities associated with the compressor. Data related to the compressor, such as oil pressure, temperature, motor current can be fed from integrated sensors into an electronic controller such as shown Fig. 4.12. This type of equipment not only eliminates mechanical switches and external pressure connections but also allows integration with other system information. Additional functionalities can include historical information. Manufacturers' web sites can give details. Fig. 9.2 shows some of the frequently used circuit components. In addition, a safety pressure-relief device may be required for any pressure vessel in the circuit, depending on its size.

Many controls are now performed by electronic controllers, but the functionalities are sometimes best described by examining the working of mechanical devices. Some components such as filters, oil separators and relief valves are physical components in the system, although certain aspects of their behaviour can be monitored electronically.

9.2 MAIN CONTROL FUNCTIONS

Since the primary purpose of a refrigeration, air conditioning, or heat pump system is to maintain temperature, a thermostat is usually fitted to stop the equipment or reduce its capacity when the required condition is reached. The following types are in use:

1. Movement of a bimetallic element
2. Expansion of a fluid

Refrigeration, Air Conditioning and Heat Pumps
http://dx.doi.org/10.1016/B978-0-08-100647-4.00009-7

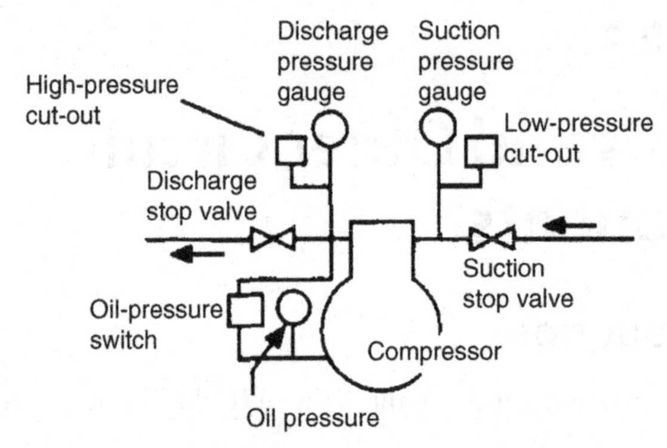

Figure 9.1 ***Components associated with compressor.***

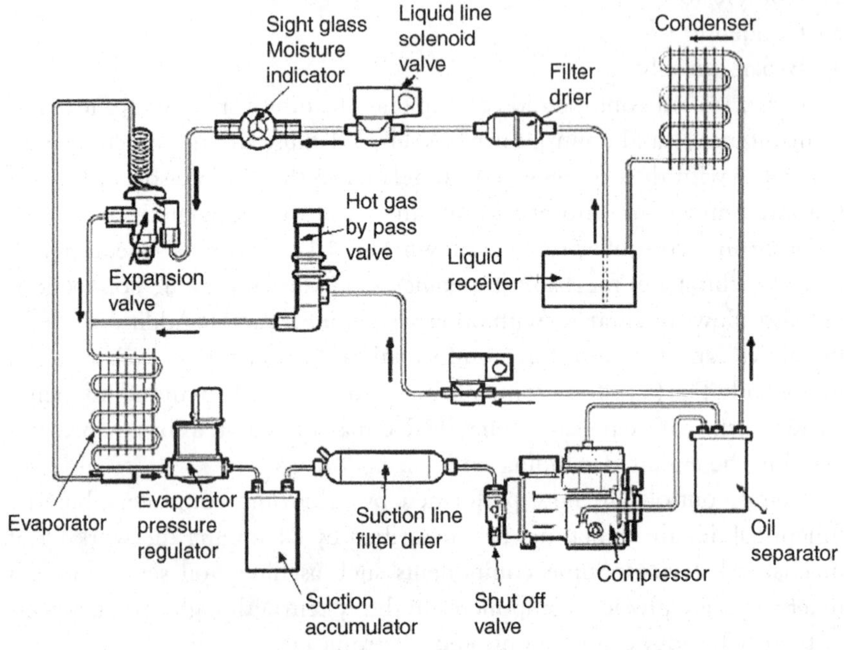

Figure 9.2 ***Components for direct-expansion circuits.***

3. Vapour pressure of a volatile fluid
4. Electric resistance
5. Electronic – various types

The first three aforementioned types produce a mechanical effect which can be used directly to operate an electric switch or modulate the pressure of an air jet (pneumatic system). These last two produce an electric signal which must be measured and amplified to operate the controlled device.

In order to prevent excessive cycling, a differential of say 2 K should be incorporated.

Where the equipment is required to maintain a predetermined level of humidity, a humidistat may be used instead of, or in addition to, a thermostat. The function will normally be to operate an electrical switch.

Mechanical humidistats employ materials which change dimension with humidity, such as animal hair, plastics, cellulosics, etc. These can work a switch directly. Electronic humidistats generally depend on the properties of a hygroscopic salt. The signal has to be measured and amplified.

A low-pressure switch can also be used in conjunction with a thermostat and a solenoid valve in the *pump-down circuit.* In this method of control, the thermostat does not stop the compressor but de-energises the liquid line solenoid valve to stop the supply of refrigerant to the evaporator. The compressor continues to run and pumps down the evaporator until stopped by the low-pressure switch. When the thermostat again calls for cooling, it opens the solenoid valve, liquid enters the evaporator and the low-pressure switch will close again to restart the compressor. This method is used to ensure that the evaporator is kept clear of liquid when the plant is off. If there is any leak at the solenoid valve, it will cause the compressor to restart periodically to remove the surplus liquid from the coil (see Fig. 9.3). Pressure switches are also made in miniature, encapsulated versions, mainly pre-set for use in integrated control circuits.

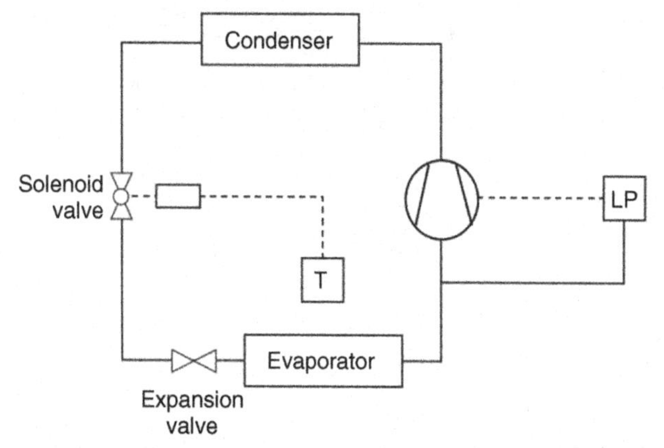

Figure 9.3 ***The pump down circuit – the solenoid valve is thermostat controlled and compressor is low-pressure controlled.***

9.3 SAFETY AND PROTECTION DEVICES

The pressure in the condenser side of the system must always be limited to a maximum allowable value, and a pressure control is used to stop the compressor if necessary.

High-pressure cut-outs (Fig. 9.4) are fitted to all but the smallest of systems. Where a compressor is fitted with a shut-off valve, the pressure connection should be upstream of the valve. The compressor outlet pressure is brought to one side of a bellows or diaphragm, and balanced by an adjustable spring. A scale on the control indicates the pressure setting to commercial accuracy and is checked on commissioning the system.

If the spring pressure is overcome, the switch will open and stop the compressor. The cut-out can also operate a warning. The cut-out point only needs to be some 2 bar higher than the maximum expected operating pressure but there is a tendency to set such controls much higher – sometimes as much as 8 bar above usual maximum pressures. At this setting, the user will not get a warning of abnormal running until the fault has reached serious proportions.

Since excess pressure indicates malfunction of part of the system – usually a condenser fault or incorrect closure of a valve – the high-pressure switch should be reset manually, not automatically.

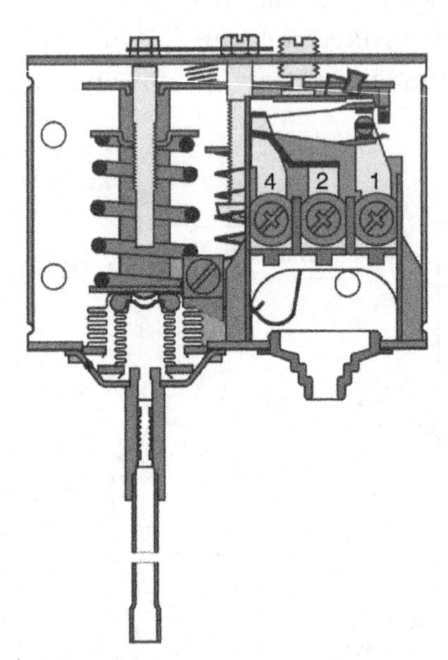

Figure 9.4 ***Pressure switch (Danfoss).***

Where the refrigeration system is providing an essential service which should not be interrupted, one high-pressure switch may be set at a warning level and operate an alarm, without stopping the compressor. A second switch, set somewhat higher, will stop the equipment if this warning is ignored and if excessive pressures are reached. All high-pressure cut-outs should be checked at least once a year, for correct setting and operation.

Abnormally low suction pressures will lead to high discharge temperatures, owing to the high compression ratio, and possible malfunction of other components. Air-cooling coils may frost excessively, or water chillers freeze.

A *low-pressure cut-out* switch is usually fitted to stop the compressor under these circumstances. Settings may be 0.6–1.0 bar below the design evaporator pressures, but depend very much on the type of system. The cut-out setting should be above atmospheric pressure if possible to avoid the ingress of air through any leaks.

Abnormally low pressure may not be an unsafe condition and the low-pressure switch may be automatic reset, closing again at a pressure corresponding to a temperature just below that of the load.

If a plant has been shut down long enough for all pressures to equalise and is then restarted, the suction pressure will pull down below normal until the liquid refrigerant has begun to circulate. Under such circumstances the low-pressure switch may operate. This is a normal occurrence, but may require the addition of a delay timer to prevent frequent starting of the compressor motor.

Under several possible conditions of malfunction, high pressures can occur in parts of the system and mechanical relief devices are advised or mandatory. The requirements which must be observed are detailed in European Safety Standard EN378 (2008) and the Institute of Refrigeration Safety Codes. The standard form of relief valve is a spring-loaded plunger valve (Fig. 9.5). Relief valves must be adequately sized and sizing details can be found in manufacturers' data sheets. The recommended method for selection is given in EN13136.

Plunger-type relief valves, if located outdoors, should be protected from the ingress of rain, which may corrode the seat. Steel valves, when installed, should have a little oil poured in to cover the seat as rust protection.

To prevent over-pressure within a compressor, a relief valve or bursting disc is often fitted between the inlet and discharge connections.

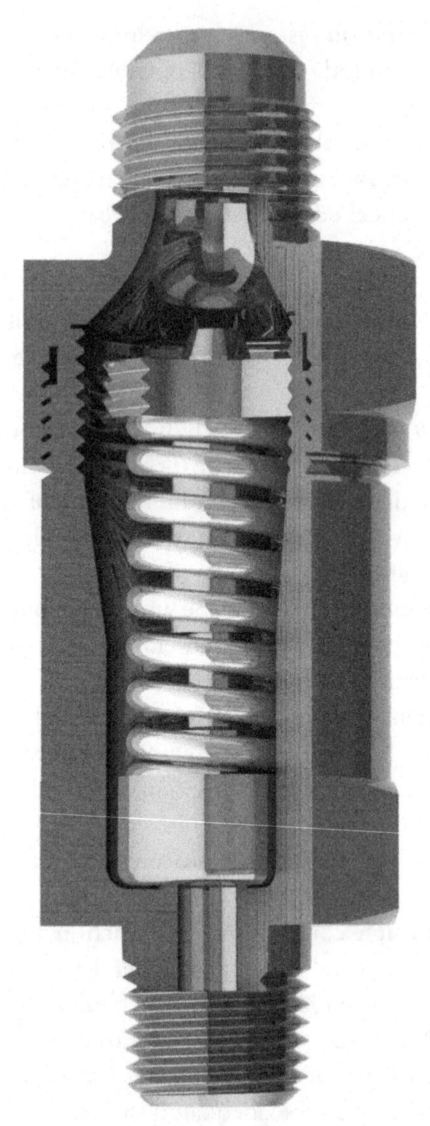

Figure 9.5 ***Spring-loaded plunger relief valve (Henry).***

9.4 OIL-PRESSURE SWITCHES

Many compressors have oil pumps to provide mechanical lubrication and will fail if the oil pressure falls because of a pump fault or oil shortage. Where an oil pump is used a cut-out, which will stop the compressor in the event of inadequate oil pressure, is provided. This takes the form of a differential-pressure switch with a starting time delay.

Since the oil pump inlet is at sump (suction) pressure, a pressure gauge on the pump discharge will indicate the total pressure at that point above atmospheric, that is suction (gauge) plus pump head. Any detection element for true oil pump pressure must sense both suction and pump outlet pressures and transduce the difference. Oil pressure cut-outs have connections to both sides of the oil pump and two internal bellows are opposed to measure the difference.

Since there will be no oil pressure at the moment of starting, a time delay must be fitted to allow the oil pressure to build up. This timer may be thermal, mechanical or electrical.

Operation of the oil cut-out indicates an unsafe condition and such controls are made with hand reset switches. Contacts on the switch can be used to operate an alarm to warn of the malfunction.

Several compressor manufacturers offer electronic oil protection systems which provide more functionality, whilst retaining the option of hand re-set.

9.5 PRESSURE AND TEMPERATURE INDICATORS

Direct indication of the operating conditions of a compressor is by pressure gauges at suction, discharge and oil delivery. Such gauges are mounted on or near the compressor.

Since the pressure losses along the discharge and suction lines are comparatively small on most systems, these pressures will also approximate to the conditions in the condenser and evaporator, and the equivalent saturation temperatures will be the condensing and evaporating temperatures. To indicate these temperatures for the common refrigerants, pressure gauges will have further calibrations showing these equivalent temperatures (see Fig. 9.6).

Gauge mechanisms are mostly of the bourdon tube type, having a flattened tube element, which distorts under pressure change. Gas pulsations from the compressor will be transmitted along the short connecting tubes and may lead to early failure of the needle mechanism. These can be damped by restricting the tube with a valve or orifice, or oil filling the gauge as shown in Fig. 9.6, or both. Gauge needles should not be allowed to flicker noticeably from gas pulsations. Pressure transducers are now used for integrated control circuits.

Many systems incorporate electronic controllers that receive data from sensors at key positions, with a display panel where pressures and temperatures

Figure 9.6 ***Refrigeration pressure gauge (Star Instruments).***

can be read. More about this is provided in the chapter: Control Systems. Temperature sensors are commonly platinum resistance or thermocouple types. Pressure transducers can provide electrical outputs corresponding to absolute or gauge pressure.

9.6 SOLENOID VALVES

Electrically operated shut-off valves (Fig. 9.7) are required for refrigerant and other circuits. These take the form of a plunger operated by a solenoid and working directly on the valve orifice or through a servo. The usual arrangement is to energise the solenoid to open the valve and de-energise to close. Sizes up to 50 mm bore tube connections are made. Beyond this, the solenoid acts as a pilot to a main servo (see Fig. 9.8).

Solenoid valves are used in refrigeration and air-conditioning systems for refrigerant lines, oil-pressure pipes (to control oil return and capacity reducers), and water and compressed air lines.

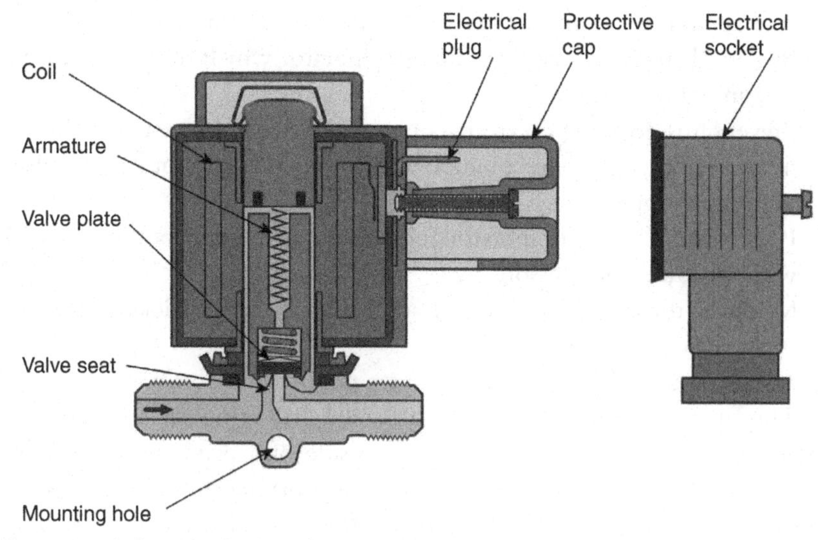

Figure 9.7 *Solenoid valve (Danfoss).*

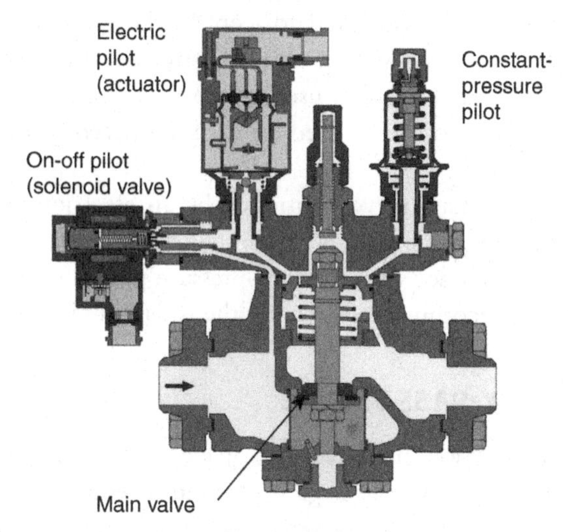

Figure 9.8 *Evaporator pressure regulator valve (Danfoss).*

9.7 EVAPORATOR PRESSURE REGULATION VALVES

Evaporator pressure regulation (EPR) valves (Fig. 9.8) can be used in the suction line, and their function is to prevent the evaporator pressure falling below a predetermined or controlled value, although the compressor suction pressure may be lower.

The application of an EPR valve is to do the following:

1. Prevent damage to a liquid chilling evaporator which might result from freezing of the liquid.
2. Prevent frost forming on an air-cooling evaporator, where this is close to freezing point, or where a temporary malfunction cannot be permitted to interrupt operation.
3. Permit two or more evaporators, at different load temperatures, to work with the same compressor.
4. Modulate the evaporator pressure according to a varying load, controlled by the load temperature.
5. Act as a solenoid valve, controlled by a pilot solenoid valve.

The simplest EPR valve is spring-loaded, balancing the thrust of the spring, plus atmospheric pressure, on one side of a diaphragm or piston, against the inlet or evaporator pressure. For working pressures below atmospheric, a helper spring is fitted below the diaphragm. Slight variations will result from changes in atmospheric pressure, but these are too small to materially affect a refrigeration control system.

A service gauge is usually fitted adjacent to the valve or as part of the valve assembly, to facilitate setting or readjustment. Above about 40 mm pipe size, the basic EPR valve is used as a pilot to operate a main servo valve. Other pilot signals can be used on the same servo and many control functions are possible.

Fig. 9.8 shows a main servo controlled by an electric pilot which responds to a signal from a controller to maintain a constant load temperature to a high level of accuracy. The constant-pressure pilot prevents the evaporator temperature becoming too low and the solenoid pilot is for on/off.

9.8 HOT GAS BY-PASS VALVES

Where a compressor does not have any capacity reduction device and on–off switching will not give the degree of control required by the process, the cooling capacity can be regulated by injecting discharge gas back into the suction (see Fig. 9.2). It has the effect of keeping the evaporator pressure constant, regardless of the load, and can have a wide range of capacity reduction, down to 10% of full load. It is a constant pressure valve, balancing the suction pressure against a pre-set spring.

They may be either direct or pilot operated. They are often equipped with an external equaliser which works just like an externally equalised expansion valve. Injecting the hot gas upstream of the evaporator enables the expansion valve to open and admit cooler refrigerant which maintains gas velocity in the evaporator and aids oil return.

Note the solenoid valve in the hot gas by pass line in Fig. 9.2. If the system is set up with a pump down, then this solenoid must be wired in parallel with the pump down solenoid in order to achieve pump down.

This control method is wasteful of energy.

9.9 SHUT-OFF VALVES

Manual stop valves are required throughout a circuit to permit isolation during partial operation, service or maintenance (see Figs 9.9 and 9.10).

Small valves which are to be operated frequently have a packless gland, either a diaphragm or bellows, and a handwheel.

Valves of all sizes which are only used occasionally will be sealed with 'O' rings. As a safeguard against leakage, they have no handwheel fitted and the stem is provided with a covering cap which is only removed when the valve is to be operated. The stem will have flats for operation by a spanner. Most such valves can be back-seated to permit changing the 'O' rings.

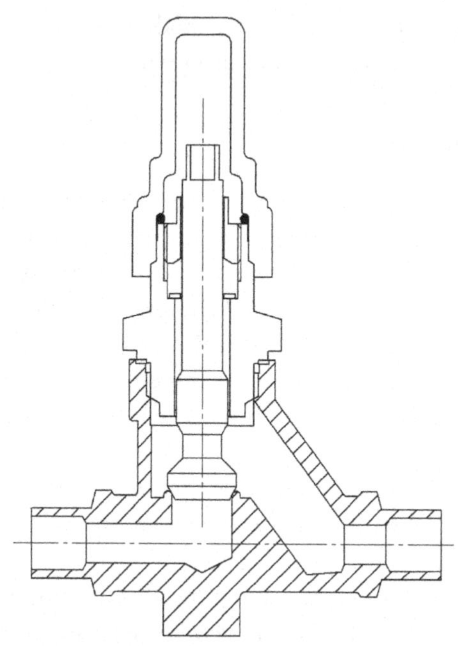

Figure 9.9 ***Seal cap shut-off valve (Henry Technologies).***

Figure 9.10 ***Ball valve (Henry Technologies).***

Valves should not be installed with the stem downwards, as any internal dirt will fall into the spindle thread.

Under low-temperature conditions, ice will form on the spindle and will be forced into the gland if the valve is operated quickly. Under such circumstances, the spindle should be well greased, or the ice melted off first.

Service stop valves on small compressors will usually carry a connection for a pressure cut-out or gauge, or for the temporary fitting of gauges or charging lines when servicing. The valve backseats to close off this port whilst gauges are being fitted. Valve seats are commonly of soft metal or of a resistant plastic such as PTFE.

Ball valves (Fig. 9.10), are sometimes used for secure isolation of sections of circuits and they give less pressure drop when open.

9.10 FILTER-DRIERS

With the halocarbons, it is essential to reduce the water content of the refrigerant circuit to a minimum by careful drying of components and the fitting of drying agents in the system. The common form of drier is a capsule charged with a solid desiccant such as activated alumina or zeolite (molecular sieve), and located in the liquid line ahead of the expansion valve. These capsules must have strainers to prevent loss of the drying agent into the circuit, and so form an effective filter-drier to also protect the valve orifice from damage by fine debris (Fig. 9.11).

Large driers are made so they can be opened, and the spent drying agent removed and replaced with new. Small sizes are throwaway. Driers may also be used in the suction line.

Suction line filter-driers are a temporary installation to clean up a system after repairs, and should be removed if the pressure drop becomes 1 K greater than when new.

Figure 9.11 *Filter drier (Henry Technologies).*

9.11 SIGHT GLASSES

Sight glasses can be used to indicate whether gas is present in a pipe which should be carrying only liquid. The main application in refrigeration is in the liquid line from the receiver to the expansion valve. Close to the expansion valve is best because observation here can determine liquid presence at the valve, and being downstream of the filter drier, it can be used to indicate the need for filter change. If the equipment is running correctly, only liquid will be present and any gas bubbles seen will indicate a refrigerant shortage (see also chapters: Installation and Construction; Commissioning and Maintenance).

Sight glasses for the halocarbons are commonly made of brass, and should have solder connections (Fig. 9.12). For ammonia, they are made of steel or cast iron.

Figure 9.12 *Sight glass/moisture indicator (Henry Technologies).*

Since the interior of the system can be seen at this point, advantage is taken in most types to insert a moisture-sensitive chemical which will indicate an excess of water by a change of colour. When such an indication is seen, the drier needs changing or recharging, and the colour should then revert to the 'dry' shade.

9.12 SUCTION ACCUMULATORS

Suction line accumulators are sometimes inserted in halocarbon circuits, to serve the purpose of separating return liquid and prevent it passing over to the compressor. Since this liquid will be carrying oil, this oil must be returned to the compressor. The outlet pipe within the separator collects gas from the top of this vessel, and dips to the bottom where there is a small bleed hole to suck the oil out (see Fig. 9.13). A version with a heating coil through which hot gas is passed can speed the evaporation of trapped liquid (Fig. 9.13b).

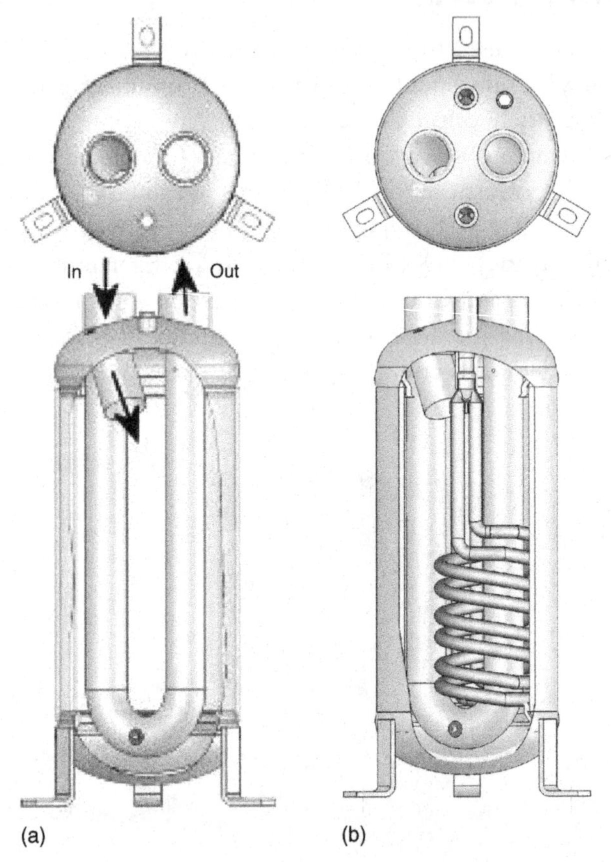

Figure 9.13 ***(a) Suction line accumulator, (b) with heating coil (Henry Technologies).***

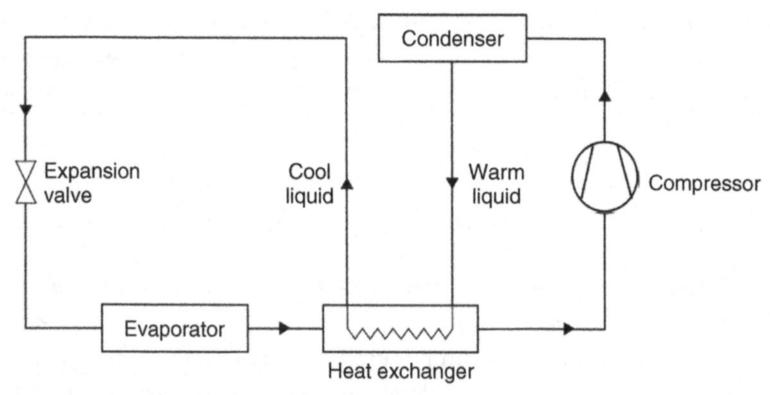

Figure 9.14 ***Suction-to-liquid heat exchanger circuit.***

9.13 SUCTION-TO-LIQUID HEAT EXCHANGERS

Cold gas returning from the evaporator to the compressor can be used to pre-cool the warm liquid passing from the condenser to the expansion valve, using a suction-to-liquid heat exchanger (Fig. 9.14). In cooling the liquid and reducing its enthalpy, a greater refrigerating effect will be obtained. This gain is offset by the superheating of the suction gas and the resultant reduction of mass flow into the compressor. The overall effect of fitting a suction-to-liquid heat exchanger in terms of thermodynamic efficiency will vary with the refrigerant and the operating conditions.

The suction-to-liquid heat exchanger will supply the suction superheat necessary for safe operation of a direct expansion evaporator, and the coil superheat may be less, giving more efficient use of the evaporator surface. The phial should be located before the heat exchanger, in which case the superheat setting is reduced. It can be located after the heat exchanger, but an external equaliser is then necessary to allow for the gas pressure drop through the exchanger.

9.14 CONDENSER PRESSURE REGULATORS

Systems are normally designed to work satisfactorily during maximum ambient conditions, and the condenser will be sized for this. In colder weather, the condensing temperature and pressure will fall and the resulting lower pressure difference across a thermostatic expansion valve may lead to malfunction. A drop of pressure difference to half the normal figure may reduce mass flow below that required, and it will be necessary to prevent the condenser pressure from falling too low.

With air-cooled condensers and water-cooling towers it is possible to reduce the air flow by automatic dampers, fan speed control or switching

Table 9.1 Indicative running cost for constant load at varying condensing temperatures

Condensing temperature (°C)	Coefficient of performance	Weekly electricity costs (£, @ 10 p/unit)
35 (summer maximum)	3.41	512
30	4.00	438
25 (probable minimum)	4.73	368

off fans, where two or more are fitted. The control should work from pressure but can be made to work from temperature.

Water-cooled condensers can be fitted with a directly controlled water-regulating valve operated by condenser pressure, or may have a three-way blending valve in the water circuit.

A condenser-pressure regulator can be in the form of a pressure-operated bleed valve in a bypass across the condenser, to divert hot gas to the receiver. The valve diaphragm is balanced by a pre-set spring and will open the bypass if the condensing pressure falls. A similar effect can be obtained by a pressure-operated valve between the condenser and the receiver, to restrict the flow and allow liquid to accumulate in the condenser, reducing its efficiency. For operating economy, it is important that such valves are not set at too high a pressure.

Where evaporative condensers and water-cooling towers have only one fan (or fan drive motor), coarse control can be effected by on–off switching. The time lag will then depend on the mass of water in the circuit, and the sensing element needs to have a wide differential to prevent frequent motor starts. Towers should have thermostatic control of the fan to prevent water freezing on the packing in winter.

An integrated control circuit with an electronic expansion valve can be arranged to permit the condensing pressure to fall, providing the valve can pass the refrigerant flow required to meet the load. This gives lower compressor energy costs.

In all forms of condenser pressure control, the minimum maintained pressure should be the lowest which will give satisfactory operation, in the interests of running economy. An indication of the relative electricity costs for a 350 kW air-conditioning plant is given in Table 9.1.

9.15 STRAINERS

Piping circuits will usually contain a small quantity of dirt, scale and swarf, no matter what care is taken to keep these out. A strainer can be fitted in the compressor suction to trap such particles before they can enter the machine.

Such strainers are of metal mesh and will be located where they can be removed for cleaning. In some configurations two strainers may be fitted.

As an extra safeguard, on new compressors a fabric liner may be fitted inside the mesh strainer to catch fine dirt which will be present. Such liners must be removed at the end of the running-in period, as they create a high resistance to gas flow.

Oil strainers may be of metal mesh and within the sump, in which case the sump must be opened for cleaning. Self-cleaning disc strainers are also used, the dirt falling into a drain pot or into the sump itself. There is an increasing tendency to provide replaceable fabric oil filters external to the compressor body, following automobile practice.

9.16 CHARGING CONNECTION

In order to admit the initial refrigerant charge into the circuit, or add further if required, a charging connection is required. The safest place to introduce refrigerant is ahead of the expansion valve, which can then control the flow and prevent liquid reaching the compressor. The usual position is in a branch of the liquid line, and it is fitted with a shut-off valve and a suitable connector with a sealing cap or flange. A valve is needed in the main liquid line, just upstream from the branch and within reach.

9.17 CHECK VALVES

Non-return or *check valves* can be found in the following positions:

1. On reversible heat pump circuits, to prevent flow through expansion valves which are not in service on one cycle
2. On hot gas circuits, to prevent the gas entering another evaporator
3. Where several compressors discharge into one condenser, to prevent liquid condensing back to an idle compressor
4. Where two or more evaporators work at different pressures, to prevent suction gas flowing back to the colder ones

9.18 LIQUID REFRIGERANT PUMPS

Where separate evaporators are used in low-temperature coldrooms, and blast freezers, for example, liquid pumps can be used to circulate refrigerant from the suction separator (or 'surge drum'), through the evaporator(s) and back. In the separator, remaining liquid falls back and is re-circulated,

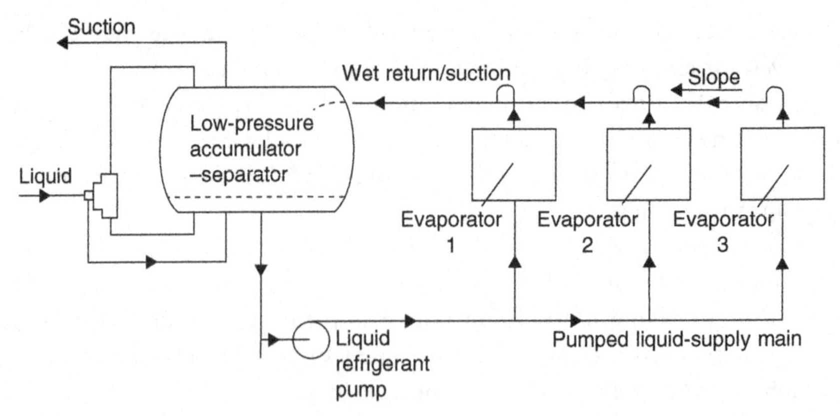

Figure 9.15 ***Pumped liquid circuit.***

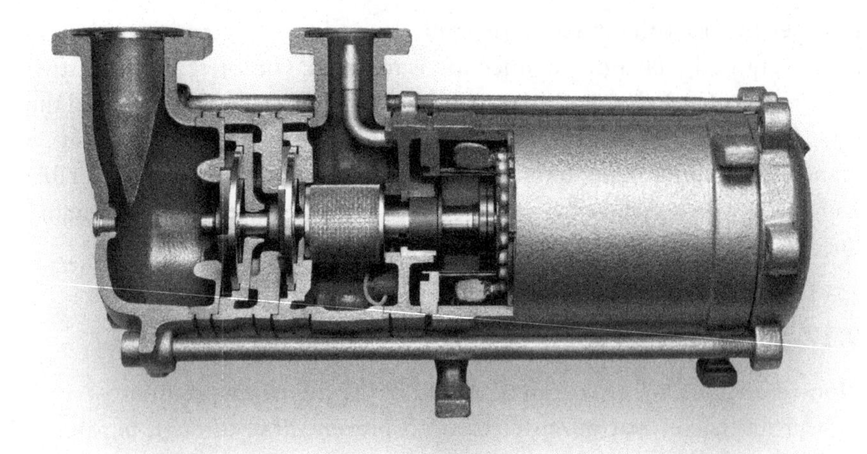

Figure 9.16 ***Witt hermetic liquid refrigerant pump (Titan).***

whereas vapour goes to the compressor (see Fig. 9.15). Liquid pumps are also used for spray chillers (see chapter: Evaporators). The liquid pump shown in Fig. 9.16 has a strainer at inlet and a second mesh to prevent particles from entering the shaft drillings which feed liquid to the bearings. These pumps are used mainly for ammonia systems, and are also supplied for CO_2.

CHAPTER 10

Component Selection and Balancing

10.1 INTRODUCTION

Selection and balancing the major system components so that they can operate over a specified range of conditions is a necessary part of the design process. Some fundamental approaches are included in order to provide an understanding, for many systems specifications will predetermine some of the conditions and constrain the options. For example, a supermarket chain may specify the evaporating, condensing temperature and refrigerant to be used. These constraints simplify the task and make comparisons easier. Moreover excellent selection software available from manufacturers and other sources can provide component and package equipment data for any desired operating conditions.

The information is based on refrigeration applications, but the same principles apply to air conditioning and heat pumps. Certain parameters such as air humidity can play a more dominant role in air conditioning.

10.2 CRITERIA AND OPTIONS

The four main components of a vapour compression refrigeration circuit – the evaporator, the compressor, the condenser and the expansion valve – must be selected to give a balanced system.

Each of these items must

1. be suitable for the application;
2. be correctly sized for the duty;
3. function as required in conjunction with the other components.

A characteristic of refrigeration system design is that there are almost always several solutions to the problem, some of which may be equally cost-effective. Individuals and companies often have their own preferred approach on the basis of experience. The system designer must examine the options that may be available in order to determine a best selection with reference to the following criteria, and there may be others.

Refrigeration, Air Conditioning and Heat Pumps
http://dx.doi.org/10.1016/B978-0-08-100647-4.00010-3

1. *Capital cost*: Obviously this criterion is of major importance, but it should be noted that reducing capital cost will almost certainly result in increased running cost and carbon emissions.
2. *Running cost*: The cost of electricity usually represents the major cost of ownership. Other costs may include other fuels, water, spares and operating and maintenance labour. It is probable that a small extra expenditure on some items, especially heat exchangers, will reduce running costs.
3. *Environmental impact*: This will be related to running cost, because the major environmental effect is associated with electricity consumption. TEWI or life cycle analyses can assist in comparing various design options.
4. *Installation time*: Installation of a new plant may cause serious disruption of the user's ongoing business, and the extent of this disruption should be determined prior to work commencing. Apart from the installation of the equipment itself, there is the associated builders' work and the temporary disconnection of other services. The use of factory-built packaged equipment helps to keep installation time to a minimum.
5. *Operational requirements*: Most systems are now automatic in operation, but users must be aware of the control system and have facilities to run on manual control, as far as this may be possible, in the event of a control failure. Operators must understand the function of the system. If not, they will not have the confidence to work on or with it, and the plant will not be operated at its best efficiency. Also, if it breaks down for any reason, they will be unable to put it right.
6. *Maintenance*: Most modern equipment is almost maintenance-free but the user must be aware of the maintenance functions required (see chapter: Commissioning and Maintenance), and to what extent the maintenance can be done in-house. Where maintenance is contracted out, it is preferable that the supplier should carry this out, at least for the warranty period.
7. *Life expectancy*: This is normally 15–20 years for refrigeration systems, and somewhat less for small packaged equipment. Where the need is for a shorter period, such as a limited production run or for a temporary building, rental or second hand equipment could be considered.

When considering the options an analytical approach should be adopted to ensure correct selection. The principles to be applied are those of value analysis – to start with the basic need and no pre-conceived method, to consider all the different methods of satisfying the need, and to evaluate each of these objectively before moving towards a choice.

The following guidelines should be taken as an indication of the factors to be considered, rather than as an exhaustive list:

1. What is the basic need?
   - a. to cool something
     - A dry product
       - (I) In air
         - (i) Temperature?
         - (ii) Humidity?
         - (iii) Maximum air speed?
     - Other solid product
     - A liquid
       - (I) What liquid?
       - (II) Temperature range?
       - (III) Viscosity?
   - b. to keep something cool
     - A solid product
       - (I) Conditions?
     - An enclosed space
2. What is the load and temperature?
   - a. If at ambient, can it be done without mechanical refrigeration?
   - b. Product cooling load?
   - c. Heat leakage, sensible and latent?
   - d. Convection heat gains, sensible and latent?
   - e. Internal heat gains?
   - f. Time required?
3. Constraints
   - a. Reliability?
   - b. Position of plant?
   - c. Automatic/manned?
   - d. Refrigerant?
   - e. Same type of equipment as existing?
4. Possible methods
   - a. Direct expansion?
     - Indirect – what medium?
     - Part by tower water or ambient air?
     - Thermal storage?
     - Existing plant spare capacity?

5. Location
   a. Plantroom?
   b. Adjacent space?
   c. Within cooled space?
   d. Maintenance access?
6. Condenser
   a. Inbuilt:
      - Water?
      - Air?
   b. Remote?
      - Availability of cooling medium?
   c. Maintenance access?

If these steps have been carried through in an objective manner, there will be at least three options for most projects, and possibly as many as five. Enquiries can now go out for equipment to satisfy the need, on the basis of the options presented. No attempt should be made to reach a decision until these have been evaluated.

10.3 EVAPORATING TEMPERATURE

The first consideration is the evaporating temperature. This will be set by the required load condition and the appropriate temperature differential (ΔT) across the evaporator. In the context of evaporator selection, the ΔT used is the difference between the evaporating refrigerant and the temperature of the fluid entering the cooler, not the log mean temperature difference.

In systems where the evaporator cools air, the air itself becomes the heat transfer medium and its temperature and humidity must be considered in relation to the end product. Where the product cannot suffer dehydration, the ΔT may be high, so as to reduce the size and cost of the coil, but the lower the evaporating temperature falls, the lower will be the capacity and COP of the compressor. In these circumstances, a first estimate might be taken with a ΔT of 10–12 K and cross-checked with alternatives either side of this range. In each case, the 'owning' cost, that is taking into account the running costs, should be considered by the user. Table 10.1 illustrates this for a cold store comparison, running 8760 h per year. The data should be used for comparative purpose only.

Unsealed products will be affected by low humidity of the air in the cooled space and may suffer dehydration. Conversely, some food products such as

Table 10.1 Comparison of running costs for various cooler sizes

			Annual electricity costs		
Cooler size	**Cost (£)**	**ΔT**	**Fans**	**Compressor**	**Total**
65	627	11.7	58	2140	2198
85	845	10.0	69	1970	2039
120	982	8.2	110	1820	1930

fresh meat will deteriorate in high humidity. Since the dew point of the air approaches the fin surface temperature of the evaporator, the inside humidity is a function of the coil ΔT. That is to say, the colder the fin surface the more moisture it will condense out of the air, and the lower will be the humidity within the space. Optimum conditions for all products likely to be stored in cooled atmospheres will be found in the standard reference books, or may be known from trade practice. The following may be taken as a guide:

Products that dehydrate quickly, such as most fruits and vegetables	$\Delta T = 4$ K
Products requiring about 85% saturated air	$\Delta T = 6$ K
Products requiring 80% saturation or drier	$\Delta T = 8$ K
Materials not sensitive to dehydration	$\Delta T = 10$ K upwards

A further consideration may be the possibility of reducing ice build-up on the evaporator, whether this is in the form of frost on fins or ice on the coils of a liquid chilling coil. Where temperatures close to freezing point are required, it may be an advantage to design with an evaporator temperature high enough to avoid frost or ice – either for safety or to simplify the defrost method.

10.4 EVAPORATOR

Once the evaporating temperature has been provisionally decided, an evaporator can be selected from catalogue data or designed for the purpose. The rating of an evaporator will be proportional to the temperature difference between the refrigerant and the cooled medium. Since the latter is changing in temperature as it passes over the cooler surface, an accurate calculation for a particular load is tedious and subject to error.

To simplify the matching of air-cooling evaporators to condensing units, evaporator duties are commonly expressed in basic ratings (see Fig. 10.1), in units of kilowatts per Kelvin. This rating factor is multiplied by the ΔT between the entering air and the refrigerant.

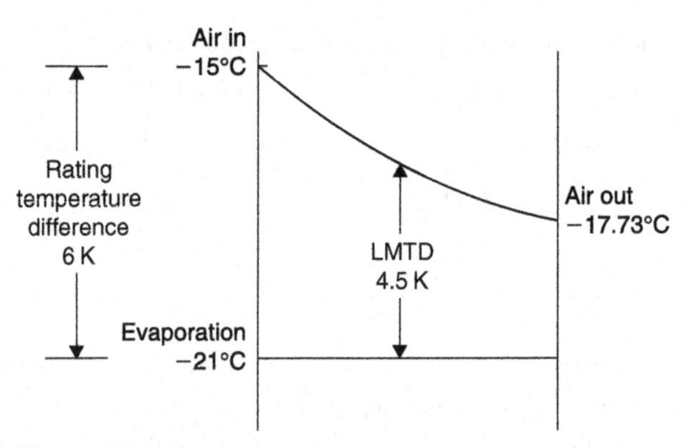

Figure 10.1 ***Basic evaporator rating and LMTD.***

This factor, the *basic rating,* is assumed constant throughout the design working range of the cooler and this approximation is good enough for equipment selection. The basic rating will change with fluid mass flow and, to a lesser extent, with working temperature. It may change drastically with fluids such as the glycol brines, since the viscosity and hence the convection heat transfer factor alter at lower temperatures. In unusual applications, the supplier should be consulted.

Example 10.1

An air-cooling evaporator has a mass air flow of 8.4 kg/s and a published 'rating' of 3.8 kW/K. What will be its rated duty at −15°C cold room temperature with refrigerant at −21°C? What is the true LMTD?

$$\text{Entering air temperature} = -15°\text{C}$$

$$\text{Refrigerant temperature} = -12°\text{C}$$

$$\text{'Rating' } DT = 6\,\text{K}$$

$$\text{Rated duty} = 3.8 \times 6 = 22.8\,\text{kW}$$

$$\text{Reduction in air temperature} = \frac{22.8}{1.006 \times 8.4} = 2.73\,\text{K}$$

$$\text{Air leaving temperature} = -15 - 2.73 = -17.73°\text{C}$$
$$\text{LMTD} = 4.5\,\text{K}$$

There would be an error at other conditions since the basic rating is only accurate at one point, so this short-cut factor must only be used within the range specified by the manufacturer.

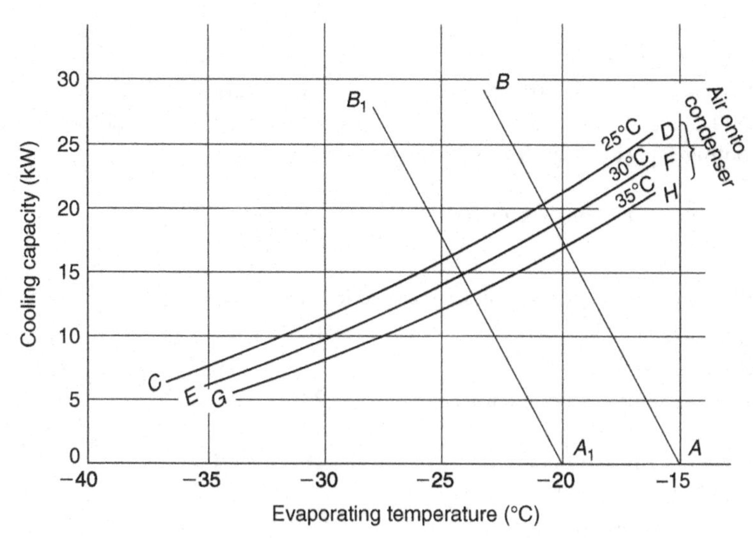

Figure 10.2 ***Graphical balance of evaporator with a condensing unit.***

A graphical method of balancing such an evaporator with a condensing unit is shown in Fig. 10.2. The condensing unit capacity is shown as cooling duty against evaporator temperature, line CD. The coil rating is plotted as the line AB, with A at the required cold room (or 'air-on') temperature, and the slope of the line AB corresponding to the basic rating. The intersection of this line with the condensing unit curve CD gives the graphical solution of the system balance point. Similar constructions for higher condenser air conditions (EF, GH) or different room temperatures (A1B1) will show balance points for these conditions.

The graph also indicates the change in evaporating temperature and coil duty when the ambient is lower or higher than the design figure. This will show if there is any necessity to control the evaporating temperature in order to keep the correct plant operation.

Frequently, coil data will be available for a design air flow, but the system resistance reduces this flow to a lower value. There is a double effect: the lowering of the LMTD and lower heat transfer from the coil by convection.

The outer surface coefficient is the greatest thermal resistivity (compared with conduction through the coil material and the inside coefficient), and a rough estimate of the total sensible heat flow change can be made on the basis of:

$$h = \text{constant} \times (V)^{0.8}$$

Example 10.2

An air-cooling coil extracts 45 kW sensible heat with air entering at 24°C and leaving at 18°C, with the refrigerant evaporating at 11°C. Estimate the cooling capacity at 95, 90 and 85% mass air flow.

$$\text{Design mass air flow} = \frac{45}{1.02 \times (24 - 18)} = 7.35\,\text{kg/s}$$

An approximate analysis gives the following result:

Air flow (%)	100	95	90	85
Mass air flow (kg/s)	7.35	6.99	6.62	6.25
Air temperature on coil (°C)	24	24	24	24
ΔT for 45 kW (K)	6	6.3	6.7	7.1
Air temperature off coil (°C)	18	17.7	17.3	16.9
LMTD, refrigerant at 11°C (K)	9.7	9.5	9.2	9.0
h, in terms of design (from V0.8) (%)	100	96	92	88
Capacity (45 × h × LMTD)/9.7 (kW)	45	42.3	39.3	36.7

With all calculations involving convective heat transfer, it must be remembered that the figures are predictions based on previous test data, and not precise.

10.5 COMPRESSOR

It is usually possible to identify several compressor types for a particular job, and at least two alternatives should be considered before making a final selection.

Compressor capacities may be published as tables or curves as in Fig. 10.3 and will be for a given refrigerant and a range of condensing pressures. Similar curves for input power and current (amps) will usually be provided. Limitations of application range are also normally given. The compressor data should state the superheat and sub-cooling reference and this should correspond to a standard vapour compression cycle or transcritical R744 cycle as defined in European Standard EN12900. The standard for the sub-cooling is zero. Several superheat options are given in the standard, and these are intended to represent conditions close to the normal running condition of the compressor. Smaller compressors can be rated at a constant suction gas return temperature which is applicable to remote installation where the

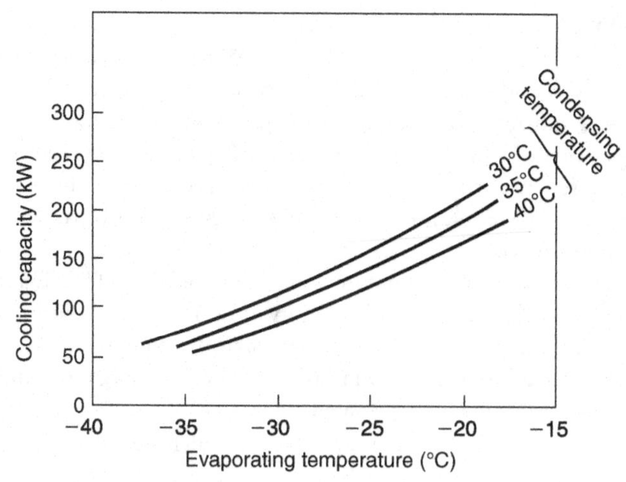

Figure 10.3 ***Compressor-capacity ratings in graphical form.***

gas temperature can be expected rise on its way to the compressor. Ammonia compressors are rated at 5 K superheat, corresponding to flooded evaporator applications.

The reason for these differences is to minimise any error arising from superheat corrections. The data will need to be corrected for the actual superheat, sub-cooling and any non-useful heat pick-up. Changing the superheat can affect the volumetric efficiency and for large changes a correction based solely on change in gas density may be inadequate. The correction for sub-cooling is a straightforward enthalpy ratio which can be made with the aid of a *P–h* chart or computer. If there are significant pressure drops between the compressor and either the condenser or the evaporator, these may also need to be accounted. These corrections can usually be made with the aid of manufacturers' software. Example 10.3 can be used to gain an understanding of the processes involved.

The rating conditions refer to the compressor refrigerant inlet and outlet (suction and discharge). There is no need for the user to consider change of conditions as the gas passes over an enclosed motor. All processes occurring within the machine are accounted in the published ratings. For enclosed types the input power refers to electrical input and for open compressors it refers to shaft power. For inverter driven variable speed enclosed types the power is normally stated at the inverter input.

Example 10.3

A semi-hermetic compressor is rated at 30 kW for R134a when evaporating at −12°C, condensing 45°C at standard conditions of 20°C suction gas temperature and zero sub-cooling. The application will be for 0°C suction gas temperature of which 5 K superheat is usefully obtained in the evaporator and the remainder is non-useful heat gain in the pipes. Sub-cooling is 5 K. Estimate the compressor capacity and the evaporator capacity.

1. Mass flow correction
   a. Compressor inlet temperature is changed from 20 to 0°C. This means the incoming gas is higher density (lower specific volume). From R134a properties using *P–h* chart, tables or computer, the ratio of specific volumes is 123.67/113.65 = 1.088. Assuming the compressor pumps the same volume flow rate:

      $m_2/m_1 = 1.088$, that is the mass flow rate increases by 8.8%
2. Correction for enthalpy at the compressor inlet
   a. From R134a properties using *P–h* chart, tables or computer,
   b. Suction gas enthalpy at 20°C and 0°C is 418.87 and 401.51 kJ/kg, respectively
   c. Liquid enthalpy at the expansion valve inlet at 45°C is 264.75 kJ/kg
   d. Evaporator enthalpy difference at rating condition = 418.87 − 264.75 = 154.12
   e. Evaporator enthalpy difference with 0°C suction = 401.51 − 264.75 = 136.76
   f. Enthalpy difference ratio = 136.76/154.12 = 0.887
   g. Compressor capacity corrected for suction temperature change = 30 × 1.088 × 0.887 = 28.96 kW
   h. The reduction in suction gas temperature increases the mass flow rate, but because the enthalpy at the compressor inlet is reduced, the effect of the change in suction gas temperature alone would be to cause a small overall reduction in compressor capacity.
3. Correction for enthalpy at the compressor inlet and expansion valve inlet:
   a. With 5 K sub-cooling, liquid enthalpy at the expansion valve inlet at 40°C is 257.15 kJ/kg
   b. Evaporator enthalpy difference at application condition = 401.51 − 257.15 = 144.36
   c. Enthalpy difference ratio = 144.36/154.12 = 0.9366
   d. Actual compressor capacity = 30 × 1.088 × 0.9366 = 30.57 kW
   e. The capacity reduction due to reduction of suction temp is more than offset by a decrease in liquid enthalpy due to sub-cooling.
4. To find the evaporator capacity consider the useful enthalpy difference
   a. Suction gas enthalpy at evaporator outlet, −7°C (5 K superheat) = 395.54 kJ/kg
   b. Useful evaporator enthalpy difference = 395.54 − 257.15 = 138.39 kJ/kg
   c. Enthalpy difference ratio = 138.39/154.12 = 0.898
   d. Actual evaporator capacity = 30 × 1.088 × 0.898 = 29.31 kW

There is a loss of capacity due to gas temperature rise in the suction line. Possible ways to reduce this loss include insulation (to reduce compressor inlet temperature), bring compressor closer to cooler. The magnitude of these changes is refrigerant and conditions dependent, and so whilst the changes in capacity for R134a calculated previously are quite small, this may not be the case with other refrigerants, particularly at low evaporating temperatures. A suction-to-liquid heat exchanger may be added to raise suction temperature and thus reduce heat pick-up.

The enthalpy differences applicable to Example 10.3 are labelled in Fig. 10.4 where two cycles are shown. The broken lines denote the standard cycle with a compressor inlet temperature of 20°C and zero sub-cooling. The standard compressor capacity is based on enthalpy difference 'A'. The solid lines indicate the desired cycle with a useful cooling capacity based on enthalpy difference 'B', the non-useful heat gain being based on 'C'. It is easy to see that B is less than A, however the mass flow rate is higher with B because of the lower specific volume of the vapour at the compressor inlet. The lines of constant specific volume have been omitted for clarity.

The condenser heat rejection is based on enthalpy difference 'D' in Fig. 10.4. The compression lines shown are for a 70% isentropic compression. The power value shown in a compressor datasheet enables the enthalpy at the compressor exit to be calculated. As a close approximation the power will not change with suction temperature.

As previously mentioned, the published compressor capacity refers to conditions at the compressor suction and discharge. If the pressure drop in the lines is significant, the evaporating and condensing temperatures at the compressor suction and discharge pressures must be considered, rather than those in the heat exchangers, when referring to compressor capacity data.

When selecting the compressor, a first guess must be taken for the condensing temperature, and this might be 15 K above the summer dry bulb for an air-cooled condenser or 12 K above the wet bulb temperature in the case of water or evaporative cooling. The balance condition between the evaporator and the compressor can be visualised in a graphic solution, superimposing the basic rating of the cooler on the compressor curves (see Fig. 10.5).

Although it is usual to consider only the balance at the maximum summer ambient, but most of the running time is likely to be at more moderate conditions. If the compressor(s) are then unloaded the condensing temperature can be reduced, taking advantage of the now proportionately larger heat exchanger surface. If the balance conditions are not favourable to the product, some average choice may be made, or an evaporator-pressure-regulating

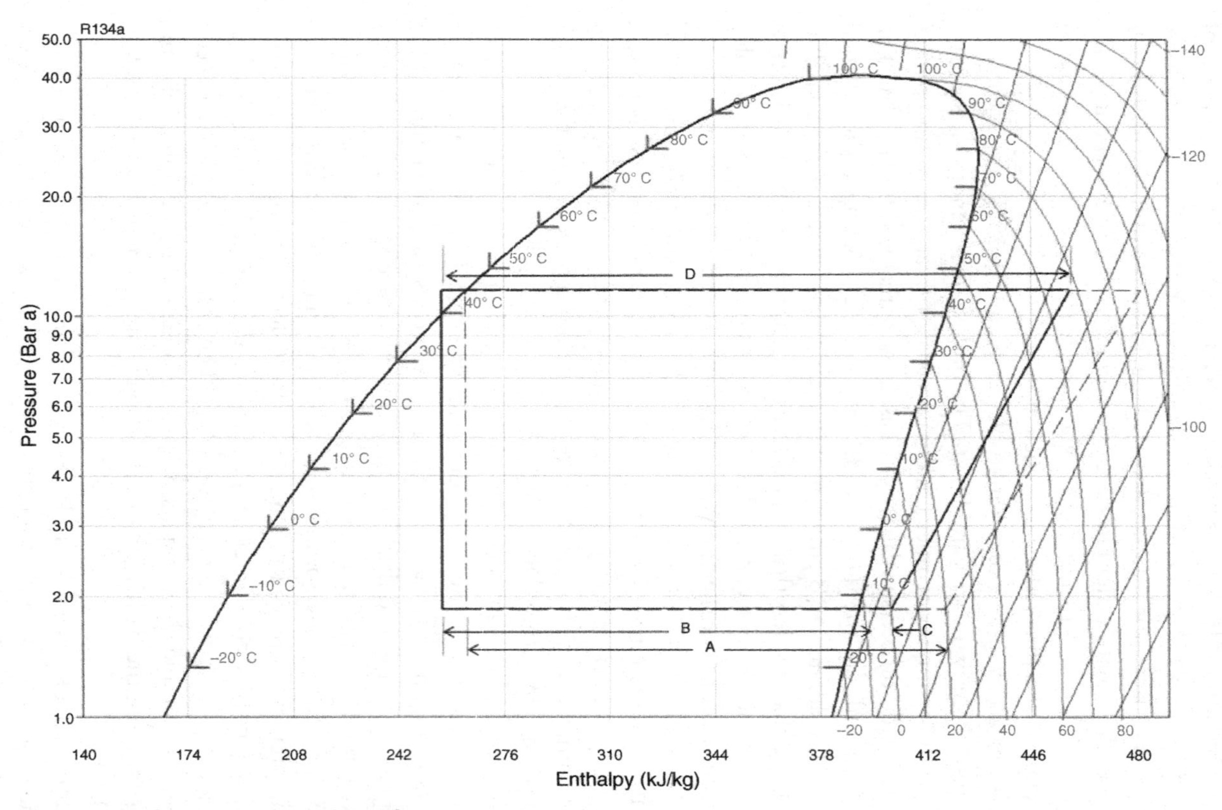

Figure 10.4 ***Compressor capacity and evaporator capacity.***

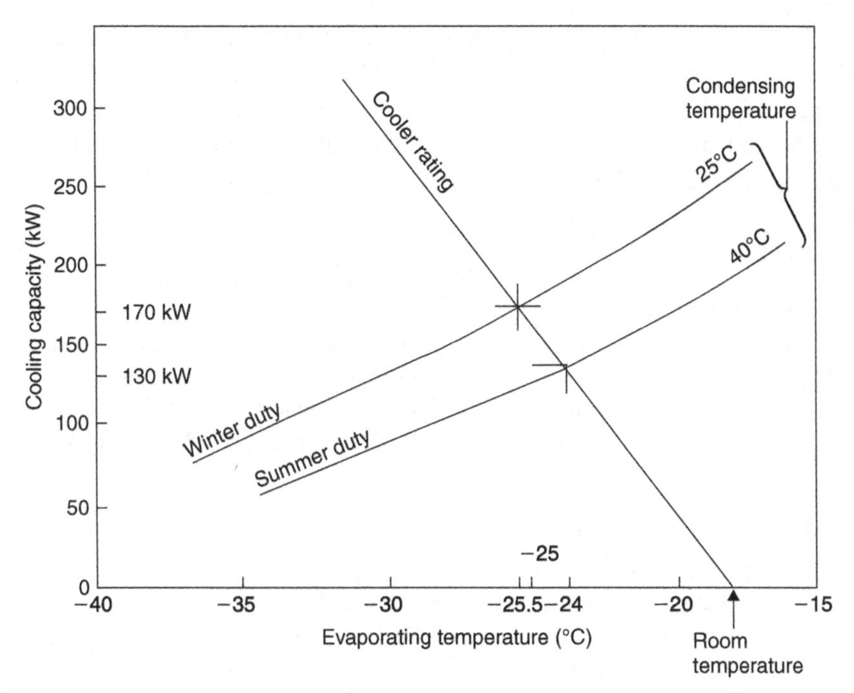

Figure 10.5 ***Balance condition between compressor and evaporator.***

valve inserted to prevent the evaporating temperature dropping too low (see Fig. 9.2). A different set of conditions will occur if the compressor has capacity control. A compressor with 50% capacity control may be connected to two equal evaporators, and one of these shut off at half load.

10.6 CONDENSER

A first guess of a condensing temperature has already been taken as a guide. Users should be aware of the large difference in owning costs arising from the choice of condenser, so the options should be compared on this basis. Certain machines, such as the centrifugal compressor, are very sensitive to high condensing conditions, and the correct choice (in this case, of a cooling tower) can give a considerable gain in COP.

Catalogue ratings show heat rejected at a stated condensing temperature and related to the following:

- Ambient dry bulb temperature for air-cooled condensers
- Available water temperature for water-cooled condensers or
- Ambient wet bulb temperature for evaporative types

For hydronic heat pumps the condensing temperature needs to be as low as possible as explained in chapter: Condensers and Cooling Towers (Fig. 6.12).

A condenser selected on first cost only will almost certainly be undersized and operate at a high head pressure.

Example 10.4

An application requires a cooling capacity of 218 kW and the running time is 2000 h/year at an electricity cost of 8 p/(kWh). In order to achieve the condensing temperature of 30°C the condenser would cost £14000, while a smaller condenser for a temperature of 35°C would cost £8500. Estimate the pay back time if the larger condenser is fitted.

Condensing temperature	**30°C**	**35°C**
Rated capacity of plant (kW)	242	218
Running time for 218 kW × 2000 h	1,802	2,000
Compressor electrical input power (kW)	60	70
Electricity cost per year (£)	11,533	14,933
Electricity saving per year (£)	3,400	

$$\text{Break-even time} = \frac{14000 - 8500}{3400} = 1.6 \text{ years}$$

This is a very approximate calculation, based on direct capital cost and not on interest rates, and needs to be analysed in terms of the general plant economics. Also variations due to seasonal air temperature changes are not accounted. It should also be borne in mind that this is based on illustrative electricity costs, and a greater saving will be made as fuel costs rise. Tendering contractors and prospective users should make themselves aware of alternatives of this sort.

In most climates the wet bulb temperature is well below the dry bulb temperature and there can be an advantage in using water or evaporative cooling for larger plant. Maintenance costs need to taken into account as outlined in chapter: Condensers and Cooling Towers. Concern over spray-borne diseases may also indicate a preference for air cooling. Table 10.2, shows condensing temperatures based on tentative temperature differences of 15 K and 12 K. For example, if it is decided to use an air-cooled condenser, there will be considerable economy in

Table 10.2 Typical condensing temperatures for air-cooled and evaporative condensers in various locations

	Air-cooled		Evaporative	
Climate	**Dry bulb (°C)**	**Condenser (°C)**	**Wet bulb (°C)**	**Condenser (°C)**
South United Kingdom	27	42	21	33
Scotland	24	39	18	30
Mediterranean	32	47	24	36
Desert	47	62	24	36
Tropical humid	33	48	28	40

sizing the condenser for a reduced condensing temperature, from a first guess of 62°C down to, possibly, 56°C.

The maximum design condensing temperature will only apply when the ambient is at its maximum, and full advantage should always be taken to allow this temperature to drop at cooler times, down to its minimum working limit. Cooling systems should allow condensing temperature to drop to at least 25°C when the cooling medium permits this, and some systems can go lower. A true estimate of total owning cost should take this into account.

The performance of condensers with a compressor–evaporator system can be shown graphically as in Fig. 10.6. The curves are the rejected heat from the compressor, that is cooling duty plus compressor power. These are plotted against the basic performance of the condenser. A development of graphical methods for matching of components is given by Sulc (2007). Some condenser manufacturers provide rating curves based on the cooling capacity of the compressor and using typical factors for the power. Where there is significant non-useful cooling as in Example 10.3, the heat rejection load will include this additional capacity. The sub-cooling may occur after the condenser in which case this load can be deducted.

Air cooled condensers require a large air flow for a given heat rejection, and the ability to locate them where this air flow can be obtained without re-circulation may limit their use. Water or evaporative cooling should always be considered as a possibility for large equipment.

10.7 EXPANSION VALVE

The expansion valve is a passive orifice through which the liquid refrigerant is forced by the pressure difference between the condensing and the evaporating conditions. If a system could always operate at fixed conditions,

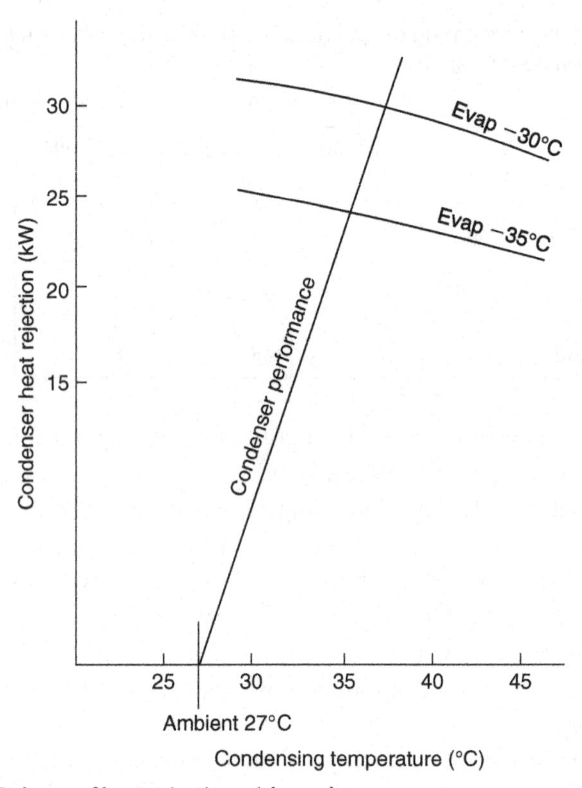

Figure 10.6 ***Balance of heat rejection with condenser.***

correctly charged, a simple restriction would fulfil the requirement. In practice this is never the case and control is necessary. Expansion devices in general use are the following:

1. Capillary tubes, for small hermetic systems. These are factory selected and cannot be adjusted.
2. Solenoid valves with liquid level sensors or liquid level valves for flooded evaporators.
3. High-pressure float valves plus handset throttle valves for some flooded and low-pressure receiver circuits.
4. Thermostatic expansion valves or electronic expansion valves for direct expansion circuits.

Capacity ratings are given in the catalogues of manufacturers and suppliers. Troubles can arise with the selection of thermostatic expansion valves of the type generally used in custom-built systems. It is usual to select a thermostatic expansion valve for the maximum duty and at the summer condensing condition, taking into account the pressure drop through a liquid

Table 10.3 Typical expansion valve capacities for specific evaporating temperature

Pressure difference (bar)	2	4	6	8	10	12	14
Valve duty (kW)	0.77	0.95	1.08	1.16	1.22	1.24	1.26

distributor in the case of a multiple-feed coil. Valve ratings are given for a range of pressure differences, that is for a range of condensing conditions for a specific refrigerant and evaporating condition, as in Table 10.3. It might be thought that the duty varies with pressure difference according to fluid flow laws, but this will not be the case because phase change occurs as the pressure falls. This means that the valve may be able to pass more liquid at low condensing pressures. Conversely, if the valve is selected at a lower pressure difference (possibly corresponding to a condensing condition in the United Kingdom of 20–25°C), the valve will not be grossly oversized at the maximum summer condition. Although no standards exist for valve rating conditions it is conventional to state a nominal capacity at 38°C condensing, 4°C evaporating with 1 K sub-cooling and 3–4 K superheat. Sometimes the data takes the form of nominal capacity and tabulated correction factors.

Unless a thermostatic expansion valve is very tightly rated, the system will operate satisfactorily at a lower condensing condition in cool weather, with a gain in compressor duty and lower power input. A growing awareness of energy economy is leading to more careful application of this component. Suppliers are ready to help with advice and optimum selections.

A greater difficulty arises where the compressor may go down to 33 or 25% capacity and the thermostatic expansion valve is called upon to control a much reduced flow. Under such conditions, the thermostatic expansion valve may be unstable and 'hunt', with slight loss of evaporator efficiency. Since the required duty is less, this may appear to be of no great importance but should avoided. It is possible to fit two expansion valves in parallel and isolate one at part load, but this arrangement is not usually necessary.

Low condensing-pressure operation should present no problem with float or electronic expansion valves, since these can open to pass the required flow of liquid if correctly sized.

10.8 SIZING PIPE AND OTHER COMPONENTS

Refrigeration system pipes are sized to offer a low resistance to flow, since this reflects directly on compression ratio, commensurate with economy of pipe cost and minimum flow velocities to ensure oil return.

Pressure losses due to pipe friction can be calculated from the basic formulas established by Reynolds and others. However, as with the calculation of heat transfer factors, this would be a time-consuming process and some of the parameters are not known accurately. Recourse is usually made to simplified estimates, tables published in works of reference, or software for this purpose.

Pressure drops on the high-pressure side will usually be small enough to have little effect on the performance of the complete system. Pressure losses in the suction pipe and its fittings, especially if this is long, should be checked, and a correction made for the actual compressor suction pressure. For low-temperature applications, pipe sizes may have to be increased to avoid excessive frictional losses at these low pressures; oil return must always be considered.

Flow-control valves, such as evaporator pressure regulators, will not necessarily be the same nominal size as the pipe in which they are fitted. Manufacturers' data for selection of their products is usually very comprehensive, and their guidance should be sought in case of any doubt.

10.9 RE-CHECK COMPONENTS

In the course of carrying through an equipment selection of this sort, several options may be tried. It is essential to make a final check on those selected to ensure that the correct balance has been achieved. Predicted balance figures should be noted, to guide the final commissioning process and subsequent operation.

CHAPTER 11

Installation and Construction

11.1 INTRODUCTION

Many larger systems require on-site installation although here is a trend towards more factory-assembled systems or packages where assembly and testing can be done under controlled conditions. The contents of this chapter relate mainly to installation of medium/large equipment in buildings.

Successful site work demands coordination of the following:

1. Site access or availability
2. Supply on time, and safe storage of materials
3. Availability of layout drawings, flow diagrams, pipework details, control and wiring circuits, material lists and similar details
4. Availability at the correct time of specialist trades and services – builders, lifting equipment, labourers, fitters, welders, electricians, commissioning engineers, etc.

Where site work is carried out by sub-contractors representing specialist trades, it is essential that authority and executive action are in the hands of a main contractor and that this authority is acknowledged by the sub-contractors. If this is not so, delays and omissions can occur, with divided responsibility and lack of remedial action.

The controlling authority must, well before the start of work, draw up a material delivery and progress chart and see that all sub-contractors (and the customer) are in agreement and that they are kept informed of any changes.

11.2 MATERIALS

Materials used in the construction of refrigeration, air-conditioning and heat pump systems are standard engineering materials, but there are a few special points of interest:

1. Compressor bodies are generally of grey cast iron, although some are aluminium or fabricated from mild steel. Hermetic types have steel shells.
2. Compressor pistons are of cast iron or aluminium, the latter following automobile practice.
3. Piping for the smaller halocarbon installations is usually of copper, because of the cleanliness and the ease of fabrication and jointing.

Refrigeration, Air Conditioning and Heat Pumps
http://dx.doi.org/10.1016/B978-0-08-100647-4.00011-5

4. Some stainless steel pipe is used, mainly because of its cleanliness, although it is more difficult to join.
5. Most other piping is mild steel. For working temperatures below −45°C, only low-carbon steels of high notch strength are used. A leak to atmosphere with R744 can cause very low temperatures and thus, low-temperature steels must be used. Where water and air contamination could cause formation of carbonic acid, stainless or half hard copper is preferable.
6. Aluminium tube is used to a limited extent, with the common halocarbons and also with ammonia.
7. Copper and its alloys are not used with ammonia.
8. Sheet steel for ductwork, general air-conditioning components and outdoor equipment is galvanised.
9. Elastomer seals must always be checked for compatibility with the refrigerant and oil.

Specific guidance on materials and their application may be had from various works of reference.

11.3 PRESSURE SAFETY AND CONTAINMENT

Refrigeration systems contain pressurised fluid and there are certain safety standards and legal requirements that must be adhered to. Under the European Pressure Equipment Directive (PED) and the UK Pressure Equipment Regulations the main duties are placed on the user/owner of the system. They are a clear and practical means of legislating for safe practices in refrigeration. Responsible contractors and users will always use such safe procedures. In addition to the regulations themselves the HSE has published 'Safety of pressure systems – Approved Code of Practice' which is a clear and helpful. The regulations apply to vapour compression refrigeration systems incorporating compressor drive motors, including standby compressor motors, having a total installed power exceeding 25 kW.

Factory-built equipment will be constructed to the relevant standards and will be pressure-tested for safety and leaks prior to shipment. In cases of doubt, a test certificate should be requested for all such items. Under the PED, vessels, including compressors, are categorised, depending on the refrigerant and volume. Those falling into certain categories will be CE marked and for smaller ones, not categorised, a statement of sound engineering practice can be obtained from the manufacturer.

It is necessary to hold a Safe Handling of Refrigerants Certificate to work with refrigerants. This can be obtained through short training courses. Maintenance engineers must keep themselves updated on safety procedures and training requirements.

Site-erected pipework, once complete, must be pressure tested for safety and leak tightness. The pressure test should be carried out in accordance with the current safety standard BS EN378. The test pressure requirement is dependent on the category under the PED 97/23/EC, currently between 1.1 and 1.43 times the maximum allowable pressure, PS. The Institute of Refrigeration's Codes of Practice provide guidance.

Factory-built components and pressure vessels which have already undergone test should not be retested unless they form part of the circuit which cannot be isolated, when the test pressure must not exceed the original figure. Site hydraulic testing is considered unnecessary, owing to the extreme difficulty of removing the test fluid afterwards. However, it must always be appreciated that site testing with gases is a potentially dangerous process, and must be governed by considerations of safety. In particular, personnel should be evacuated from the area and test personnel themselves be protected from the blast which would occur if a pressure vessel exploded.

Systems should be pressure tested with dry (oxygen-free) nitrogen (OFN) or high-purity nitrogen. Nitrogen is used from standard cylinders, supplied at about 200 bar, and a proper reducing valve must always be employed to get the test pressure required. A separate gauge is used to check the test pressure, since that on the reducing valve will be affected by the gas flow.

If the high side is being tested, the low side should be vented to the atmosphere, in case there is any leakage between them that could bring excessive pressure onto the low side. It may be necessary to remove relief valves. Other valves within the circuit will have to be open or closed as necessary to obtain the test pressure. Servo-operated valves will not open on a 'dead' circuit, and must be opened mechanically.

The test pressure should be maintained for at least 15 min. If the pressure has not significantly reduced in that period the nitrogen is slowly vented until the pressure in the system has reduced to the pressure test (leak test) pressure. To determine if leaks exist, new equipment may be left pressurised at leak test pressure overnight or for longer periods, and any pressure drop noted. Pressure will change with temperature, and so this must be taken into consideration. Another option is to leave the equipment under vacuum for a period. A traditional way of finding leaks is to use soapy water. Many people discount it, but for finding leaks it is possibly the most effective method. It can be used to find very small leaks. All leaks must be fixed before equipment is put into service. Electronic leak detectors are should be checked for their suitability for different refrigerants. It is important to use a detector of sufficient sensitivity; it should be capable of detecting a leak of 5 g/year.

Reference should be made to the codes of practice and guidance notes published by the Institute of Refrigeration (see Bibliography). Leak testing is covered in chapter: Commissioning and Maintenance.

11.4 PIPE-JOINING METHODS

Steel pipe should be welded, except for joints which need to be taken apart for service, which will be flanged. It is essential that welding is carried out by competent craftsmen and is subject to stringent inspection.

Flanges for ammonia (and preferably, also, for other refrigerants) must be of the tongued-and-grooved type which trap the gasket.

Copper tube should use brazed joints. It can be bent to shape in the smaller sizes and the use of bending springs or formers is advised to retain the full bore.

Where fittings are required, these should be of copper or brass to give a correct capillary joint gap of not more than 0.2 mm, and joined with brazing alloy.

The brazing of copper tube in air will leave a layer of copper oxide inside, which may become detached and travel around the circuit. It is therefore necessary to pass nitrogen into the pipe before heating, and to maintain a flow until brazing is complete. The use of special grades of oxygen-free or moisture-free nitrogen is not necessary.

11.5 PIPING FOR OIL RETURN

The sizing and arrangement of suction and discharge piping for the halocarbons must ensure proper entrainment of oil, to return this to the compressor. Pipes for these gases usually have a higher velocity at the expense of a greater pressure drop than those for ammonia. Pipe sizes may only be increased in runs where the oil will be assisted by gravity to flow in the same direction as the gas.

Horizontal pipes should slope slightly downwards in the direction of flow, where this can be arranged. If a suction or discharge line has to rise, the size may be decreased to make the gas move faster. In the case of a lift

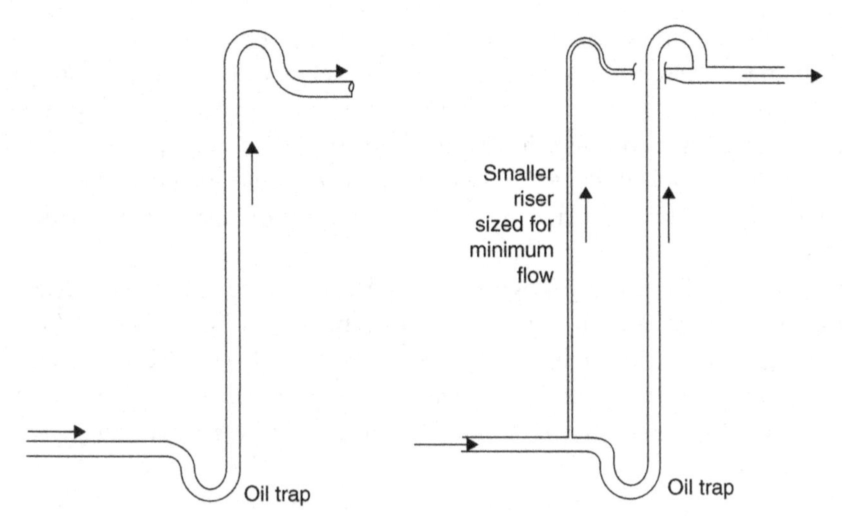

Figure 11.1 ***Gas risers for oil return.***

of more than 5 m, a trap should be formed at the bottom to collect any oil which falls back when the plant stops.

Suction and discharge risers (Fig. 11.1) will normally be sized for full-compressor capacity, and velocities will be too low if capacity reduction is operated. In such installations, *double risers* are required, the smaller to take the minimum capacity and the two together to carry the full flow. Traps at the bottom and goosenecks at the top complete the arrangement. At part capacity, any oil which is not carried up the main riser will fall back and eventually block the trap at the bottom, leaving the smaller pipe to carry the reduced flow, with its quota of oil. When the system switches back to full capacity, the slug of oil in this trap will be blown clear again.

11.6 SITING, ACCESS AND PIPE SUPPORTS

Air movement is always a consideration when siting units that depend on air flow to function correctly. Air-cooled condensers must be spaced from walls and each other to allow free movement of air without recycling. Indoors, close proximity of air distribution units to walls and ceiling can cause mal-distribution in the conditioned space and even result in the air flow to cause unsightly marking.

Access for maintenance is another consideration when siting equipment. Sufficient room should be allowed to facilitate compressor or heat exchanger removal. Piping must be properly supported at frequent intervals to limit stress and deflection. Supports must allow for expansion and contraction which will occur in use, special care must be taken with long straight runs which can expand significantly when hot. This has been known to cause distortion, high stresses and equipment failure. Pipework which might form a convenient foothold for persons should be protected from damage by providing other footholds and guarding insulation.

Stop valves, especially those which might need to be operated in a hurry (and this means most, if not all, of them), should have easy access. Where they are out of reach, reliance should not be placed on moveable ladders, which may not be there when needed, but permanent access provided. Chain-operated wheels can be fitted to the larger valves, to permit remote operation.

Emergency stop valves must not be placed in tunnels or ducts, since personnel may be subject to additional danger trying to operate them.

11.7 INSTRUMENTS

Provision should be made for temperature measurements at compressor suction and discharge, and at expansion valve liquid entry. Likewise, pressure measurements should be possible at compressor suction and discharge. Even if sensors are permanently fitted to relay information to control systems, provision should be made for connecting external instruments. This can be considered as a minimum requirement and will enable a log to be kept during initial commissioning and also during the life of the plant (see chapter: Commissioning and Maintenance). The advent of the electronic probe thermometer has simplified commissioning work, and the fitting of thermometer wells is less important. Even so, such facilities are worth considering when large pipes are being erected, and it will be necessary with insulated pipes if true temperatures are to be taken without damaging the insulation.

Wells should slope downwards into the pipe, so that they can be part filled with liquid to provide better thermal contact. Where a pipe temperature is a critical factor in the operation of a system, it is usually worth fitting a permanent thermometer.

The monitoring of temperatures for electronic control systems is now mainly by thermocouples or resistance thermometers and these can be secured onto the outside of the pipe using a thermal paste and aluminium tape and the pipe then insulated over.

Pressure gauges should always be fitted on the discharge side of liquid pumps, to check performance and give warning of a possible drop in flow resulting from dirty strainers. Manometer pressure gauges are required across air filters.

11.8 RISING LIQUID LINES

If liquid refrigerant has to rise from the condenser or receiver to an expansion valve at a higher level, there will be a loss of static head, and the refrigerant may reach its boiling point and start to flash off. Under such circumstances, bubbles will show in.

Example 11.1

R717 condenses in a circuit at 32°C and is subcooled to 30°C before it leaves the condenser. How much liquid lift can be tolerated before bubbles appear in the liquid line?

Saturation pressure at 32°C = 12.37 bar
Saturation pressure at 30°C = 11.68 bar
Permissible pressure drop = 0.69 bar (69 000 Pa)
Liquid density = 590 kg/m^3
Possible loss in static head = 69,000/(9.81 × 590) = 12 m approximately

Ammonia has low density, and the safe liquid lift for other fluids may be much less. Where a high lift cannot be avoided, the liquid must be sub-cooled enough to keep it liquid at the lower pressure. Sub-cooling can be accomplished by fitting a sub-cooling coil to the condenser, a water-cooled sub-cooling coil, a suction-to-liquid heat exchanger before the lift, or a refrigerated sub-cooler.

To reduce the risk of 'flash gas' forming, the condenser should always be higher than the evaporator, if this can be arranged.

The same effect will occur where the liquid line picks up heat on a horizontal run, where it may be in the same duct as hot pipes, or pass through a boiler house. If the liquid in the sight glass flashes even with the addition of refrigerant, the possibility of such extra heat gain should be investigated. To cure this, insulate the pipe.

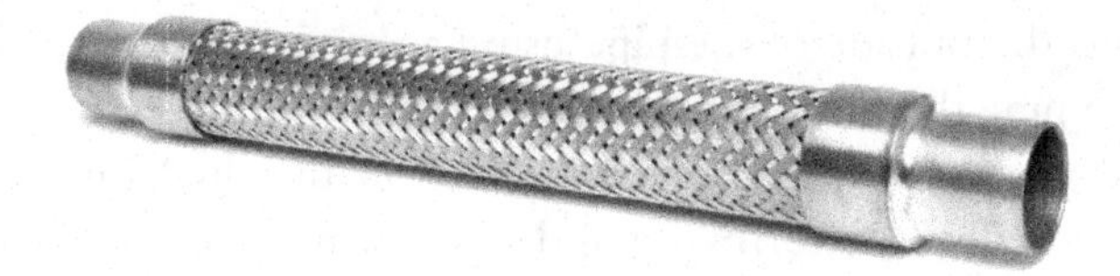

Figure 11.2 ***Vibration eliminator (Henry Technologies).***

11.9 VIBRATION

Compressors and pumps will transmit vibration to their connecting pipework. Mounting methods will be recommended by manufacturers and should be used. Where vibration eliminators (Fig. 11.2) are inserted in suction and discharge lines, they should be in the same plane as the compressor shaft, that is horizontal for semi-hermetic and open reciprocating types and vertical for all vertical scrolls.

Water and brine pumps may be isolated with flexible connectors. For small-bore pipes, these can be ordinary reinforced rubber hose, suitably fastened at each end. For larger pipes, corrugated or bellows connectors of various types can be obtained. In all cases, the main pipe must be securely fixed close to the connector, so that the latter absorbs all the vibration. Flexible connectors for the refrigerant usually take the form of corrugated metal hose, wrapped and braided. They should be placed as close to the compressor as possible.

A great deal of vibration can be absorbed by ordinary piping up to 50 or 65 mm nominal bore, providing it is long enough and free to move with the compressor. Three pieces, mutually at right angles and each 20 diameters long, will suffice. At the end of these vibration-absorbing lengths, the pipe must be securely fixed.

In all instances of anti-vibration mounting of machinery, care must be taken to ensure that other connections – water, electrical, etc. – also have enough flexibility not to transmit vibration.

11.10 CLEANLINESS OF PIPING

Only perfectly clean and corrosion-free tubing should be used. All possible dirt should be kept out of pipes and components during erection. Copper pipe will be clean and sealed as received, and should be kept plugged at all times, except when making a joint. Use the plastic caps provided with the tube – they are easily seen and will not be left on the pipe. Plugs of paper and rag tend to be forgotten and left in place. Steel pipe will have an oily coating when received, and it is important that this should be wiped out, since the oil will otherwise finish up in the sump and contaminate the proper lubricating oil. If pipe is not so cleaned, the compressor oil should be changed before the plant is handed over.

Avoidable debris includes loose pieces of weld, flux and the short stubs of welding rod often used as temporary spacers for butt welds. Pipe should only be cut with a gas torch if all the oxidised metal can be cleaned out again before closing the pipe.

It should be borne in mind that all refrigerants have a strong solvent effect and swarf, rust, scale, water, oils and other contaminants will cause harm to the system, and possible malfunction, and shorten the working life.

11.11 EVACUATION

Following pressure testing and prior to evacuation, all air and moisture should be removed, as for as possible, by draining, purging with oxygen-free nitrogen and possibly using heated air. Significant quantities of water in a system cannot be removed in practice by a vacuum pump as this would take far too long. The principle of evacuation is to reduce the pressure of any moisture left in the piping to a saturation temperature well below the ambient temperature, shut off the vacuum pump and observe any pressure rise which indicates a leak or large quantity of moisture in the system which needs to be boiled off. Then the vapour pressure can be lowered as much as possible. The vapour pressure of water, expressed in a number of units, all as absolute pressures, is shown in Table 11.1 (see also units of measurement in Appendix).

Table 11.1 Vapour pressure of water at low temperature

	Vapour pressure of water, abs			
Temperature (°C)	Pa	mb	mm Hg torr	μm Hg micron
−60	1	0.01	0.01	7.5
−50	4	0.04	0.03	30
−40	13	0.13	0.10	96
−30	38	0.38	0.29	285
−20	103	1.03	0.77	775
−10	260	2.60	1.95	1950
0	611	6.11	4.5	4585

The vacuum pump should be connected to draw from all parts of the circuit. This may require two connections, to bypass restrictions such as expansion valves, and all valves must be opened within the circuit, requiring electrical supplies to solenoid valves and the operation of jacking screws, where these are fitted.

The final vacuum pressure should be lower than the water-vapour pressure at the lowest part of the system, and is therefore dependent on the system temperature at its coldest point. The process may be speeded by using a triple evacuation procedure that involves breaking the vacuum with oil-free nitrogen to help absorb the moisture and re-evacuating.

On small systems, such as factory packages, a typical final pressure of 50 μmHg (7 Pa) should be reached. Vacuum pumps of good quality, which can tolerate refrigerant gas, must be used. Evacuation of a large system may take a couple of days. During this time, checks should be made around the pipework for cold patches, indicating water boiling off within, and heat applied to get this away.

11.12 CHARGING WITH REFRIGERANT

Refrigerant may be charged as a liquid through the connection shown in Fig. 11.3. The cylinder with dip tube is connected as shown and the connecting pipe purged through with a little of the gas to expel air from it. It will be necessary to invert the cylinder if there is no dip tube. For small charges, the bottle may be supported on a weighing machine, or a calibrated charging cylinder may be used.

The compressor must not be started whilst the system is under vacuum, so refrigerant is admitted first up to cylinder pressure. At this point, the compressor may be started, assuming that all auxiliary systems (condenser fan, pump, tower, cooler fan, etc.) are running. A liquid-line valve upstream of the charging connection is partially closed to reduce the line pressure at this point below that of the supply cylinder, and the refrigerant will continue to flow in. Although the refrigerant can be safely admitted in this way, the system is not running normally, since the throttle valve is reducing the pressure across the expansion valve. At intervals during charging, the cylinder valve must be closed and a throttle valve opened fully. Only under these conditions can correct running be observed. When fully charged, the sight glass will be clear.

As an alternative method, the cylinder pressure can be increased by gently heating it. Any heating of this sort should only be done using a bath

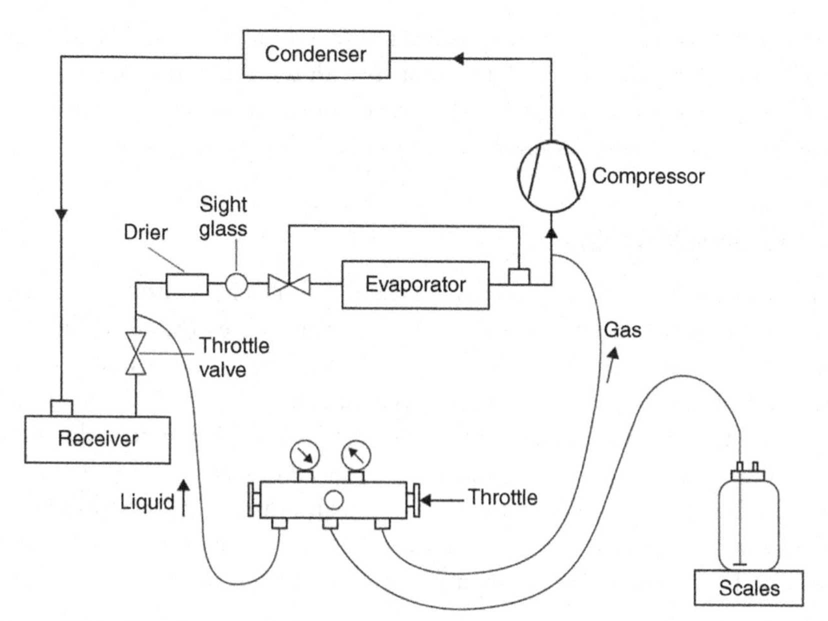

Figure 11.3 ***Charging connection.***

of hot water at less than 40°C or a thermal jacket whilst keeping a careful watch on the cylinder pressure. Raising the cylinder pressure in this way avoids the use of the throttle valve and the charging process is much quicker.

If a receiver is in circuit, this should be about one-sixth full under normal running conditions. Refrigerant may also be charged as gas into the suction line. This is usually done when the system is running and is being topped up. The liquid must be evaporated before it goes into the compressor. This can be done by charging through a manifold set as shown in Fig. 11.3 and throttling its low side valve. With the valve just open, the liquid is turned into gas before going into the system.

Records of refrigerant quantities added to systems should be kept, and the cylinder is placed on scales for this purpose.

Systems having high-pressure float expansion valves, and those without sight glasses, must be charged gradually, observing the frost line or using a contact thermometer to measure superheat.

Small packages will have the charge marked on the nameplate and must be carefully charged to this weight, which will be critical.

Systems may need to have further lubricating oil added, to make up for that which will be carried around with the refrigerant. In the absence of any firm guidance from the supplier, the crankcase must be topped up gradually during normal running, until it is level with the middle of the sight glass under operating conditions. This is not so with the small hermetic systems, where there is usually sufficient oil in the compressor to supply the needs of the circuit, and which, in many cases, have no oil level sight glass.

11.13 INSULATION

Pipework and other components should be insulated after the safety pressure test, but usually before prolonged running of the plant, since it is very difficult to remove water and frost once it has formed. For cooling systems the low-pressure piping is insulated, where it does not form part of the evaporator, that is after the expansion valve, where this may be outside the cooled space, and from the evaporator back to the compressor. For heating systems the liquid line and discharge lines should be insulated to conserve superheat and secure any condenser sub-cooling benefit. Discharge lines may in any case benefit from insulation for safety reasons.

Insulants for pipework and curved pressure vessels can be obtained ready shaped, so that they fit tightly to the surface. All surfaces should be quite dry before the material is applied, even if the adhesive is a water-based emulsion, and the water or other solvent must be given ample time to dry or set before any outer wrapping is applied.

Any air spaces within the insulation should be avoided, since this air will contain moisture, which will condense and freeze, leading to early deterioration of the insulant.

The essential part of the insulation system is the vapour barrier, which must be complete and continuous over the outer (warm) surface.

The application of insulating materials is a specialist trade and justifies careful supervision and inspection.

Much use is made of flexible foamed plastic material, which can be obtained in tubular form for piping up to 114 mm diameter and in flat sheets of various thicknesses for tailoring onto other shapes. This material has a vapour-tight outer skin, but must be sealed at all joins and the ends. The manufacturers are helpful in advising users.

11.14 WATER CIRCUITS

Water and other fluid circuits are pressure tested for safety and leakage, using water at a pressure of 1.5 times the working pressure, or as required.

The opportunity is taken, whilst filling for testing, to ensure that the circuits can be filled without airlocks. Air vents at high points of the circuit may be automatic or manual. Although the pipes are full, pumps should be run if possible to dislodge any dirt before draining down and cleaning the strainers. If a lot of dirt is found, the pipework should be filled again and re-flushed. In any case, the pumps should be run at the earliest opportunity and the strainers cleaned out.

Fluids, if other than water, are not put in until this pressure testing and flushing has been carried out.

11.15 NON-CONDENSABLE GASES

Other gases, mainly ambient air, may enter a refrigeration system as a result of incomplete evacuation before charging, opening of parts for maintenance or repair or inward leaks on circuits operating below atmospheric pressure. These gases will be circulated with the refrigerant vapour until they are all in the condenser and receiver. They cannot move further around the circuit because of the liquid seal at the outlet to the expansion valve.

Within the confines of the condenser and receiver, the gases will diffuse together and will exist in the same proportions throughout. The non-condensable may therefore be removed through purge valves on either vessel, but such valves are commonly fitted on or near to the hot gas inlet to the condenser. The presence of non-condensable gas will be shown as an increase of condenser pressure (Law of Partial Pressures) and may be detected during normal operation if the running log is accurate. The effect of this higher condenser pressure is to increase the compression ratio and so reduce the volumetric efficiency and increase the power. There will also be the effect of the gas blanketing the condenser surface, reducing heat flow.

Where the presence of such gas is suspected, a cross-check can be made, providing the high-pressure gauge is of known accuracy. The method is to switch off the compressor after a short running period, and so stop the flow of thermal energy into the condenser, but continue to run the condenser until it has reached ambient conditions. The refrigerant vapour pressure can then be determined from the coolant temperature. A value above the coolant temperature indicates non-refrigerant gas in the system.

The bleeding of gas from the purge valve will release a mixture, the content of which can be estimated from the total pressure.

Example 11.2

A system containing R407C is cooled to an ambient temperature of 20°C and the condenser gauge then indicates 11.70 bar. What is the partial pressure of the non-condensable gas, and how much R407C must be lost to purge 10 g of this gas assuming that it is air?

Because R407C is a zeotropic mixture its vapour pressure is dependent on the proportion of vapour phase in the vessel. Assuming that this is small compared to the mass proportion of liquid in the condenser, the pressure will be the bubble point pressure at 20°C.

$$\text{Vapour pressure of R407C at } 20°\text{C} = 10.34 \text{ bar abs}$$
$$\text{Observed pressure} = 11.70 \text{ bar abs}$$
$$\text{Partial pressure of non-condensible gas} = 1.36 \text{ bar abs}$$

R407C consists of R32/R125/R134a in proportions 23/25/52%. Therefore its molecular mass is:

$$0.23 \times 52 + 0.25 \times 120 \times 0.52 \times 102 = 95$$

Note: The vapour composition in the condenser will differ slightly from the mass proportions given, but this will be ignored.

Gas	Proportion by pressure	Molecular mass	Proportion by weight	Weight ratio
Air	1.36	28.97	39.40	1
R407C	10.34	95	983.25	25.0

So 250 g of R407C must be wasted to purge 10 g of non-condensable gas.

Purging must only be allowed if absolutely necessary, and must be carefully controlled.

Ammonia has a much lower molecular mass and the proportion by weight in this example would only have been approximately 40 g of ammonia lost. Also, ammonia is much cheaper than R407C!

Wastage of refrigerant can be reduced by cooling the mixture of gases and thus reducing the ratio. By means of a refrigerated purge device, which cools the mixture down to the evaporator temperature (eg, −35°C in a blast freezer plant), the ratio would become

$$\text{Vapour pressure of R407C at } -35°\text{C} = 1.5 \text{ bar}$$
$$\text{Partial pressure of non-condensible gas} = 10.2 \text{ bar}$$

The ratio now becomes 10 g of R407C lost per kilogram of air.

Automatic gas purgers comprise a collection vessel for the gas mixture with an inbuilt cooling coil connected to the main suction, or with its own refrigeration system. Condensed refrigerant returns to the condenser, and any gas remaining in the vessel will be non-condensable and can be vented by an inverted bucket trap.

Purging of gases must always be to the open air. The release of ammonia–air mixture is usually made through a flexible tube into a container of water. The water will absorb the ammonia, and any bubbles seen to rise to the surface will be other gases.

CHAPTER 12

Distributed Cooling and Heating

12.1 INTRODUCTION

The generated cooling or heating effects must be applied where and when they are needed. The distribution of the heating effect of burning gas using hot water radiators is familiar and similar hydronic systems can be used for heat pumps. A building or process having a large number of separated cooling loads could have a refrigeration system for each of these loads but it is usually more convenient to concentrate the cooling into one plant. The cooling effect of a central refrigerating system can be distributed by a heat-transferring liquid or secondary refrigerant. These are sometimes termed *indirect systems*. Where the working temperatures are always above 0°C, such as in air-conditioning, and water is commonly used. At temperatures below this, non-freezing liquids are used.

Thermal storage is used to store generated heat or cool until it is needed. Domestic applications of thermal storage are very familiar. Ice cubes are used to cool drinks. Plastic containers holding a phase change material are frozen by placing them in the domestic deep-freeze cabinet. Once frozen, they can be used for the short-term storage of cold foods and drinks.

In commercial use, thermal storage is used to smooth cooling or heating loads or take advantage of off-peak electricity tariffs. It can be used to make available large capacities with limited plant size and typical applications, which include the following:

1. To make ice over a period of several hours and then use ice water for the cooling of a batch of warm milk on a dairy farm. This is also used at main creameries, to reduce peak electricity loads. The available water is very close to freezing point, which is the ideal temperature for milk cooling.
2. To run the refrigeration or heat pump system at night, or other times when electricity is cheaper. It is also in use in areas where the electricity supply is unreliable. Where the cold water is to be used at a higher temperature, such as in air conditioning, the circuit requires three-way blending valves.
3. As hold-over cooling plates in transport (see chapter: The Cold Chain – Transport, Storage, Retail).

Refrigeration, Air Conditioning and Heat Pumps
http://dx.doi.org/10.1016/B978-0-08-100647-4.00012-7

12.2 REFRIGERANT CIRCULATION

This is the simplest distribution method, when there is no additional liquid pump it is termed a direct expansion or DX system. The evaporators or condensers are placed locally with refrigerant metered to each on demand. A combination of heating and cooling loads in buildings can be supplied by VRF systems (see chapter: Air Conditioning Methods and Applications). In centralised supermarket systems refrigerant is sent to each cabinet which has its own expansion valve and evaporator. This usually transfers cooling to the air that finally chills the product. Heat or cool can be distributed by ducted air systems to individual locations. Ammonia can be circulated by pump to cold room evaporators.

The disadvantages of DX are relatively large refrigerant charge, pressure drops in refrigerant lines and the need for maintenance expertise in occupied zones. Regulations on charge limits for flammable refrigerants can be found in BS EN378, and R717 cannot be tolerated in occupied areas. These factors drive the use of alternative indirect systems that carry the penalties of an additional heat transfer barriers and secondary pumps.

12.3 CHILLED WATER

The preferred secondary refrigerant is water, if this can possibly satisfy the temperature requirement, that is if the load temperature is sufficiently above 0°C that water can be circulated without the risk of freezing.

The greatest demand for chilled water is in air-conditioning systems. For this duty, water is required at a temperature usually not lower than 5°C and, for this purpose the evaporator operates with refrigerant temperatures close to freezing point. A very wide range of factory-built package chillers is available with either water or air-cooled condensers (see chapter: Packaged Units). Other types may have remote air-cooled or evaporative condensers, and so require refrigerant pipe connections on site to these condensers. Sizes range from 14 to 35,000 kW, most installations being within the range 100–1,500 kW.

At water temperatures close to freezing, and with evaporators which are vulnerable to ice damage, it is important to have adequate safety controls, to check the calibration of these frequently and to avoid interference by unauthorised persons. Nearly all troubles from packaged water chillers arise from a failure of safety controls. Several types of controls are in use, frequently three or more on the same equipment, but there should never be less than two of the following:

1. Water-flow switch, to stop the machine if flow stops in the chilled water circuit

2. Refrigerant low-pressure cut-out to protect against loss of charge
3. Water outlet low-temperature cut-out
4. Evaporator pressure-regulation valve
5. Hot gas bypass valve, to maintain the evaporating temperature above freezing point, not for normal capacity control

Most packaged water chillers are large enough to have capacity-control devices in the compressor. The main control thermostat will unload the compressor as the water temperature approaches a lower safe limit, so as to keep the water as cold as possible without the risk of freeze damage.

In all but the smallest installations, two or more chillers are used, or one chiller with two separate circuits. This arrangement gives some continuity of the service if one machine is off-line for maintenance or other reason, gives better control and provides economy of running when loads are light.

If water is required below 5°C, the approach to freezing point brings considerable danger of ice formation and possible damage to the evaporator. Some closed systems are in use and have either oversize heat exchange surfaces or high-efficiency-type surfaces. In both of these, the object is to improve heat transfer so that the surface in contact with the water will never be cold enough to cause ice layers to accumulate.

12.4 BAUDELOT COOLERS AND ICE BANK COILS

Water can be cooled safely to near freezing point, using evaporators which have the refrigerant inside, with space for ice to form on the outside of the surface without causing damage. Following two types are used:

1. *Baudelot coolers (see chapter: Evaporators).* The evaporator stands above a collection tank, and the water runs down the outside surface in a thin layer. Evaporator construction can be pipe coils or embossed plates. The latter are usually made of stainless steel to avoid corrosion. Evaporators may be flooded or direct expansion. During operation, a Baudelot cooler may sometimes build up a thin layer of ice, but this does no damage to the evaporator, and should melt off again when the load changes.
2. *Pipe coils within a water tank (see chapter: Evaporators).* Both flooded and direct expansion evaporators are in use. Water is circulated by pumps and/or special agitators. This type of water chiller may be operated without formation of ice, or ice may be allowed to accumulate intentionally (see further).

Water chillers of these two types are not usually made as single packages with their condensing unit, owing to the bulk of the system and subsequent difficulty of installation.

12.5 ICE MANUFACTURE

Ice storage goes back to the earliest days of low-temperature food preservation when stored winter ice was used. Mechanical refrigeration was subsequently introduced to make the ice and it would be transported to where the cooling effect was required. The refrigeration energy available in this way is mainly its latent heat of melting, 334 kJ/kg, as it changes back to water. One of the largest plants in UK, capable of 1000 tonnes ice per day, was installed at Grimsby (Lester and Hundy, 2014)

Today ice is used in certain industries such as fishing vessels, and as a thermal storage medium. Small icemakers generate ice for drink and food displays. Ice can be made as thin slivers on the surface of evaporator drums, and removed mechanically when the correct thickness has been formed – either the drum or the scraper may rotate. This is a continuous process and the ice flakes fall directly onto the product or into a storage bin below the machine. Smaller units are made as packages with the bin integral and cooled by a few turns of the suction line or by a separate evaporator. Small pieces of ice can be formed in or on tubes or other prismatic shapes made as evaporator tubes, arranged vertically. Water is pumped over the surface to freeze to the thickness or shape required. The tube is then switched to 'defrost' and the moulded section thaws sufficiently to slide off, possibly being chopped into short pieces by a rotating cutter. The machine itself is made as a package, and the smaller sizes will include the condensing unit.

12.6 SECONDARY COOLANTS

Where a secondary refrigerant fluid is to be circulated, and the working temperatures are at or below 0°C, then some form of non-freeze mixture must be used. Aqueous solutions of sodium chloride and calcium chloride were the first types of sub-zero secondary fluids, and this is why the collective term brine is sometimes used. Propylene glycol–water mixtures are the most common of the glycols used as secondary coolants in refrigeration installations and are permissible where contact with food is possible. Ethylene glycol is another type and there are many other products on the market, some of which do not contain water. The thermal properties of Propylene glycol–water mixtures are shown in Fig. 12.1.

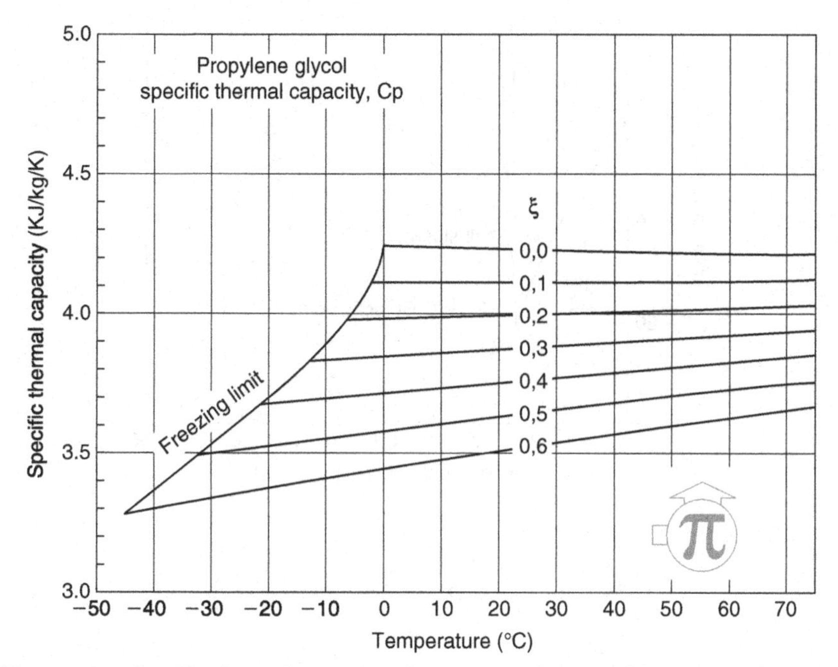

Figure 12.1 ***Specific thermal capacity of aqueous solutions of propylene glycol for various concentrations (M Conde).***

With any solution, there will be one concentration that remains liquid until it reaches a freezing point, and then it will freeze solid. This is the eutectic mixture, and its freezing point is the eutectic point of the solute (see Fig. 12.2). At all other concentrations, as the solution is cooled it will reach a temperature where the excess water or solute will crystallise out, to form a slushy suspension of the solid in the liquid, until the eutectic point is reached, when it will all freeze solid. For economy of cost, and to reduce the viscosity (and so improve heat transfer), solutions weaker than eutectic are normally used, provided there is no risk of freezing at the evaporator.

The concentration of a solute has a considerable effect on the viscosity of the fluid and the surface convective resistance to heat flow. There is little published data on these effects, so applications need to be checked from basic principles.

Brine may be pumped to each cooling device, and the flow controlled by means of shut-off or bypass valves to maintain the correct temperature (see Fig. 12.3). The brine pump is usually in the return line to the chiller as shown, so the pumping rate is based on the return temperature density.

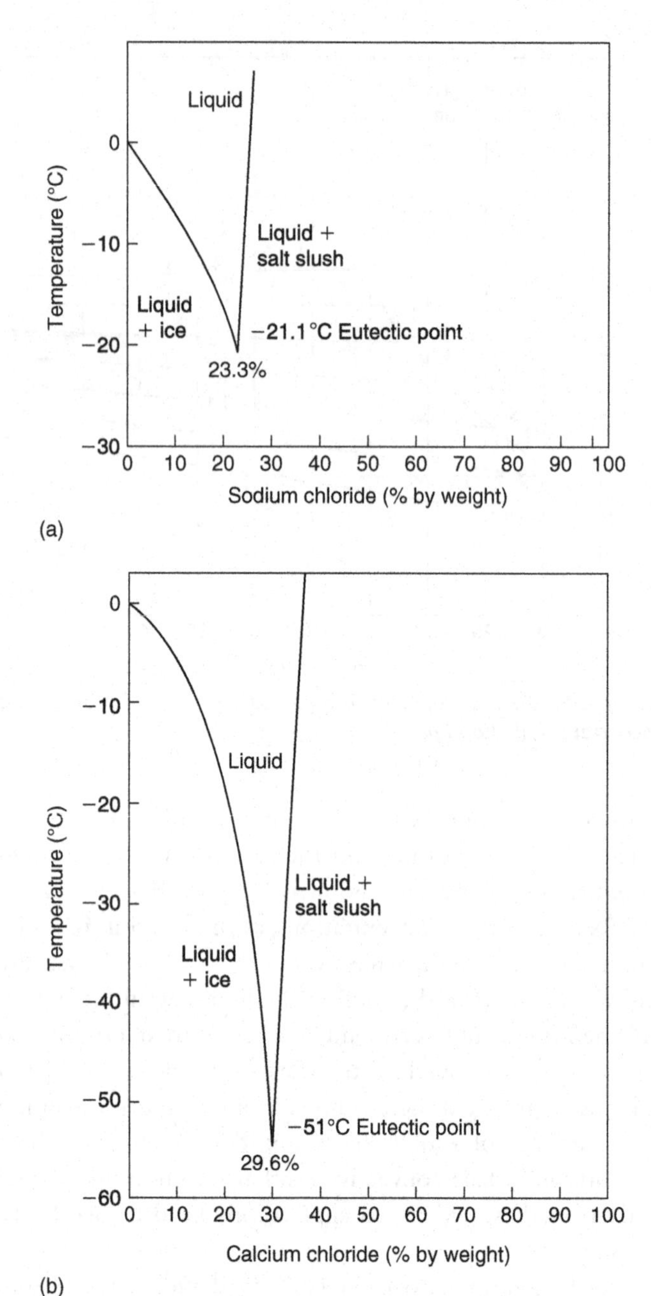

Figure 12.2 ***Eutectic curves. (a) Sodium chloride in water; (b) calcium chloride in water.***

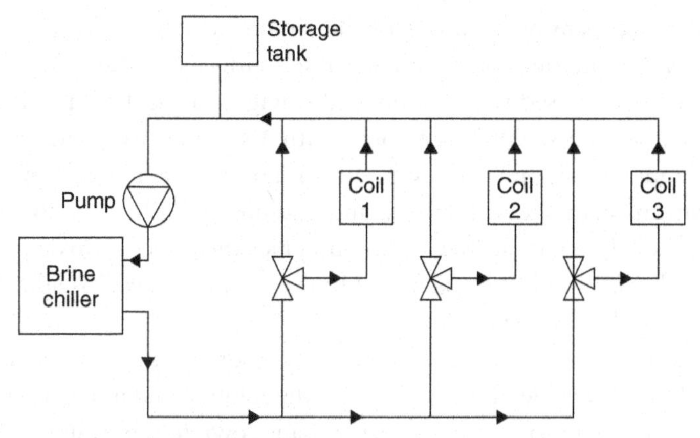

Figure 12.3 ***Brine circuit for separate rooms.***

Where a brine system services a multiple-temperature installation such as a range of food stores, the coolant may be too cold for some conditions, causing excessive dehydration of the product. In such cases, to cool these rooms the brine must be blended. A separate three-way blending valve and pump will be required for each room (see Fig. 12.4).

Air and moisture must be kept out of the system to avoid corrosion, and use of a closed system rather than an open one is preferable where possible. Pressure-controlled dry nitrogen can be applied above the expansion tank or storage tank. The preferred brine circuit is that shown in Fig. 12.3, having

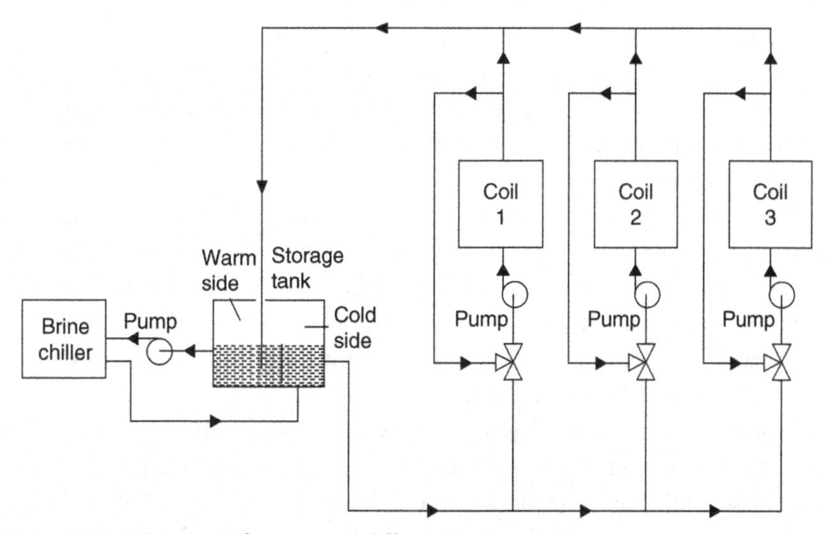

Figure 12.4 ***Brine circuit for rooms at different temperatures.***

the feed and expansion tank out of the circuit, which is otherwise closed. This avoids entrainment of air and too much surface exposure. The same arrangement can be used with the divided storage tank as shown in Fig. 12.4, except that the tank will be enclosed with a separate feed and expansion tank. To reduce the effects of corrosion inhibitors are added, typically sodium chromate in the salt brines, and sodium phosphate in the glycols. These are alkaline salts and help to counteract the effects of oxidation, but periodic checks should be taken, and borax or similar alkali is added if the pH value falls below 7.0 or 7.5.

Brines are hygroscopic and will weaken by absorbing atmospheric moisture. Checks should be made on the strength of the solution and more salt or glycol is added as necessary to keep the freezing point down to the required value.

12.7 PHASE CHANGE MATERIALS AS SECONDARY COOLANTS

A pumpable ice slurry can be used instead of chilled water or brine. The composition of these fluids is quite simple, consisting of water with ice crystals mixed with another fluid such as glycol, ethanol, ammonia and NaCl. Only a fraction of the water is transformed into small ice crystals (around 1 mm). This ensures a uniform solution whilst enhancing cold transfer. Ice slurry can take advantage of the high latent heat of freezing whilst retaining the ability to pump the fluid to the coolers. Optimisation is directed towards finding the least viscous secondary refrigerant from within the available range, which has the highest energy efficiency.

The ice crystals can be produced by scraped surface generators that may be quite costly. Alternative generation methods are being investigated. Various methods, products and applications are given by Paul and Jahn (1997) and Paul (2002).

The properties of carbon dioxide make it very suitable for circulation as a volatile secondary coolant. Liquid carbon dioxide at approximately −10°C and 25–26 bar pressure can be pumped to coolers where a fraction evaporates. The two-phase mixture returns to a reservoir where it is separated. A conventional refrigeration circuit maintains the pressure at an appropriate saturation temperature.

Liquid carbon dioxide has a very low viscosity; and because of its high latent heat, the volume to be circulated is very small compared to any water-based solution. Carbon dioxide exhibits a high-boiling heat-transfer coefficient that favours the design of the coolers. This type of system is starting to be applied in supermarkets for medium temperature refrigeration. It also has significant potential in air conditioning applications (Pearson, 2004).

12.8 THERMAL STORAGE

Phase change materials (PCMs) for thermal storage are available in a number of forms to suit various applications. In its simplest form variations in cooling load can be provided from the latent heat of melting of ice or a frozen eutectic. Ice can be formed by allowing it to build up on the outside of evaporator coils in a tank. Brines are more normally held in closed tanks or plates, again with evaporator coils inside, the outside of the tank forming the secondary heat exchange surface. Eutectics can be formulated according to the temperature required.

An example of alternative PCM is a paraffin-based material. The specific heat capacity of latent heat paraffins is about 2.1 kJ/(kg K). Their melt enthalpy lies between 120 and 160 kJ/kg, which is very high for organic materials. The combination of these two values results in an excellent energy storage density. Consequently, latent heat paraffins/waxes offer four to five times higher heat capacity by volume or mass than water at low operating temperature differences. They have a low heat conductivity, but for thermal storage this can be overcome by introducing sufficient surface.

A variation is to have a pumpable fluid such as one of the glycols, and to contain a PCM within capsules in a storage tank. The capsules are in the form of plastic balls and the PCM within may be formulated to suit any required thermal storage temperature.

Bound PCMs incorporate the PCM into an existing structure. When the PCM changes into a liquid within the supporting structure, the combined influence of capillary forces of the supporting material and the special cross-linking additives ensures that the PCM, even when in the liquid phase, remains within the structure. This means that one always works with a 'dry' material and liquid handling problems are eliminated. Free spaces within bound PCMs allow for volume expansion of the PCM to take place within the structure.

A study of the use of a PCM in a domestic refrigerator is given by Marques et al. (2013). Water is used as the PCM to provide cooling during long off cycle of several hours and a large high-capacity, efficient compressor can recharge with moderate on-cycle times. In another innovative application a wax with melting point 15°C is used to store heat derived from liquid sub-cooling to provide energy for a rapid defrost (Davies et al. 2014).

CHAPTER 13

Packaged Units

13.1 INTRODUCTION

Factory-built self-contained systems are very familiar in the form of domestic appliances, and retail units such as vending machines, drinking water chillers and display cabinets and counters. Although larger systems may need to be finally assembled in situ, groups of components can be conveniently delivered as pre-assembled units, and it is the exception for a system to be custom built from individual components, except in the case of large industrial installations.

The main benefit of packaging is reduction in cost because a high proportion of the total cost of a refrigeration or air-conditioning system is made up of work which can be carried more quickly, efficiently and under better control within a factory rather than on the installation site. There are other advantages as well including the following:

1. Correct selection and balance of components
2. Factory control of system cleanliness, leak tightness and, in some cases, charging
3. Inspection and testing of the complete unit before it leaves the factory
4. Delivery to the site, complete and in working order, so avoiding site delays for materials
5. Simplified site installation, with a minimum of disruption, inconvenience and cost.

A possible disadvantage is that the size range of available packages may be limited, but bespoke packages are frequently possible. There is a risk of misapplication, and care should be taken to understand the functionality of the package and its limitations. Comprehensive application data is normally available for all marketed packaged units to allow designers or sales engineers to make the correct selections for their purposes. Errors in application stem mainly from a lack of understanding of the requirement and a tendency to purchase at the lowest price without the protection of a clear specification. Once the application is fully understood, assessed and specified, the possibility of error is reduced.

Refrigeration, Air Conditioning and Heat Pumps
http://dx.doi.org/10.1016/B978-0-08-100647-4.00013-9

Transport applications use specialised systems, and these are covered in chapter: The Cold Chain – Transport, Storage, Retail, as are refrigerated display cabinets. Heat pump package units can be found in chapter: Commissioning and Maintenance.

13.2 CONDENSING UNITS

A *condensing unit* is a single package comprising the compressor, the condenser (either air- or water-cooled) mounted on a base plate or frame, and all connecting piping, together with the necessary wiring and controls to make the set functional (Fig. 13.1). Condensing units generally include a liquid receiver and are ready for site connection to an evaporator. Accessories such as pressure cut-out switches, liquid-line drier, sight glass and fan speed control may be included, and the specification should be carefully checked. They are built in sizes from less than a kilowatt to more than 100 kW cooling capacity and may incorporate more than one compressor. Air-cooled types may have two or more fans. Sometimes the condenser is separate in which case the packaged part is termed a compressor unit. Units intended for outdoor application have weatherproof housings and may be suitable for either wall or floor mounting (Fig. 13.2).

Figure 13.1 ***Air-cooled twin fan condensing unit, semi-hermetic compressor, suitable for several refrigerants (Bitzer).***

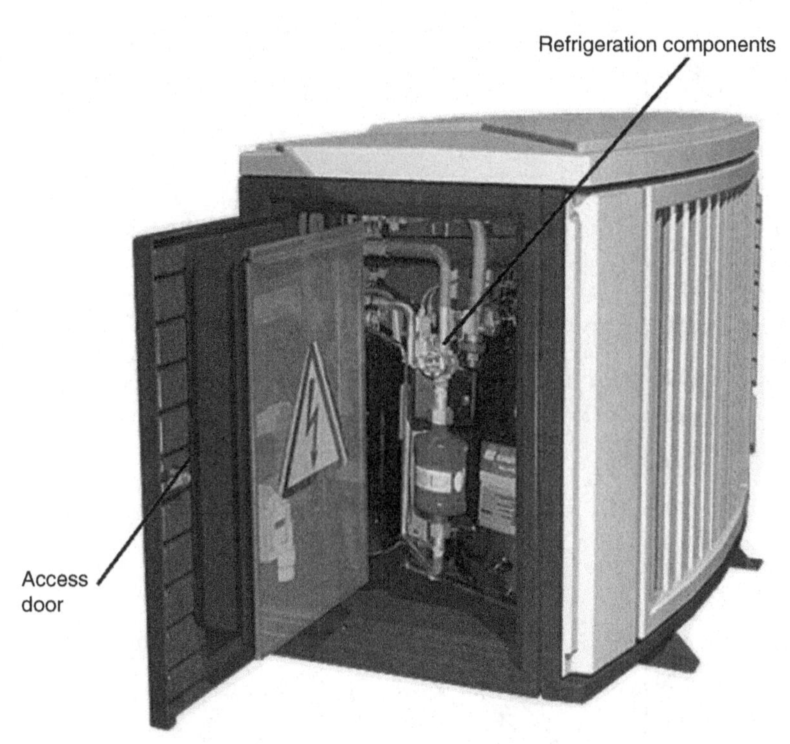

Figure 13.2 ***Outdoor air-cooled condensing unit, scroll compressor, suitable for several refrigerants (Emerson Climate Technologies).***

The condenser is matched to the compressor, and cooling capacity data refer to air or water temperature onto the condenser as detailed in chapter: Condensers and Cooling Towers. Since compressor and condensing units do not include an evaporator, they are not complete systems and are not charged with refrigerant, but have a holding charge of dry nitrogen, or a little of the refrigerant gas to maintain a slight positive pressure for transit. Suction and liquid interconnecting lines and electrical connection is installed on site.

Ratings for condensing units refer to entering temperatures of the condensing medium – air or water; traditionally presented as curves (Fig. 13.3) and now most commonly found from manufacturers' computer selection programmes. The standard conditions for rating condensing units are given in European Standard EN13215, 2000. An estimate of condensing unit capacity at an ambient condition different to the one

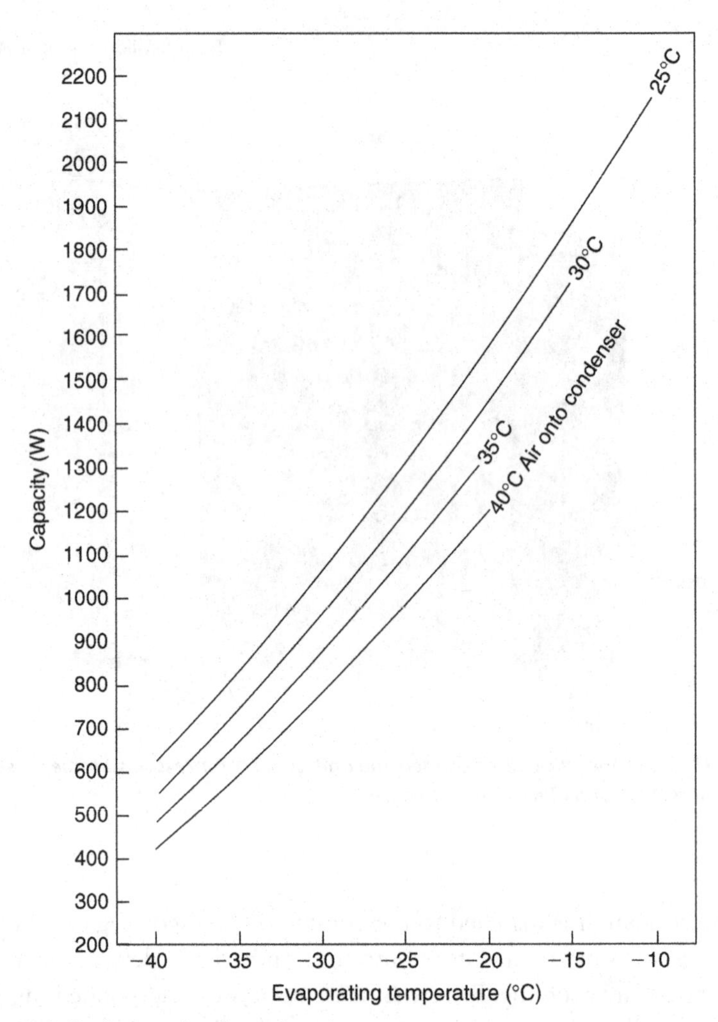

Figure 13.3 ***Typical condensing unit rating curves.***

provided can be made graphically as indicated in chapter: Component Selection and Balancing.

Some manufacturers of air-cooled condensing units offer a range of condenser sizes for each compressor, and these should be closely compared in terms of refrigeration capacity and running costs. A smaller condenser will operate at a higher condensing temperature. This can be visualized in Figure 10.6, where the condenser performance line would be less steep moving the balance point to the right. The refrigeration capacity is reduced, in accordance with the compressor curve and the power input is increased.

13.3 COMPRESSOR PACKS

This term is used to describe an assembly of several compressors mounted on a frame, complete with liquid receiver, suction and discharge headers, oil separation and oil return piping and controls (Fig. 13.4). They are widely used in centralised supermarket systems and the compressors serving the low-temperature evaporators and those serving the chill loads may be mounted on the same frame and piped accordingly. Compressor packs provide an efficient means of dealing with a wide span of refrigeration load. There can be up to 10 small compressors, either scroll or semi-hermetic, and this offers a large number of capacity steps by switching compressors. Alternatively a smaller number of uneven compressors may be used. Compressor packs are usually custom built to individual supermarket requirements

Figure 13.4 ***Scroll compressor pack (Climate Center).***

Figure 13.5 ***Scroll compressor pack within rooftop condenser (Searle).***

although manufacturers will probably use standardised design and construction methods.

The concept has been extended to factory-built rooftop units where the compressor pack is installed within the overall condenser housing (Fig. 13.5).

13.4 CHILLERS AND AIR COOLING PACKAGES

These are true packaged units in the sense that all the parts of the refrigeration system and its controls are factory assembled and tested in the complete state. There are four configurations:

- Air cooling, air-cooled
- Air cooling, water-cooled
- Liquid cooling, air-cooled
- Liquid cooling, water-cooled

Air cooling packages are limited by the volume of cooled air that needs to be in close proximity to the machinery. Examples of fully packaged air coolers are room air conditioners of the integral 'through the wall' type (chapter: Air Conditioning Methods and Applications) and refrigerated vehicle units (chapter: The Cold Chain – Transport, Storage, Retail).

Liquid cooling packages, frequently termed 'chillers', are self-contained systems and may be either air-cooled or water-cooled (Figs 13.6 and 13.7). Because chilled water is a highly flexible and efficient means of distributing cooling in many types of building, the range of chillers

Figure 13.6 ***Air-cooled water chiller, nominal 200-1775 kW capacity with two R134a/R1234ze refrigerant circuits, single shell and flooded evaporator, and capacity controlled centrifugal compressors (Airedale).***

available on the market is very large. They can be found with compressors of all types – reciprocating, scroll, screw and centrifugal together with heat exchangers of the shell and tube type or plate heat exchangers. Two refrigerant circuits are commonly used because this gives more capacity flexibility and offers some redundancy. Air conditioning is the main application, but they are also used for process cooling where the range of temperatures can be extended downward by use of brine instead of water. Water-cooled chillers are most frequently linked to cooling towers and this improves the chillers' overall thermodynamic effectiveness as compared to air-cooled chillers because heat rejection is at or near the ambient air's wet-bulb temperature rather than the higher, sometimes much higher, dry-bulb temperature. Many chillers are designed to work in reversed mode to provide heating when required (see chapter: Heat Pumps and Integrated Systems.

Chiller ratings, in accordance with EN14511, refer to the entering and leaving chilled fluid temperatures and either the ambient air condition (air-cooled) or the cooling water entering and leaving temperatures (water-cooled).

Figure 13.7 ***Water-cooled water chiller, nominal 250 kW capacity with two R407C refrigerant circuits, brazed plate heat exchangers (one evaporator, two condensers) and multiple scroll compressors (Airedale).***

13.5 SPLIT PACKAGES

To avoid the constraint of having all parts in one package, the evaporator set may be split from the condenser and the compressor going with either (see Fig. 13.8).

The unit is designed as a complete system but the two parts are located separately and connected on site. On some small units, flexible refrigerant

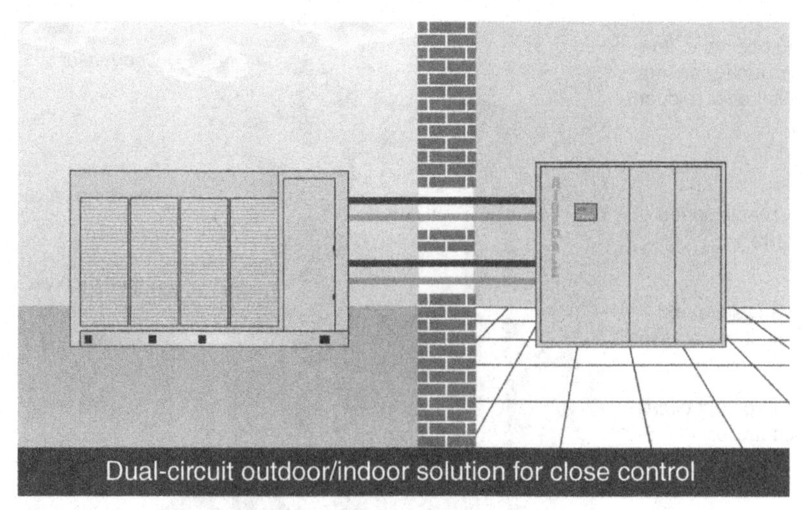

Dual-circuit outdoor/indoor solution for close control

Figure 13.8 ***Dual-circuit split-system package, outdoor air-cooled condensing unit with scroll compressors matched with indoor evaporator unit, typically 50–80 kW capacity (Airedale).***

piping may be provided. Larger split packages must be piped on site by normal methods, and then processed and charged as an open plant. The height of a split unit evaporator above the condenser may be limited and the supplier's instructions should be followed. See also *multi*-splits, chapter: Air Conditioning Methods and Applications.

Evaporator sets, as supplied as part of a split package or for application with a condensing unit, are of three main types:

1. *Air conditioning*: Having the air-cooling coil with drip tray under, expansion valve, fan and motor, air filters, inlet and outlet grilles. They may also include dampers and duct connections for return and fresh air, heaters, humidifiers and various controls.
2. *Cold store evaporators*: Having the coil with drip tray under, fans, and possibly the expansion valve.
3. *Cold store evaporators*: For use below 12°C with defrost elements.

Except for large industrial installations, heat pumps are invariably supplied as packages, commonly a split system similar in principle to that in Fig. 13.8, but the outdoor unit containing the evaporator and the condenser located indoors. The compressor may be in either unit. Examples of heating packages can be found in chapter: Heat Pumps and Integrated Systems. A new concept is a heating module package consisting of an indoor unit containing all the matched components and all necessary

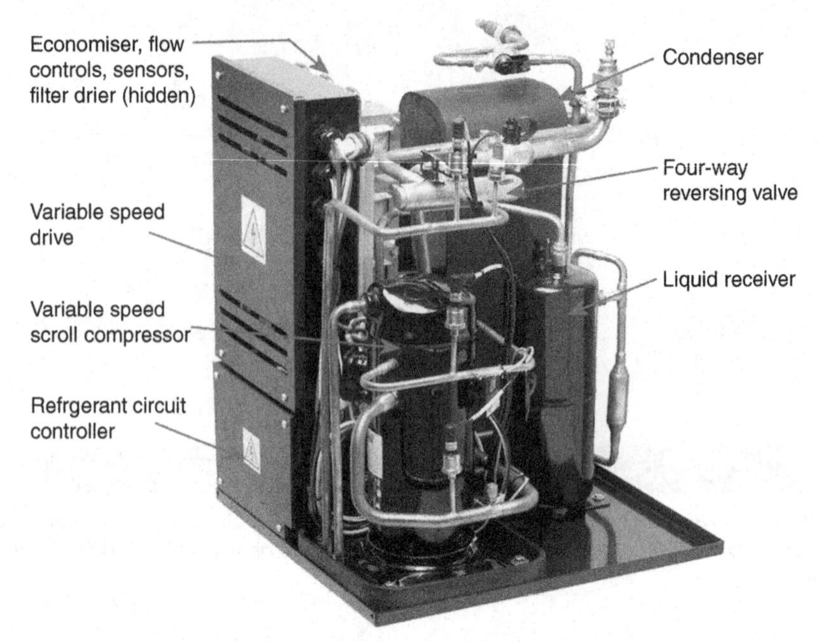

Figure 13.9 ***Heating-module package containing variable-speed compressor, condenser and all necessary controls (Emerson Climate Technologies).***

controls for economiser and compressor speed to match the load and outdoor conditions, (Fig. 13.9). The installer has only to link it to a suitably sized outdoor evaporator unit, water circuit, and provide a heating/ cooling demand signal.

13.6 TESTING OF PACKAGED UNITS

Manufacturers' test procedures for packaged units may include some of the following:

1. Safety-pressure tests and leak tests should always be included (see also Section 11.3).
2. Rating test, from a representative unit which forms the basis for the published capacity and application leaflets.
3. Type tests, which verify the product as designed will function properly when applied within the required operating envelope, and fail-safe when subjected to certain abnormal operating modes.
4. Rating check tests on a proportion of production units, to verify that standards are being maintained.

5. Function tests on all production units, to verify correct operation of components.
6. A short-running test at normal conditions to check for reliability of operation plus, possibly, an approximate capacity check.
7. Run at maximum operating conditions.

Factory test procedures can be costly, but are much more efficient than performing the work in the field. Records will be kept of all such tests and, in the case of larger units, manufacturers will, if asked, provide a certified copy of the test on the equipment supplied.

13.7 OTHER PACKAGES

A very large variety of smaller self-contained refrigeration and air-conditioning packages are made, mainly for the consumer durable market and small domestic applications. They include the following:

- Domestic refrigerators and freezers.
- Retail display cold and freezer cabinets and counters.
- Cooling trays for bottles (beer, soft drinks, wines).
- Instantaneous draught beer coolers. These usually comprise a tank of constantly chilled water, through which the beverage flows in stainless steel piping.
- Ice makers – cubes and flakes.
- Cooled vending machines.
- Soft ice-cream freezers.
- Dehumidifiers, in which air is passed first over the evaporator to remove moisture and then over the condenser to re-heat and lower the humidity (see chapter: Practical Air Treatment).
- Drinking water chillers.

CHAPTER 14

Food Refrigeration and Freezing

14.1 INTRODUCTION

The present-day food industry is almost totally dependent on refrigeration in one form or another, to manufacture, preserve, store and bring the product to the point of sale. Chapter 'Food Refrigeration – Product by Product' gives some examples for specific food products, whilst this chapter gives an overview of cooling methods.

The use of low temperatures for food preservation has been known and practised for many thousands of years, but it was not until the mid-19th century that Pasteur and others determined the bacteriological nature of food spoilage and the beneficial effect of cooling which slows chemical reactions and breakdown by bacteria. Mechanical refrigeration made it possible to provide the extra food required by the growing urban populations. A large international trade was built up, starting with the transport of frozen meats to Europe in 1873 and 1876 from Australasia and South America.

As a general rule, foods which are not to be frozen are handled and stored at a temperature just above their freezing point, providing this does no damage (exceptions are fruits such as bananas and lemons). Produce which is to be frozen must be taken down to a temperature low enough to significantly reduce the amount of free moisture and hence bacterial activity. Until the temperature is reduced below the minimum temperature for growth, some microorganisms can potentially multiply.

A distinction must be drawn between the cooling process and the subsequent storage. Careful control of temperature and humidity is needed when cooling warm produce since evaporative cooling plays a part in both product temperature and weight loss. Considerable research has been carried out to find optimum methods for different foodstuffs, especially meats, for cooling and for short-term and long-term storage.

Cooling and freezing cannot improve a fresh vegetable, fruit or meat product, and the best that can be achieved is to keep it near to the condition in which it entered the cooling process. This means that only the best produce should be used, and this should be as fresh as possible. However sometimes preservation in cold stores is essential to prevent wastage, regardless of the quality of the crop.

Refrigeration, Air Conditioning and Heat Pumps
http://dx.doi.org/10.1016/B978-0-08-100647-4.00014-0

14.2 PRE-STORAGE TREATMENT

All foods must be clean on entry. Vegetables and fruits should be dirt-free and some, such as fish, leaf vegetables and some fruits, may be washed and left wet. Fish will tend to dry out and lose its fresh appearance, so it is packed wet or given a sprinkling of ice chips to keep the surface moist. For meat and poultry the degree of surface microbial contamination is critical.

The shelf life of meat is dependent on the initial numbers of spoilage bacteria on carcasses. With higher numbers, fewer doublings are required to reach a spoilage level. Contamination of carcasses may occur at virtually every stage of slaughtering and processing, particularly during flaying and evisceration of red-meat animals and scalding, and mainly affects the surface of the carcass. Sources of contamination have been reviewed by James et al. (1999). The adoption of good production practices throughout the slaughtering system and hygienic handling practices should ensure that bacteria counts on the finished carcass are at an acceptable level. Decontamination methods are also sometimes applied (James and James, 1997).

Potatoes will start to sprout after a long period in storage. This can be checked by spraying the freshly lifted tubers with a chemical sprout depressant. Certain fruits, notably grapes and dates, may have some surface contamination or infestation when first picked, and they are fumigated with sulphur dioxide or some other gas. Chlorine washing is also used. They must, of course, then be thoroughly ventilated before going into storage.

The techniques of this processing will be known to the user or can be found in sources from the particular branch of the food industry.

Handling conditions must be hygienic. Some types of food, such as milk, can be kept sealed within the processing system. If the food will be exposed to the air during handling, the conditions of the surrounding air – in terms of temperature, humidity and cleanliness – must be the best that can be maintained. This is especially the case with fresh meats.

14.3 PRE-COOLING

If warm produce is taken into a cold store, moisture will evaporate from its surface and this may result in excessive humidity and condensation on the cold produce already there. This will be of no consequence with wet products such as fish and leaf vegetables. Meat and poultry are pre-cooled in a separate room under controlled conditions so that the product is reduced to near-final storage temperature.

Pre-cooling can be achieved by allowing produce to stand in ambient air, especially at night. For example, apples and pears picked in the daytime at 25°C may cool down to 12°C by the following morning, halving the final refrigerated cooling load.

Wet products can be pre-cooled in chilled water or by the addition of flake ice. Ice is also used with fish and leaf vegetables to help maintain freshness in transit to storage. Leaf vegetables can be cooled by placing them in a vacuum chamber and so evaporating surface water at low pressure.

14.4 FREEZING

Storage in the frozen state enables products to be kept for longer than maintaining chilled conditions. Freezing reduces bacterial degradation reactions to a very low level but causes structural change in the product due to the formation of ice crystals. The cells of animal and vegetable products contain a water solution of salts and sugars. When this solution starts to freeze, surplus water will freeze out until the eutectic mixture is reached (see Section 12.6). If freezing is not carried out quickly, the ice crystals will grow and pierce the cell walls; then when the product thaws out, the cells will leak and the texture will be spoiled. This is of no great consequence with the meats, whose texture is changed by cooking, but fresh fruit and vegetables need to be frozen quickly.

The texture and moisture content of the product after thawing will differ from that of the fresh product, and for some products it also results in weight loss in the form of 'drip loss'. Different freezing methods are used to minimise these effects.

As a general rule, any product which will be eaten without cooking, or only very brief cooking (such as green peas, strawberries and beans), should be quick-frozen in a blast-freezing tunnel or similar device. Other foodstuffs need not be frozen so quickly, and placing food items in large refrigerated rooms is the most common method of freezing. For meat and poultry there is no clearly defined optimum freezing rate. Many factors such as final product quality (tenderness, flavour), weight loss, drip loss, and uniformity of texture have been investigated. A comparison (Sundsten et al., 2001) revealed some commercial advantages of fast freezing, but no quality advantages. During industrial processing, frozen raw material is often thawed or tempered before being turned into the final product which is subsequently frozen. Meat-based products, that is pies, convenience meals, burgers, etc., often include meat which has been frozen twice.

Frozen confections such as ice cream rely on speed of freezing to obtain a certain consistency and texture, and they require special treatment (see chapter: Food Refrigeration – Product by Product).

14.5 QUICK FREEZING

The speed of freezing is a relative matter, but 'quick frozen' produce is generally frozen in 5–10 min in an air blast, and depending on the thickness this can be somewhat quicker if immersed. Various methods have evolved, depending on the available resources, the product concerned and the premium value it might earn in an improved frozen state.

Where the product shape is irregular, the only way to extract its heat will be by using a cold fluid surrounding it. Air is the obvious choice – it is economical, hygienic and relatively non-corrosive to equipment. The air temperature will be of the order of −40°C and the air speed over the product will be high, to get good heat transfer. Circulation of air around stacked product in a cold room is used for batch processes. This requires good air distribution and an optimum value exists between the decrease in freezing time and the increasing power required to drive the fans to produce higher air speeds.

Discrete pieces of product, such as peas, slices of carrot, beans and items of this size, can be conveyed on a perforated belt. This may be a fluidised bed where cold air is directed upwards through the mesh belt and the food particulates begin to tumble and float. This exposes all sides of the food to the cold air and maximises heat transfer. Flat pieces of product, such as fish fillets, would suffer a change in shape in a free air blast and are better on a flat moving belt. Here, some of the heat goes directly to the cold air and some by conduction to the belt, which is usually of stainless steel. The tunnel can be designed to absorb much less fan power which improves efficiency, because fans are located in the cooled space and the heat generated by their motors adds to the cooling load.

Linear tunnels are restricted by the length of belt necessary to achieve the cooling time required and on the space available. Conveyors wound into a spiral shape and contained within a cold-room with an air blast coil offer a more compact arrangement (see Figs 14.1 and 14.2). Spiral freezers are very good for larger items, such as tubs of ice cream which take a long time to harden and where a straight conveyor would be too long for convenience.

14.6 CONTACT FREEZING

Modern plate cooling systems differ little in principle from the first contact freezer patented in 1929 by Clarence Birdseye. Products in regular-shaped packages, such as ice cream in flat cartons, are pressed between horizontal, flat, refrigerated plates (see Fig. 14.3). These can be opened apart slightly to admit the product and are then closed by hydraulic rams to give close thermal contact. When freezing is complete, the plates open again to remove the packs.

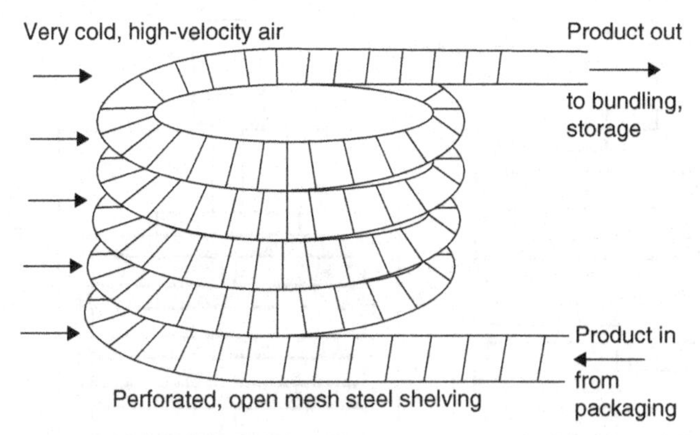

Figure 14.1 *Arrangement of spiral freezer.*

Figure 14.2 *View of spiral freezer (Star Refrigeration).*

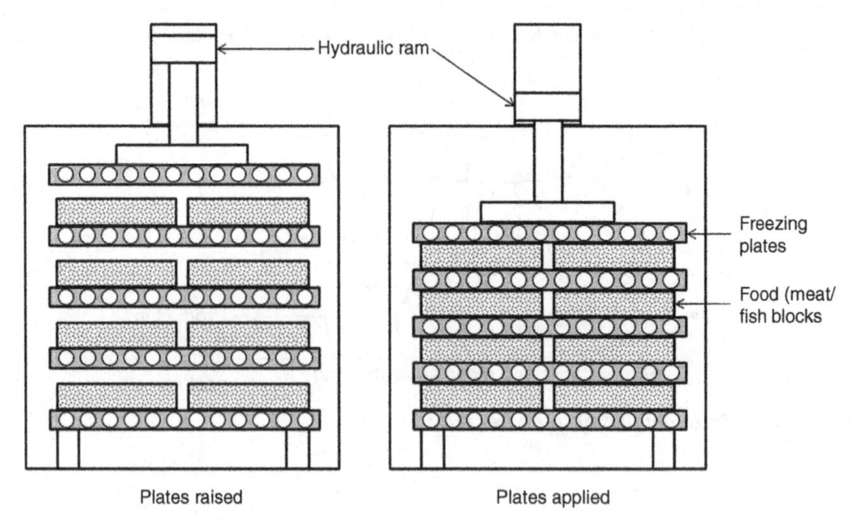

Figure 14.3 ***Arrangement of plate freezer (RD&T).***

A horizontal plate freezer is shown in Fig. 7.13a. The vertical plate freezer (Fig. 7.13b) is used for a loose product such as wet fish, which is packed into the gaps between the plates. When the freezing is complete, the product is removed as a solid block and may be of 75 or 100 mm thick (Fig. 14.4).

Contact freezers are less costly to operate because they do not use fans for air movement. The cooling is accomplished by direct contact of product with a surface, which in turn is in direct contact with the refrigerant or secondary coolant.

Material to be frozen can be fully immersed in a cold liquid such as a brine. This is only suitable for wrapped product. Sodium chloride and glycol brines are not cold enough to get complete freezing, so this may be the first pre-cooling stage before a final air blast. Alternatively, liquid nitrogen (−196°C) or carbon dioxide (−78.5°C) can be sprayed onto the surface. This is termed *cryogenic freezing*.

14.7 FREEZE DRYING

Certain products cannot be kept in the liquid form for an appreciable time and must be reduced to dry powders, which can then be kept at chill or ambient temperatures. The water must be removed to make them into powders, but any heating above ambient to boil off the water would lead to rapid deterioration. The water must therefore be removed at low temperature, requiring low pressures of the order of 125 Pa.

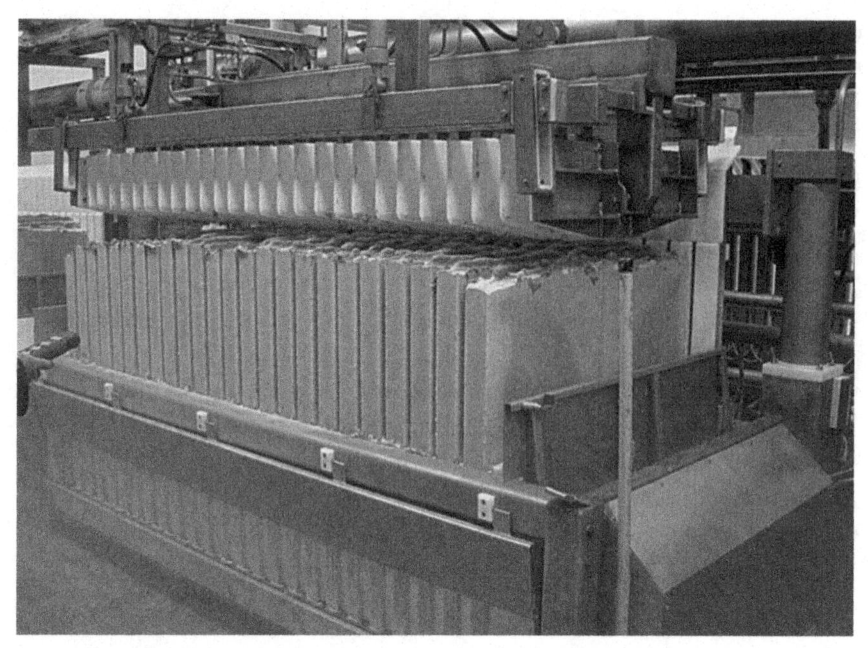

Figure 14.4 ***Blocks of chicken at end of freeze cycle (Star Refrigeration).***

The process is carried out in a vacuum chamber fitted with refrigerated contact freezing plates, heaters and a vacuum pump. Between the chamber and the pump may be a refrigerated separator to prevent too much of the moisture entering the pump. The product is placed in containers on the plates and frozen down to about −25°C, depending on the product, but sometimes as low as −50°C. The vacuum and, at the same time, a carefully controlled amount of heat, is then applied to provide the latent heat of sublimation (ice to vapour) without allowing the temperature to rise. As the water is driven off, the product collapses to a dry powder. This is extremely hygroscopic and must be packed in air-tight containers as quickly as possible on completion of the cycle.

This process was developed for the preservation of antibiotics, but is now in widespread use for other products such as 'instant' coffee, tea, soup, etc.

14.8 POST-STORAGE OPERATIONS

As a general principle, products leaving cold storage for ultimate consumption may be allowed to rise slightly in temperature. They should be kept as close to the storage temperature as long as possible down the chain of delivery. This requires prompt handling and the use of cooled vehicles up to the final retail outlet (see chapter: The Cold Chain – Transport, Storage, Retail).

Some products require special treatment, for which provision should be made, for example:

- Frozen meat coming out of long-term storage to be sold chilled must be thawed out under controlled conditions. This is usually carried out by the retail butcher, who will hang the carcass in a chill room (−1°C) for 2 or 3 days. On a large scale, thawing rooms use warmed air at a temperature below 10°C.
- Potatoes and onions coming out of storage will collect condensation from the ambient air and must be left to dry or they will rot.
- Fruits of various sorts are imported in a semi-green state and must be ripened off under the right conditions for sale.
- Some cheeses are frozen before they have matured. On thawing out for final distribution and sale, they need to mature.

CHAPTER 15

Cold Storage and Refrigeration Load Estimation

15.1 INTRODUCTION

Cold storage is an essential link in the cold chain, and has seen many improvements in technology during the period from 1945 to the present day, resulting in better quality and performance of insulation and door design. The older types of insulation tend to be prone to deterioration, resulting in loss of efficiency and increased running costs in maintaining required temperatures. Early stores were generally small and of low height with multi-chambers allowing different products to be stored at different temperatures. In the early 1970s, there was an increase in bulk long-term storage requiring larger enclosures with and relatively few doors. With the increase in demand for fresh, frozen and chill products, there has since been a move towards throughput storage with rapid turnaround of goods, requiring improved access and racking systems. EU regulations now generally require chilled enclosed loading bays.

Stores constructed generally since about 1990 have enclosed loading bays either internal or external to chambers with panel insulation, rapid closing doors with a number of ports and lorry docking facilities. They are designed to suit purpose-built static or mobile racking systems. Heights are variable but usually of 10 m or more with recent buildings in excess of 12 m and as high as 30 m.

Smaller cold storage items such as chilled display cases, frozen food cabinets and freezing tunnels will have capacity ratings which enable their refrigeration load to be defined. The following sections indicate the procedures to be followed in cases where it is necessary to make direct calculations.

15.2 COLD STORAGE LOADS

The refrigeration load comprises some or all of the following elements:

- *Sensible*: Cooling of fluids or solids
- *Latent*: Change of state – freezing
- *Heat of respiration*: From vegetables, fruits, etc.
- *Insulation*: Heat gain through the walls, ceiling and floor

Refrigeration, Air Conditioning and Heat Pumps
http://dx.doi.org/10.1016/B978-0-08-100647-4.00015-2

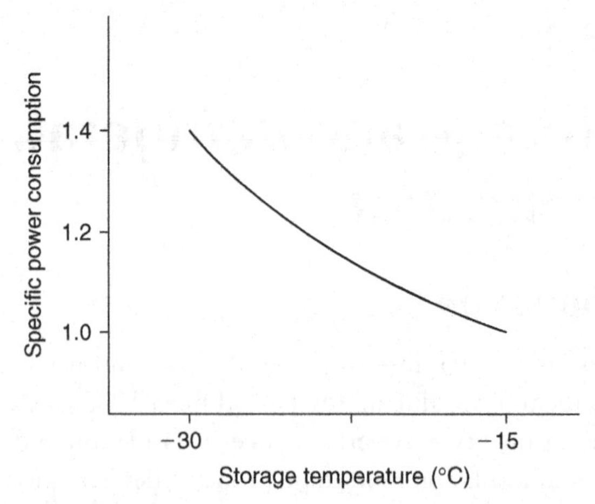

Figure 15.1 ***Typical specific power consumption variation with storage temperature.***

- *Infiltration*: Heat convection from outside (air infiltration or ventilation), both sensible and latent
- *Radiation*: Direct from surroundings
- *Auxiliaries*: Evaporator fans, pumps, lighting
- *Other*: People and trucks in cold stores, etc.

If the product is at storage temperature on arrival, then the first two will be negligible. Heat penetration of the insulation and infiltration will be the major source of heat load, and both are dependent on the storage temperature. Both the efficiency and COP of the system are dependent on the temperature lift, and the lower the storage temperature, the greater the lift must be. Although low temperatures are beneficial for the product, there is a trade-off against running costs. The graph in Fig. 15.1 gives an indication of typical power consumption characteristics for the same amount of cooling at different store temperatures in medium ambient conditions. At lower store temperatures the amount of cooling is greater because heat infiltration tends to be greater. The power costs could be 50% greater at −30°C compared to −15°C. Additionally, the compressor capacity is lower at the lower evaporating temperature and larger plant will incur additional capital costs.

15.3 PRODUCT COOLING

If product cooling is required the total amount of sensible and latent heat to be removed in cooling a product is given by:

$$H = M((c_a \times \Delta T_a) + h_l + (c_b \times \Delta T_b))$$

where H = total quantity of heat to be removed (kJ)
M = Mass of product (kg)
c_a = Specific heat capacity above freezing (kJ/kg K)
ΔT_a = Temperature decrease above freezing (K)
h_l = Latent heat of freezing (kJ/kg)
c_b = Specific heat capacity below freezing (kJ/kg K)
ΔT_b = Temperature decrease below freezing (K)

Some of these components will be zero if cooling does not take place through the range of temperatures above and below the freezing point. Typical specific heat capacities, freezing points and latent heats are given in Table 15.1.

The rate of heat extraction, that is the product cooling load, will be: $Q = H/t$ where t = the time available for cooling.

Example 15.1

What is the cooling duty to freeze water from 15°C to ice at 0°C, at the rate of 20 t/day?

$$Q = \frac{20{,}000[(4.187 \times 15) + 334]}{24 \times 3600} = 92\,\text{kW}$$

Example 15.2

What duty is required to cool 8 t of lean meat (specific heat capacity 3.1 kJ/(kg K)) in 14 h from 22°C to 1°C?

$$Q = \frac{8000[3.1 \times (22 - 1)]}{14 \times 3600} = 10.3\,\text{kW}$$

The load is probably a daily batch from an abattoir and the duty will be less at night, once the meat has been cooled. The maximum capacity will therefore be 10.3 kW, plus the fans and other room losses, and the plant will run continuously only whilst the meat is being chilled.

All assumptions regarding the load and estimated cooling duty should be recorded as the design parameters of the system, and agreed with the user. There may be several unknown quantities in an estimate. A good example is milk cooling. Milk is not usually stored in cold stores but its temperature control through the cold chain is critical. A dairy farm may produce 2400 L/day (a rate of 100 L/h), but this will come from two milkings, possibly 1400 L in the morning and 1000 L in the afternoon, and the milk must be cooled in 2 h, so the peak rate is 700 L/h.

Table 15.1 Specific and latent heats of foodstuffs (typical values)

Product	Specific heat capacity above freezing	Highest freezing point (°C)	Latent heat of freezing	Specific heat capacity below freezing
Apples	3.65	−1.1	280	1.89
Bananas	3.35	−0.8	250	1.78
Beer	3.85	−2.2	–	–
Cabbage	3.92	−0.9	–	–
Carrots	3.79	−1.4	294	1.94
Celery	3.99	−0.5	–	–
Dairy products				
Milk	3.75	−0.6	–	–
Butter	1.37	Down to −20	53	1.04
Ice cream	2.95	−6	210	1.63
Cheese	2.1	−13	125	1.3
Dried fruits	1.8	−2		
Eggs, shell	3.05	−2.2	220	1.67
Fish, white	3.55	−2.2	270	1.86
Blue	2.9	−2.2	210	1.63
Meats, bacon	1.5	−2	64	1.07
Beef	3.2	−2	230	1.7
Ham	2.7	−2	188	1.55
Lamb	3	−2	215	1.65
Pork	2.6	−2.5	125	1.3
Poultry	3.3	−2.8	246	1.77
Melons	3.95	−0.9	310	2
Mushrooms	3.89	−0.9	304	1.98
Onions	3.8	−0.9	295	1.95
Oranges	3.75	−0.8	–	–
Pears	3.62	−1.6	–	–
Potatoes	3.5	−0.7	265	1.84
Tomatoes	3.98	−0.5	–	–

Many of these figures will be slightly different, according to the variety, breed or location of the product.

The milk-cooling requirement of 700 L/h is a maximum rate. There is no need to allow for any more than this, but it cannot be any less. Alternatively, this could be cooled using an ice bank, in which case the total load of 2400 L could be spread over 16 h of running time. With an allowance for water tank insulation heat gains and an ice water pump, the peak load might be reduced enabling use of a refrigeration plant one-third the size.

The entering temperature of a product may be uncertain, being warmer in the summer or after a long journey. The dwell time within the cooling system may vary, beer leaving an instantaneous cooler at 4°C when first tapped, but at 12°C if drawn off continuously. The exact product may not be known – a general foodstuffs cold store might contain bacon (sensible heat capacity 2.4) or poultry (sensible heat capacity 3.3).

Observations may need to be taken of the operation, to form an estimate of unknown figures, or the process analysed to decide representative rates. In these circumstances, the probable minimum and maximum should be calculated from the best available data and an average decided. Assumptions should be stated and agreed by the parties concerned, since these estimates are to form the basis for the selection of the required plant.

Example 15.3

A dockside frozen meat store has a capacity of 1000 t stored at −12°C, and leaving the store at a maximum rate of 50 t/day. Meat may arrive from a local abattoir at 2°C or from ships in batches of 300 t at −10°C. Estimate a product-cooling load.

Case 1

Meat goes out at the rate of 350 t/week and may arrive from local supplies. There is possibly of a 4-day week, allowing for odd holidays, and so there may be 90 t/day from the abattoir. Cooling load is 90 t/day from 2°C to −12°C. Tables give the following:

Specific heat capacity above − 1°C = 3.2 kJ/(kg K)
Freezing point of meat, average = −1.0°C
Latent heat of freezing = 225 kJ/kg
Specific heat of frozen meat = 1.63 kJ/(kg K)

$$Q_f = \frac{90,000}{24 \times 3,600}[(3.2 \times 3) + 225 + (1.63 \times 11)] = 263\,\text{kW}$$

Case 2

Shipments may come in on consecutive days (unlikely, but possible if store is almost empty):

$$Q_f = \frac{3,00,000}{24 \times 3,600}(1.63 \times 2) = 11\,\text{kW}$$

These show a wide variation. Since meat will keep for several days at 2°C, rework case 1 on the basis of a steady input of 50 t/day all coming from the abattoir.

Case 3

$$Q_f = \frac{50,000}{24 \times 3,600}[(3.2 \times 3) + 225 + (1.63 \times 11)] = 146\,\text{kW}$$

It would seem, then, that the minimum safe cooling capacity required is 146 kW, with the possible risk of 263 kW for a day or so. Most of the time the load will be much less.

A practical approach would be to install plant having a maximum product-cooling capacity of 146 kW (to which must be added the other load components of heat leakage, internal heat, and service). After an estimate of the total cooling load has been formed, this must be converted into a refrigeration plant capacity.

General practice, after having calculated the average load over a period of 24 h, is to take the absolute maximum, or allow 50% over the average, that is a plant running time of 16 h in the 24. This general rule must be assessed for the particular application

15.4 CONDUCTED AND RADIATED HEAT

Conducted heat is the heat that flows through the insulation. It enters through cold store surfaces, tank sides, pipe insulation, etc. A single value is normally used, based on an average summer temperature, probably 25–27°C for the United Kingdom, unless some other figure is known. Cold room surfaces are measured on the outside dimensions and it is usual to calculate on the heat flow through the insulation only, ignoring other construction materials, since their thermal resistance is small.

The total conduction heat load is found from the area of the walls, ceiling and floor multiplied by temperature difference and respective 'U' values.

Solar radiation may fall on outside walls or roofs, raising the skin temperature, and this must be taken into account. Most cold stores are built within an outer envelope which protects them from the elements and from direct sunshine. In cases where the insulation itself is subject to solar radiation, an allowance of 5 K higher outside temperature should be taken. Heat load must be estimated through all surfaces including piping, ducts, fan casings, tank walls, etc., where heat flows inwards towards the cooled system.

Radiant heat is not a serious factor in commercial or industrial refrigeration systems, being confined to sunshine through refrigerated display windows (which should have blinds) and radiation into open shop display cabinets from lighting.

15.5 CONVECTED HEAT

Warm air will enter from outside mainly during the opening of doors for the passage of goods. This must be estimated on the basis of the possible use of the doors, and such figures are based on observed practice. The parameters are the size of the store, the enthalpy difference between inside and outside air and the usage of the doors. The latter is affected by the existence of airlocks and curtains. Some textbooks give data on which to base an estimate, and this can be summed up as

$$Q_f = (0.7V + 2)\Delta T$$

where Q_f = heat flow
V = volume in m^3
ΔT = temperature difference between room and ambient.

This is for cold rooms up to 100 m^3 with normal service. For heavy service, that is a great deal of traffic through the doors, this figure can be increased by 20–35%.

Rooms above 100 m^3 tend to be used for long-term storage, and are probably fitted with curtains. For such rooms, the service heat gain by convection may be taken as

$$Q_f = (0.125V + 27)\Delta T$$

A more detailed calculation given in Section 15.9 shows how to account for the contribution of moisture ingress and further background information with examples is provided in chapter: Air and Water Vapour Mixtures.

15.6 INTERNAL HEAT SOURCES

The main sources of internal heat are fan motors and circulating pumps. Where the motor itself is within the cooled space, the gross energy input to the motor is liberated as heat which must be removed. Where the motor is outside, only the shaft power is taken. Defrost loads must be included.

Other motors and prime movers may be present – conveyors, lifts, fork-lift trucks, stirrers, injection pumps, packaging machines, etc. The gross power input to these machines may be read from their nameplates or found from the manufacturers.

Each person in a cooled space represents a load of approximately 120 W.

All lighting within the space must be included on the basis of the gross input. An 80 W lighting tube takes about 100 W gross. Where the lighting load heat input is seen to be a large proportion of the total, it is desirable to look at low energy options such as LED lighting systems.

Example 15.4

A cold room has 12 lighting fittings labelled 280 W. The four evaporators each have three fan motors of 660 W gross per fan and 18 kW defrost heaters which operate alternately for 15 min twice a day. The fork-lift truck is rated 80 A at 24 V and will be in the store 20 min each hour during the 8-h working day. Two packers will be present for 10 min each hour. Estimate the average and peak internal heat source loads (see Table 15.2).

Table 15.2 Heat source loads for Example 15.4

Heat load	Average over 24 h	Peak
Lighting, 12 × 280, 8 h/day	1.12	3.36
Fan motors, 12 × 660 W	7.78	7.92
Defrost heaters, 72 kW, 1 / 2 h/day	1.50	18.00
Fork-lift, 1.92 kW, 1 / 3 × 8 h	0.21	1.92
Fork-lift driver, 120 W, 1 / 3 × 8 h	–	0.12
Packers, 240 W, 1 / 6 × 8 h	–	0.24
Total	10.61	31.56

This example shows that the greatest load is the fan motors, since these run all the time, except during defrosting. There are several unknowns. For example, it is assumed that the defrosting of the evaporators will not coincide, but this may occur if badly timed, and cause a peak load which may raise the store temperature for a time. The last two items can be ignored, making the load 11 kW average. However, the greatest heat input is still the fan motors, which indicates that any reduction in this component of the load, possibly by switching off two evaporators at night, can appreciably reduce the energy requirements, in terms of both the electricity input and the cooling load to take this heat out again.

15.7 HEAT OF RESPIRATION

Certain stored foodstuffs are living organisms and give off heat as their sugar or starch reserves are slowly consumed. This is known as the heat of respiration, since the products consume oxygen for the process. The heat of respiration varies with the sugar or starch content of the product, and the variety and temperature of the product, and is between 9 and 120 W/t at storage temperatures. Typical figures are shown in Table 15.3. These figures

Table 15.3 Typical values for heat of respiration

Product	Temperature (°C)	Heat of respiration (W/t)
Apples	2	12
Pears	1	16
Bananas	13	48
Strawberries	0	45
Potatoes	1.5	9

increase with temperature, roughly doubling for every 10 K, so that fruits and many vegetables deteriorate very rapidly if they are warm, using up their food reserves and then decaying.

15.8 PACKING AND HANDLING

Cold storage packing must contain and protect the product, whilst allowing the passage of cooling air to keep an even temperature.

Once the decision of storage temperature has been made, the primary factor determining the size and layout of the storage is the type and amount of product to be held and for what duration. The final design will be based on these factors, additionally influenced by material-handling considerations including packaging; pallet stacked height and weight. Grouping of product in chill stores is also a consideration where some products may become tainted by others.

This then leads to the selection of a rack system design, and associated material handling equipment. A small store is usually less than 4500 m^3 of storage space, whilst anything larger than this must be considered as a large store where mechanical handling may be required. The selection of type(s) and size(s) of fork-lift truck affects the type of rack system, rack layout, building layout, sizes and locations of freezer doors and sizes of aisles. Fig. 15.2 illustrates wide aisle racking that requires aisles approximately 2 m wide to allow normal fork lift truck manoeuvrability. This is the simplest and most commonly used system but results in excessive space usage, resulting in poor storage factors and consequently higher capital building costs for a given storage capacity. In order to overcome these shortcomings more complex systems can be employed, details of which are given in the IOR Cold Store Code, 2015.

The overall height of the truck when collapsed, that is the minimum height with the forks just clear of the ground, is the limiting parameter for the door heights.

Figure 15.2 ***Wide aisle racking (IOR).***

15.9 FROST AVOIDANCE

Frost and ice are a hazard for people working in a cold store and can cause damage to mechanical-handling equipment as well as impairing the performance of the cooling system (Fig. 15.3).

Moisture moves due to the difference in vapour pressure. Typical vapour pressures found in cold storage application design are shown in Table 15.4.

It is the difference in vapour pressure that draws moisture into the store. This will happen even if the store is at a positive pressure. Moisture will move in the opposite direction to airflow and through the insulation if allowed to do so (see Fig. 15.4). For example, 1% moisture by volume can reduce thermal performance by up to 85%, reducing the U-value from 0.25 to 0.47 W/m^2K.

Cold storage areas operate at a negative pressure; this is due to the cooling effect of the air passing through the evaporator coil block. This

Figure 15.3 ***Iced-up evaporator (Michael Boast).***

pressure difference draws outside ambient air into the store, with its associated moisture.

Air also infiltrates into the store from door openings where high-velocity air currents can be created. These air currents increase as the height of the door and store increase. The rate of increase is not linear; it increases

Table 15.4 Properties of air (Michael Boast)

Dry bulb and relative humidity	Vapour pressure (kPa)	Air moisture content (g/kg)	Dew point (°C)	Enthalpy (kJ/kg)
−25°C, 80% rh	0.38	0.31	−27	−24.3
−22°C, 80% rh	0.5	0.42	−24	−21
5°C, 80% rh	0.7	4.33	2	15.9
32°C, 40% rh	1.96	12.31	22	63.7

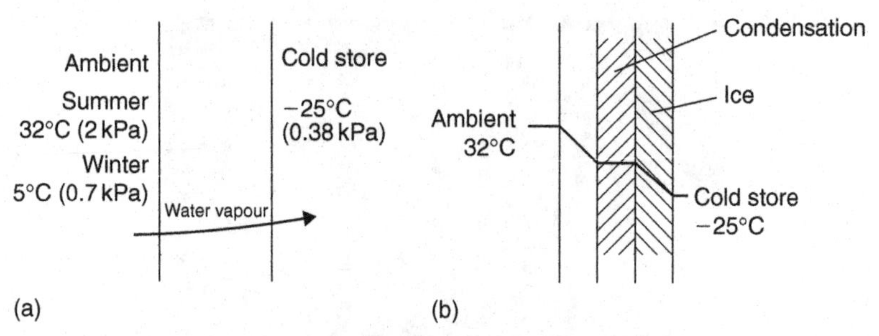

Figure 15.4 ***Section through cold room insulation where diffusion is allowed to occur. (a) Vapour diffusion; (b) Thermal gradient.***

almost by the square. The infiltration air velocity through an open cold store door will be between 0.5 and up to 2 m/s, depending on door size and cold store height.

Example 15.5

If a store has two pallet truck doors, dimensions of 2.2 × 3.2 m, each with 300 traffic movements per day each of 30 s, what is the air infiltration? What is the number of air changes per day if the store measures 50 × 70 × 10 m high?

The cross-sectional area is 7.04 m² and the door openings are 18,000 s per day. The resulting infiltration at 1.0 m/s is

$$7.04(\text{m}^2) \times 18{,}000(\text{s}) \times 1.0(\text{m/s}) = 1{,}26{,}720\,\text{m}^3/\text{d}$$

The store volume is 35,000 m³. Hence the rate of air change is 3.6 per day.

For the typical store described previously, the plant capacity required would be about 500 kW. Of this 80 kW (16%) is for infiltration if the frozen area has a temperature-controlled loading facility. If the air is infiltrating from ambient then the total load rises to 680 kW of which 200 kW (29%) is the infiltration load.

Frost-free operation requires the following:

1. Dehumidification of the air which enters the store. Desiccant wheel dehumidification equipment will remove almost all the moisture from an air stream.
2. Appropriate door design and traffic management is essential.

Example 15.6

For the store in Example 15.5, assuming the air enters at 5°C what is the infiltration load without dehumidification, and with dehumidification using two air locks, each supplied with 600 m^3/h of dehumidified air at 20°C, and all of this air is assumed to enter the cold store? (Take air specific volume = 0.8 m^3/kg)

Without dehumidification the air change is

$$35{,}000\,(m^3) \times \frac{3.6\,(\text{air change})}{0.8\,(m^3/kg)} = 1{,}57{,}500\,kg/day$$

The cooling load is

$$1{,}57{,}000\,(kg/day)\,\frac{15.9-(-24.3)}{86{,}400} = 84.5\,kW$$

With dehumidification the air change is

$$600\,(m^3/h) \times 2\,(\text{doors}) \times \frac{24}{0.8\,(m^3/kg)} = 36{,}000\,kg/day$$

The cooling load is

$$36{,}000\,(kg/day) \times 1.0\,(kJ/kg) \times \frac{20-(-25)}{86{,}400} = 18.75\,kW.$$

To this the energy required to regenerate the desiccant wheel has to be added.

For more detail on frost control please refer to Boast (2003).

15.10 INSULATION AND VAPOUR BARRIERS

The purpose of insulation is to reduce heat transfer from the warmer ambient to the store interior. Many different materials have been used for this purpose but most construction is now with the following:

1. *Mineral fibre*: It is a high-quality non-combustible bonded Rockwool with a vertical fibre structure, which ensures a high insulation value, excellent fire resistance and acoustic performance. Rockwool is water repellent, free of (H)CFCs and fully recyclable solution for wall and ceiling applications in hygienic chill and ambient environments where internal temperatures are above 0°C.
2. *Foamed polyurethane*: The basic chemicals are mixed in the liquid state with foaming agents, and swell into low-density foam, which sets by

polymerisation into a rigid mass. As the swelling material will expand into any shape required, it is ideal for the core of sandwich panels, and the sheet material skins may be flat or profiled. When the panels are manufactured the mixture is injected between the inner and the outer skins and expands to the thickness required, adhering to the lining materials.

3. *Polyisocyanurate (PIR)*: This is amongst the most efficient insulating materials available. Its low-density rigid foam exhibits superior thermal stability and lower combustibility when compared to polyurethane foams. Polyisocyanurate insulation products have excellent thermal conductivity, do not support the growth of fungi, are resistant to moisture ingress and are (H)CFC-free.

Polystyrene has been used until quite recently; but because of fire risk, panels made from this material are no longer acceptable to insurers for new buildings and many existing ones.

The ability of an insulant to reduce heat flow is expressed as resistivity or its reciprocal conductivity. The units of the latter are watts per metre kelvin (W/(m K)). Values for these materials used are approximately as follows:

Mineral fibre 0.03–0.04 W/(m K)
Expanded polystyrene 0.035–0.45 W/(m K)
Foamed polyurethane 0.025–0.03 W/(w K)
Polyisocyanurate 0.02 W/(m K)

The U-value is used to describe the performance of particular panels or building components. It is defined as the rate of heat flow over unit area of any building component through unit overall temperature difference between both sides of the component. The term 'U' represents overall thermal conductance from the outside to the inside covering all modes of heat transfer. The units are W/m^2K.

Insulation material is generally supplied as sandwich panels with a coated steel outer casing. They must be highly impermeable to water vapour. Moisture in the air in the room will condense on the coil and, if cold enough, will freeze. This will reduce the humidity inside the room – it will tend to approach the saturation pressure at the coil fin temperature. Since this is lower than the vapour pressure of the ambient air, water vapour will try to diffuse from the hot side to the cold, through the wall (see Fig. 15.4). At the same time, heat is passing through the wall, and the temperature at any point within the insulation will be proportional to the distance through it.

At some point through the wall, the temperature will be equal to the saturation temperature of any water vapour that is allowed to pass through it, and condensation into liquid water will occur within the insulation. This process will continue and the water will travel inwards until it reaches that part of the insulation where the temperature is 0°C, where it will freeze. The effect of water is to fill the air spaces in the material and increase its conductivity. Ice, if formed, will expand and split the insulant.

It is to prevent this deterioration of the insulation that a vapour barrier is required across the warm face. This must be continuous and offer the best possible barrier to the transmission of water vapour. The use of impervious materials on both skins requires meticulous attention to the sealing of any joints. It is sometimes thought that the plastic insulants, since they do not easily absorb moisture, are vapour barriers. This is not so, and no reliance should be placed on the small resistance to vapour transmission which they may have.

Great care must be exercised at wall-to-floor junctions, door thresholds and all changes of direction of walls and ceilings. In the case of a wall-to-floor junction, this will often occur at two dissimilar types of construction, that is pre-formed wall panels to in situ floor insulation. A satisfactory continuous vapour barrier needs careful design.

Any conductive material, such as masonry and metal structural members or refrigerant pipes, which must pass through the insulation will conduct heat, and the outer part may become cold enough to collect condensation and ice. Such heat bridges must be insulated for some distance, either inside or outside the main skin, to prevent this happening. If outside, the vapour barrier must, of course, be continuous with the main skin vapour barrier.

15.11 SECTIONAL COLD ROOMS

Small cold rooms can be made as a series of interlocking and fitting sections, for assembly on site on a flat floor (see Fig. 15.5). Standard ranges are made up to about 70 m^3, but larger stores can be made on this principle. The floor section(s) is placed on a flat floor and the sides erected on this, located, sealed and pulled up together. The roof sections then bridge across the walls. Such packages are supplied complete with all fittings. They can be dismantled and moved to another location if required. Specialist site work is restricted to cutting necessary holes for pipework and fitting the cooling equipment.

Stores of this size can be built, using standard size factory-made sandwich panels, cutting these to size, joining and sealing on site. This form

Figure 15.5 ***Sectional cold room (Michael Boast).***

of construction is prone to fitting errors, with subsequent failure of the insulation, if not carried out by skilled and experienced people.

15.12 COLD STORE CONSTRUCTION

The plastic insulants are rigid, homogeneous materials, suitable as the core of constructional sandwich panels. This method of fabrication is facilitated when using foamed rigid polyurethane, since the liquids can be made to foam between the inner and the outer panel skins and have a good natural adhesion, so making a stiff structural component.

The panel-edge locking devices may be built in or applied on site. To build such a store, the floor is first prepared, bringing the vapour barrier up at the outer face. Wall sections are erected on end on the edge of the floor and locked together, making the inter-panel seal at edges and corners. Ceiling panels are fitted over the tops of the walls and sealed at the warm face of the junction.

The insulation panels are normally erected within a frame building so that panel joints are protected from the weather. Long vertical panels can be additionally braced to the structure. It is possible, with suitable construction and finishes, to erect the insulation panels around an internal supporting framework. Various systems and panel joining methods to suit all situations are available from panel suppliers.

Care must be taken regarding the method of supporting ceiling panels. Large portal framed steel buildings may provide an effective outer shell but do have a considerable amount of roof movement. Panels hung from this type of structure have support systems that can accept some movement, and wind bracing in the sides of the frame is used to prevent movement.

Careful consideration must be given to the roof design. Sunlight radiation can expand the external panel steel sheets during daytime with surface temperatures up to 80°C and may be higher in hot climates, whilst the internal steel sheet may be cooled to a temperature as low as −30°C. At night the external panel temperature could drop to 20°C. The fluctuating stress on the external sheeting of these panels going through this temperature cycle every 24 h can result in eventual vapour seal and panel breakdown. It is beneficial to use white panels on outside surfaces.

15.13 FLOORS

Heavy floor loadings and the use of ride-on electric trucks demand a strong, hard-working floor surface, which must be within the insulation envelope.

Floor construction starts with a firm concrete foundation slab about 200–250 mm below the final floor level. This is covered with the vapour barrier, probably of overlapping layers of heavy-gauge polythene sheet. On this is placed the insulation board in two layers with staggered joints; this is fitted as tightly as possible. Bitumen building paper is used to cover the surface and prevent penetration of the insulation joints by the floor slab. The concrete floor itself is reinforced with steel mesh and is laid in sections, to allow for contraction on cooling. It can be floated in strips up to 100 m long and these are subsequently saw cut when partially cured. The saw cuts are usually underneath the rack locations to avoid joints in the fork-lift truck paths. As the floor shrinks, the saw cuts open up and can finally be filled with hard setting mastic.

The need for good design and expert installation of floor finishes cannot be emphasised too strongly. The floor receives the greatest wear of all the inner linings, and once the temperature has been reduced in the store, it will usually remain low for the rest of its life. Repairs to damaged floors require the use of low-temperature epoxy resins or low-temperature cement mixes.

Where a store is to take post-pallets, or will have internal racking to store pallets, careful calculation is necessary of the load on the feet. They can have a considerable point load, having the effect of punching a hole through the floor finish.

15.14 FROST HEAVE

If floors are laid on wet ground, the vapour pressure gradient (Fig. 15.4) will force water vapour up towards the vapour seal. Given a ground temperature of 10°C in the United Kingdom, the underside slab may become as cold as 0°C after many months of store operation, and any moisture condensed under the floor insulation will freeze and, in freezing, expand. In time this 'lens' of ice under the floor slab, unable to expand downwards, will lift the floor (frost heave).

Frost heave is prevented by supplying low-intensity heat to the underside of the insulation, to keep it above freezing point. This may take several forms:

1. Low-voltage electric resistance heater cables fixed to the structural floor slab and then protected within a 50-mm thickness of cement and sand to give a suitable surface on which the floor vapour barrier can be laid. The heating is thermostatically controlled, and it is usual to include a distance reading or recording thermometer to give visual indication of the temperature of the floor at several locations below the insulation.
2. Pipes buried in the structural slab: These are connected to delivery and return headers, and glycol circulated. This is heated by waste heat from the refrigeration plant. MDPE (medium density polyethylene) is the normal piping for under floor glycol heater mats. Steel pipe should not be used under the floor unless protected against corrosion.
3. Air-vent pipes to allow a current of ambient air through the ground under the base slab. This is not very suitable in cold climates.
4. On very damp ground or where the finished floor level is in line with the deck of transport vehicles, the cold store floor can be raised above the existing ground level. This is done by building dwarf walls or extending the length of the piles, if these are used, to support a suspended floor at the required height. This leaves an air void of some 1 m under the cold store, through which air can naturally circulate; although forced ventilation may be required to reduce condensation.

15.15 FIRE RISK

A cold store may appear to be an unlikely location for a fire. However, factors such as an ultra-dry atmosphere, and the highly combustible nature of some foam insulation materials, wooden pallets and plastic wrapping present a high fire risk. Electrical faults from handling equipment, lighting or maintenance operation are potential hazards. Storage racking can affect the airflow and impede the detection and response to a fire.

Fire alarms can take the form of an aspirating 'active' smoke-detection system. These systems continuously sample the air at strategic points and lead the air to a detector placed outside the cold environment. Sprinkler systems for fire suppression are widely utilised in warehouses, and for installations in zones where temperatures are below zero, an extremely fast response (EFR) type is used. The pipe installation is dry until the sprinkler head is activated. An alternative is a system in which a low-temperature grease is used to fill the pipes within the sub-zero zone.

The emphasis of buildings regulations is shifting towards placing responsibility on to owners to prove the fire safety of their buildings rather than specifying the design details. Insurance requirements are likely to dictate the use of PIR filled panels or mineral wool, together with appropriate fire alarm and suppression systems. For this reason higher-cost PIR cores have now superceded polystyrene.

15.16 DOOR AND SAFETY EXITS

Appropriate door design and traffic management are essential. Traffic has to be segregated, firstly into pallet trucks and reach trucks, with each group having its own entry and exit facility. Having segregated the traffic, it is then necessary to introduce one-way flow. For large stores a six-door installation is a good configuration. It can handle large volume of traffic efficiently and safely and have a degree of redundancy.

Two one-way doors deal with reach trucks. Two pairs of smaller one-way doors are used for pallet trucks. All the entry facilities are through air locks about 6 m long. The doors can be vertical or horizontal in operation; often a combination of both gives the best result. Each end of the air lock has a rapid movement door. Strip curtains are used for smaller stores (see Fig. 15.6).

The latest technology for rapid movement doors will open and close the door at speeds of 3 m/s or faster. The door installation for reach trucks needs careful consideration, as these entrances may be 4.5 m high. Low-temperature rapid roll doors for door heights up to 5 m are now available.

All mechanical doors are required by law to be capable of hand operation in the event of power failure, and doors of all types must have fastenings which can be opened from either side in case an operator is shut in the store. Larger rooms must have an escape door or breakout hatch or panel at the end remote from the doors, for use in an emergency.

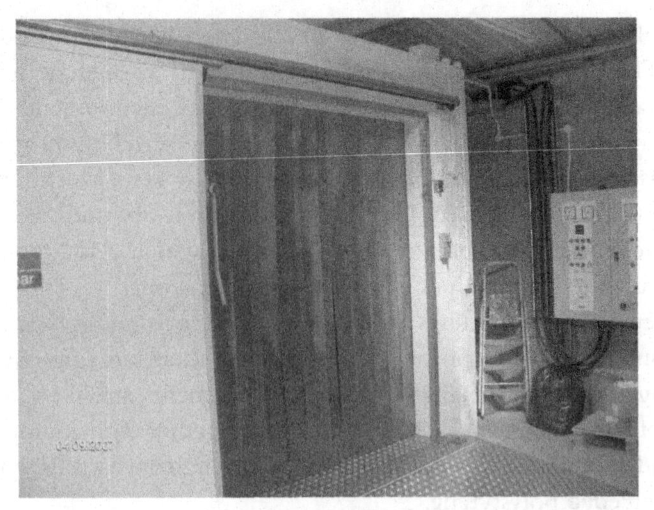

Figure 15.6 ***Cold room loading door open and strip curtain (Michael Boast).***

15.17 INTERIOR FINISH AND FITTINGS

The interior surface finish, to comply with health standards, must be rust-proof, cleanable and free from any crevices that can hold dirt. Bare timber in any form is not permitted. Most liners are aluminium or galvanised steel sheet, finished white with a synthetic enamel or plastic coating. Glass reinforced plastic (GRP) liners are also in use. Floors are of hard concrete. Protective curbs are placed once the floor has cured and shrunk.

In the past, timber dunnage battens were fixed around the walls to protect the surface from collision damage and ensure an air space for circulation of the air from the evaporators. Since timber is no longer used, dunnage may be provided in the form of metal rails. The provision of the floor curb at the walls will ensure that pallets cannot be stacked to prevent air circulation.

Lighting in higher-temperature rooms is normally by fluorescent tubes fixed to the ceiling and having starters suitable for the temperature concerned. Low-temperature stores mostly have high-pressure sodium lamps and it is possible to obtain an overall lighting intensity of 300 lux with an electrical load of 10 W/m^2 floor area. The design of efficient lighting systems merits close attention, since all energy put into the store for lighting must be removed again. Control switches are usually outside the entrance doors.

Large stores must be fitted with an emergency lighting system, battery maintained, to enable the routes to the exits to be seen clearly in the event of a mains power failure.

15.18 EVAPORATORS

Ceiling-mounted evaporators (see Fig. 15.7) should be positioned to ensure a good airflow pattern (see Fig. 15.8). Modern coolers when fitted with a short sock can achieve air throws of up to 80 m without excessive velocity by making use of the Coandă effect.

Owing to the weight, evaporators must be supported from the outer structural roof by tie-rods passing through the insulation. Access gangways may be needed in the roof void to facilitate maintenance and inspection of piping, valves and insulation.

15.19 AUTOMATED COLD STORES

The need for access by fork-lift trucks can require up to 60% of the floor area for gangways. There are two main methods of avoiding this wastage of store space.

Figure 15.7 ***Ceiling-mounted coolers (Michael Boast)***

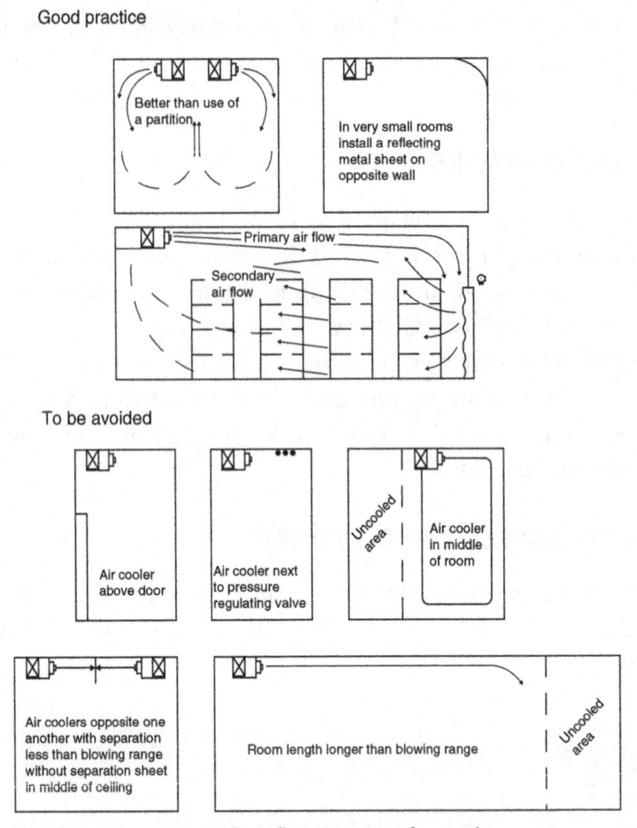

Figure 15.8 ***Preferred location and air flow patterns for coolers.***

Automatic stacker cranes were first used in a cold store in the USA in 1962 and there are now many installations throughout the world. The store height can be increased considerably, up to 30 m and possibly higher using traditional panel construction within a steel frame. The operation of such a store can be by using a crane with the operator inside the store, driving the crane from a heated, insulated cab, or can be fully automatically operated by a computer. One crane can service some 4000 pallet positions at the rate of 50 pallets per hour.

Mobile racking, where the lines of racking are on transverse rails, can be closed together when access is not needed, but rolled apart to provide an aisle for a fork-lift truck. This system is best for a limited range of products moving in rotation, since the racking will not have to be moved very often. A typical small installation might have seven mobile racks; each 25 m long by four pallets high, and require an extra 3 m width for one access aisle, plus an end access of 4 m. This results in a store of 504 pallet capacity and a floor area of 270 m^2.

The tight stacking when the racks are closed impedes the airflow around the pallets, so this system is not suitable where some cooling of the product may be required.

15.20 SECURITY OF OPERATION

The value of the produce in a large cold store may be several times the cost of the store itself, and every effort should be made to maintain the refrigeration service at all times, even if plant may be inoperative for inspection, overhaul or repair. The principle of plant security is that there should be sufficient pieces of each item of equipment and that they should have enough capacity to hold conditions as required by the produce, regardless of any one item which might be stopped.

Usual arrangements can be summarised as follows:

1. At least two compressors, either of which can keep the store at temperature. It may run continuously to hold this.
2. Two condensers or a condenser assembly having two separate refrigerant circuits and permitting repair to one circuit whilst the other is working. If there is one assembly with forced convection, there are at least two fans.
3. All circulating pumps to be in duplicate, with changeover valves.
4. At least two evaporators, to maintain conditions if one is not working.
5. Where two compressors and two condensers are installed as independent circuits, provide changeover valves so that either compressor can work with either condenser or evaporator.

Before installation, the planned system should be analysed in terms of possible component failures to ensure that it can operate as required. Commissioning running tests should include simulated trials of plant failure, and operatives should be made aware of failure drills to keep the plant running.

CHAPTER 16

Food Refrigeration – Product by Product

16.1 INTRODUCTION

Almost all perishable foods are commercially prepared with the aid of cooling processes. These pages can do no more than give a very brief outline of some of the refrigeration techniques used in the food and drink industry today.

16.2 MEAT INDUSTRY APPLICATIONS

In the meat industry, the main applications of mechanical refrigeration are the following:

- Chilling of carcasses directly after slaughter and dressing
- Cooling of meat-handling rooms such as butcheries
- Chill storage of edible meats and offal
- Meat and poultry freezing

Animals when slaughtered are at a body temperature of 39°C. The carcass cools slightly as it is being dressed, but must be put into refrigerated chambers as soon as possible. The speed of cooling depends on the thickness of the joint, so the larger carcasses are usually halved into sides. Although there is a need to remove body heat to check deterioration, if this process is too quick with beef or lamb, the resulting meat may be tough (cold shortening). A general rule for lean meat such as beef is that no part should be cooled below 10°C for at least 10 h after slaughter, although this limit may be varied by the local producer. The total time in this chiller stage will be at least 24 h for a beef side. During the initial cooling stage, the surface of the meat will be quite warm, and careful design of the coolers and their operation is needed to reduce weight loss by evaporation from the surface and good air circulation is required at a humidity level of 90–94%. Meat-cooling curves which indicate weight loss as found by Bailey and Cox (1976) are shown in Fig. 16.1. At low air speeds the weight loss is high because the process takes a long time, but higher air speeds promote rapid evaporation from the surface with corresponding weight loss and cold-shortening effect. Therefore, there is an optimum air velocity and air temperature to minimise weight loss.

Refrigeration, Air Conditioning and Heat Pumps
http://dx.doi.org/10.1016/B978-0-08-100647-4.00016-4

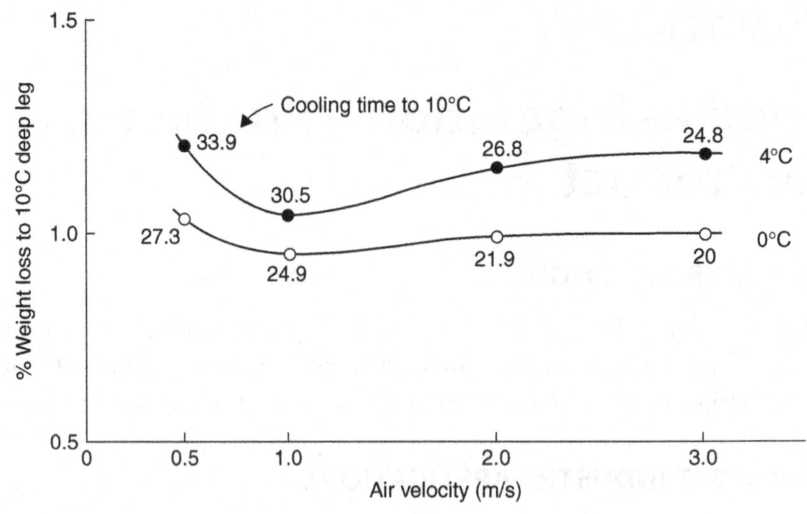

Figure 16.1 ***Effect of air velocity and temperature on the weight loss of beef carcasses (IOR).***

In order to maintain a good and steady air circulation around the carcasses at this time, they are hung from rails and the cold air is normally fan circulated.

Storage conditions in terms of air movement and humidity will be different from those used when initially chilling the carcase. In Fig. 16.2 the air is distributed using air socks. Chilled meat on the bone is stored at about 0°C, up to the point of sale (see Table 16.1). The humidity of the surrounding air is also critical in the case of fresh meats. If the air is too dry and the meat will lose weight and discolour, and if temperature is allowed to drift upwards in high-humidity conditions too slime can form on the surface.

16.3 BONED, BOXED AND PROCESSED MEATS

For processing, the meat needs to be at 0°C or just below, that is, just above the temperature at which it starts to freeze. The air temperature is usually not lower than 10°C, for the comfort of the people working in the area, but some establishments work down to 2 or 3°C. Air movement in the working area must be diffused and not too fast, to give an acceptable environment to the operators. Textile sock distribution is a good solution (see chapter: Air Conditioning Methods and Applications).

Cut meats are usually wrapped or vacuum packed directly after cutting. Reduction in temperature, called *tempering*, prior to the cutting process is

Figure 16.2 ***Post-slaughter meat chill area (Star Refrigeration).***

used to ensure that it is cleaved cleanly and with minimal distortion. The viscera, bones and other parts not going for human consumption have a by-product value and will probably need to be stored at chill temperature before disposal. Cut meats may be frozen or kept at 'chill' temperatures. If the latter is the case, the shelf life is comparatively low.

In 'protein economy' processes, parts of the carcass which are not to be sold as joints or cuts are made up in moulds into artificial joints, 'gigots' or meat loaf, in a pre-cooking operation. The made-up product must then be cooled to about 0°C and may then be sliced and vacuum packed. These operations should take place in air conditioned rooms kept at temperatures of 10°C or lower. Most such items are for 'chill' storage and immediate distribution for sale.

There are many variations in the manner of handling and processing meats, and these will be known only to specialist companies in the trade. The principles of cooling are the same for all.

Meat may be frozen on the bone, but this is not a very convenient shape for packing and handling. It is more usually boned, vacuum wrapped, boxed and then frozen. Boxed meat sizes are about 635 × 350 mm and 100, 125 or 150 mm thick, the largest of these holding some 25 kg. The freezing may be in a cold air blast, and the freezing time will depend on the thickness of the slab and the insulation effect of the box or wrapping Fig. 16.3

Table 16.1 Indicative storage conditions for foods.

Products	Temperature	Humidity	Life
Apples	1–4★	85–90	2–8 months
Bananas, green	12–14	90	10–20 days
ripe	14–16	90	5–10 days
Beer, barrel	2–12	65	3–6 months
Cabbage	0–1	95	3–5 weeks
Carrots, young	0–1	95	1–2 months
old	0–1	95	5–8 months
Celery	0–1	95	1–2 months
Cucumber	10–12	90–95	10–14 days
Dairy products, milk	0–1	–	2–4 months
Ice cream	−23 – (−28)	–	6–12 months
Cheese	1–4	65–70	6–18 months
yoghurt			
Dried fruits	0–1	Low	6 months up
Eggs, shell	−1 to 0	80–85	5–6 months
Fish, wet	1–2	90–95	5–15 days
Fruit soft (berries)	0–1	90–95	5–7 days
Grapefruit	10–14	85–90	4–6 weeks
Grapes	0–1	90–95	2–5 months
Lemons, green	14–15	85–90	1–6 months
Lettuce	0–1	90–95	1–2 weeks
Meats, bacon	1–4	85	1–3 months
Beef	−1 −(1)	85–90	1–6 weeks
ham, fresh	0–1	85–90	7–14 days
lamb, mutton	0–1	85–90	5–14 days
pork, fresh	0–1	85–90	3–7 days
poultry	−1 to 0	85–90	1 week
frozen	−12	90–95	2–8 months
frozen	−18	–	4–12 months
Melons	4–10†	85–90	1–4 weeks
Mushrooms	0	90	1–4 days
Onions	0–1	65–70	1–8 months
Oranges	0–9†	85–90	3–12 weeks
Pears	−1−(1)★	90–95	2–6 months
Pineapples	7–10	90	2–4 weeks
Plums	0–1	85–90	2–8 weeks
Potatoes, new crop	10–12	85–95	3–6 weeks
main crop	1–3	90–95	6–10 months
Tomatoes, green	12–15	85–90	3–5 weeks
ripe	10	85–90	8–12 days
Wine unfortified	8–10	–	Indefinite

For detailed guidance and advice, refer to Food Standards Agency (United Kingdom) and similar local authorities.

★ See also Section 18.2.

† Depending on variety, harvest time and other factors.

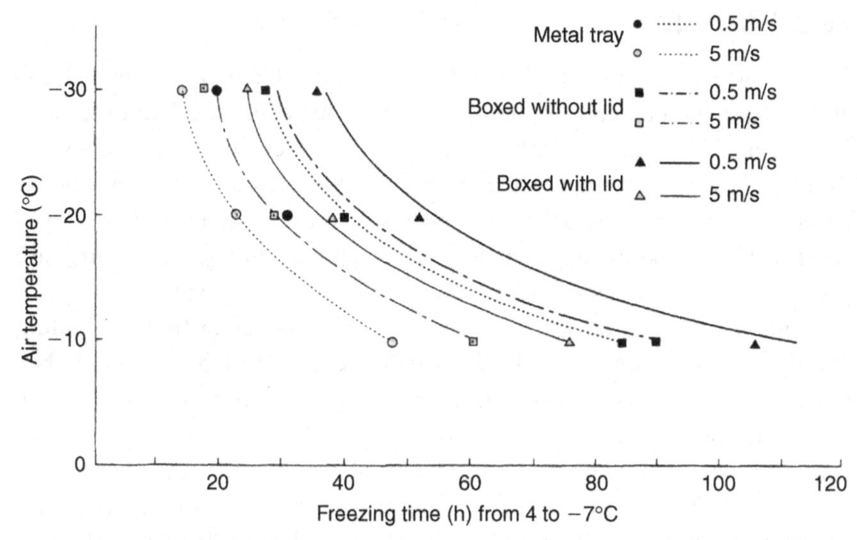

Figure 16.3 ***Freezing times at two air speeds for 150 mm wrapped boxed beef (RD&T).***

shows some experimental and predicted freezing times. The curves are predictions and the points are experimental data. Thinner pieces of meat can be frozen between refrigerated plates (see Figs 7.13a and 14.3).

16.4 PORK AND BACON

Fresh pork has a shorter shelf life than beef, but is handled in the same way and at the same chill-room temperatures. Although no latent heat of the freezing of water content will be extracted at chill temperatures, some heat will be removed when the fat 'sets' or crystallises. The quantity of heat to be removed should be estimated and may be included in the sensible heat capacity in that temperature range. For example, the sensible heat capacity of pork meat averages 2.5 kJ/(kg K), but a figure as high as 3.8 may be used for carcass cooling to allow for this factor.

A high proportion of pork is pickled in brine and smoked, to make ham or bacon. The original process was to immerse the meat in a tank of cold brine for a period. A quicker method is to inject the cold pickle with hypodermic needles into the cuts. Smoking is carried out at around 52°C, so the cured bacon must be cooled again for slicing, packing and storage. Bacon has to be tempered (part-frozen) before it can be sliced in a high-speed slicing operation. Traditionally, the desired slicing temperature was achieved in a long single-stage process. Increasingly a two-stage process of heat removal followed by temperature equalisation is used to reduce processing time and weight loss (Brown et al., 2003).

16.5 POULTRY

The carcasses are dipped into a scald tank of hot water, which helps to loosen the feathers. The carcases are then moved in to the mechanical plucking machines where the feathers are removed. After evisceration, they are thoroughly washed using potable water and chilled down to 4°C by cold air jets. Larger birds may be reduced to portions, so the flesh must be cooled to about 0°C to make it firm which is preferred for cutting. Whole birds are prepared for cooking and then vacuum wrapped for hygiene

Poultry may be chilled for the fresh chicken market, or frozen. Chilling and freezing are mainly by cold air blast. More and more birds are now being cut into portions to meet the demands of consumers for convenience and value for money. Portioning may be done by hand or by automatic-portioning machinery that has the advantage of removing contact with the carcases and of speed. Some poultry is frozen by spraying with liquid carbon dioxide. Storage of chilled poultry is at −1°C. The shelf life is relatively short, and the product will not remain in store for more than a few days.

16.6 FISH

Fish caught at sea must be cooled soon after it is taken on board and kept cold until it can be sold, frozen or otherwise processed. The general practice is to put the fish into refrigerated seawater tanks, kept down to 0°C by direct expansion coils or a remote shell-and-tube evaporator. The seawater must be clean and may be chlorine dosed. In this condition, fish can be kept for up to 4 days.

Ice is also used on board, carried as blocks and crushed when required, carried as flake, or from shipboard flake icemakers. Artisanal fishermen in hot climates may take out crushed ice in their small boats. Fresh fish is stored and transported with layers of ice between and over the fish, cooling by conduction and keeping the product moist. Fish kept at chill temperatures in this manner can travel to the final point of sale, depending on the time of the journey. Where refrigerated storage is used, the humidity within the room must be kept high, by using large evaporators, so that the surface of the fish does not dry.

Most vessels freeze their catch at sea, enabling them to stay offshore without the need to run back to a port within the limited life of the chilled product. If the fish is to be cleaned and processed later, it is frozen whole, either by air blast or, more usually, in vertical plate freezers (see Fig. 7.13b), followed by frozen storage. Some fishing vessels and the fish factory vessels will carry out cleaning, filleting and other operations on board and then freeze and store the final product.

A limited amount of fish is frozen by immersing it in a cold concentrated sodium-chloride brine. This is mainly tuna for subsequent canning, or crustaceans.

Fish which is frozen in air blast will often be dipped into clean water afterwards, resulting in a layer of ice on the surface. This glazing process protects the fish from the effects of dehydration in subsequent storage.

Some freezing of fish fillets and other processed fish is carried out between or on freezer plates, in an evaporator assembly similar to that shown in Fig. 7.13a. Flat cartons of fish and fish fillets are frozen in these horizontal plate freezers.

Health and safety requirements continue to become stricter in the maintenance of the cold chain and the latest regulations should be adhered to.

16.7 MILK AND MILK PRODUCTS

Milk is converted in the creamery and associated factories to whole or 'market' milk, skimmed milk, creams, butters, cheeses, dried milk, whey, yoghurts, butter oil, condensed milk, milk powder and ice-cream.

In the dairy industry as a whole, the following are the main needs for mechanical cooling:

- Cooling milk directly after it leaves the cow and before transport to a central creamery
- Keeping the raw milk cool after it enters the creamery
- Chilled water for use in plate heat exchangers to cool milk and milk products directly after pasteurising
- Chilled water to wash butter
- Chill-temperature stores for milk, butter, cheese, yoghurt and other liquid milk products
- Frozen storage for butter (and sometimes cheese)
- Continuous, plate and air-blast freezers for ice cream
- Low-temperature brine for freezing of ices

Milk comes from the cow at about 37°C and must be cooled within 2 h to 4°C or lower, under hygienic conditions. At this temperature, any microorganisms present will not multiply at a dangerous rate and the milk can be transported to the creamery.

Dairy farms have bulk-storage milk tanks with their own refrigeration plants. These are usually made in the form of a double-skin, insulated tank, having the evaporator in the jacket. The stainless steel pressing forming the jacket constitutes the distribution system for the evaporating refrigerant which is supplied by a condensing unit. The load is intermittent, corresponding to milking times, and the milk temperature must be rapidly reduced. To reduce the size of cooling equipment ice banks are sometimes used to precool the milk before it enters the tank. The refrigeration system runs throughout the 24 h and builds up a layer of ice on the evaporator coils when there is no milk cooling load. This stored cooling effect is available to help cool the warm milk when required (see also Section 12.4).

Bulk tanker vehicles will not collect milk which is warmer than 4°C. If milk can be picked up from the farm at this temperature in bulk tankers, and transported quickly enough to the creamery, then there is no need to have refrigeration equipment on the vehicle. On arrival at the creamery the milk is tested and transferred to bulk-storage tanks, which may hold up to 150 t each. These will be heavily insulated and may have some method of cooling, so as to keep the milk down to 4°C until it passes into the processing line.

Throughout the subsequent processes, milk and milk products will require to be re-cooled down to 4°C or thereabouts. The main method of achieving this is to use chilled water at a temperature just above the freezing point as the secondary refrigerant. Creameries have a large central water-chilling system, using Baudelot coolers or spray chillers (see chapter: Evaporators). Chilled water is piped to all the cooling loads within the plant.

Whole milk for human consumption is pasteurised at 75°C for a short time and then re-cooled to 4°C immediately. This is done by contraflow heat exchange between milk entering and leaving the process, hot water and chilled water, in plate heat exchangers (see Fig. 16.4) in the following stages:

1. Raw milk at 4°C is heated by the outgoing milk up to about 71°C.
2. This milk is finally heated by hot water up to the pasteurising temperature of 75°C (or hotter for UHT milk) and held for a few seconds.
3. The milk is cooled by the incoming milk, down to about 10°C.
4. The final stage of cooling from 10 to 4°C is by chilled water at 2°C.

Milk for other products is treated in the following ways:

1. In a centrifuge to obtain cream and skim milk.
2. In churning devices to make butter and buttermilk.
3. With rennet to make cheese (leaving whey).
4. With cultured bacteria to make yoghurt.
5. By drying, to milk powder.

Figure 16.4 ***Plate heat exchanger (Alfa Laval).***

Butter is made from cream in continuous churning machines. At stages during this process, the butter is washed in clean, cold water to keep it cold and remove surplus buttermilk. At the end of the churning stage, butter is still in a plastic state and, after packaging, must be stored at 5°C to crystallise the fat. Long-term storage of butter is at −25°C.

Cheeses may be pressed into a homogeneous block, or left to settle, depending on the type and methods of manufacture. They then undergo a period of ripening, to give the characteristic flavour and texture. The cold storage of cheese during the ripening period must be under strict conditions of humidity and hygiene, or the cheese will be damaged. Some cheeses can be frozen for long-term storage, but must then be allowed to thaw out gradually and complete their ripening before release to the market.

Other processes (except milk drying) require the finished product to be cooled to a suitable storage temperature, usually 4°C or thereabouts, and kept cool until the point of sale. Conventional-type cold stores can be used for mixed dairy products, since all of them will be packaged and sealed after manufacture.

16.8 ICE CREAM

Ice cream is a product which has been developed since mechanical refrigeration became available. Ice-cream mixes comprise fats (not always dairy), milk protein, sugar and additives such as emulsifiers, stabilisers, colourings, together with extra items such as fruit, nuts, pieces of chocolate etc., according to the particular type and flavour. The presence of this mixture of constituents means that the freezing process covers a wide band of temperatures, starting just below 0°C and not finishing until −18°C or lower. The manufacturing process is in three main stages – mixing, freezing to a plastic state and hardening

The basic mix is made up in liquid form, pasteurised, homogenised and cooled, using chilled water in plate heat exchangers. It is then 'aged' for a few hours and, for this, it will be stored at 2–3°C in jacketed tanks, with chilled water in the jacket.

The next stage is to freeze it rapidly, with the injection of a controlled proportion of air, to give it a light, edible texture. Aerated mix of about 50% air, 50% ice-cream mix by volume is passed into one end of a barrel which forms the inside of a flooded evaporator. The mix freezes onto the inside of the barrel and is then scraped off by rotating stainless steel beater blades, and passes through the barrel with a continuous process of freezing, beating and blending. The most usual refrigerant for continuous ice-cream freezers is ammonia, which will be at an evaporating temperature of −35 to −30°C. About half of the total heat of freezing is removed in this stage, and the ice cream leaves at a temperature of around −5°C, depending on the particular type of product. A continuous ice-cream freezer process is shown in Fig. 16.5.

The ice cream is still plastic as it comes from the freezer, and it is extruded into the final sales shape – carton, tub, box or stick product. It must then be hardened by cooling down to a storage temperature of −25°C or lower, during which the other half of its heat of freezing is removed. Stick products are extruded into trays and go directly into a hardening tunnel. Fig. 16.6 shows four lanes of sticks on their way to the hardening tunnel. This tunnel can handle up to 36,000 pcs/h. The extrusion head is in the top right-hand corner. Flat boxes can be hardened between refrigerated plates as shown in Fig. 7.13a, although tunnels are commonly used.

An important factor of this final freezing process is that it must be as rapid as possible, in order to limit the size of ice crystals within the ice cream. Rapid freezing implies a high rate of heat transfer and, therefore, a very low refrigerant temperature. Air blast at −40°C is common. Two-stage compression systems are used.

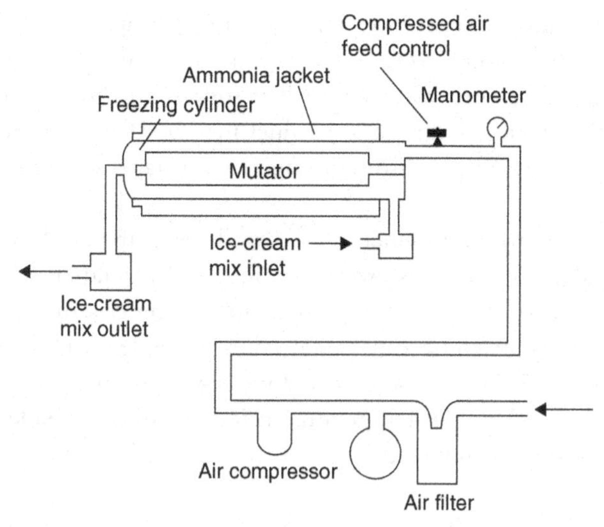

Figure 16.5 ***Continuous ice-cream freezer process.***

Figure 16.6 ***Stickline for ice-cream products (Gram).***

Ice cream must be kept at low temperature right up to the point of final consumption. If it is allowed to soften, the entrained air bubbles may escape and the original texture will be lost. If it softens and is then re-frozen, a hard, solid skin forms, making the product inedible. Ice cream must always be handled quickly when passing through transit stages from the factory to consumer.

Novelties of frozen product on a wood stick are produced in large numbers, and these products are frozen in metal moulds/trays which are submerged in a brine tank having a built-in cooling coil or shell and tube chiller. A novelty machine capable of taking a wide variety of shapes is shown in Fig. 16.7. Both ice-cream and water-ice products can be handled. Cooling is supplied by low-temperature brine or carbon dioxide. The brine tank temperature is maintained at a temperature which can be between −28

Figure 16.7 ***Novelty ice-cream stick product machine (Gram).***

and −35°C depending on the process. The moulds are made from stainless steel or nickel, and pass in rows through the brine bath. Different layers of confection may be built up by allowing one outside layer to freeze, sucking out the unfrozen centre and refilling with another mix. The sticks are inserted before the centre freezes solid. The moulds finally pass through a defrost section of warm brine to release the product from the mould, and extractor bars grab the sticks, remove the products and drop them into packaging bags.

16.9 BEER AND BREWING

The production of beers and ciders requires the fermentation of sugary fluids by the action of yeasts, and the cooling, filtration, clarification and storage of the resulting alcohol–water mixture.

The starting mix for beers is a warm brew of grain-based sugar and flavouring. This 'wort' leaves the hot brewing process and is cooled to a suitable brewing temperature – around 10°C for lagers and 20°C for traditional bitters. This was originally carried out with Baudelot coolers, but now plate-heat exchangers are mainly used, with chilled water as the coolant.

The process of fermentation gives off heat, and the tanks may need to be cooled with chilled water coils, with jackets, or by cooling the 'cellar' in which the tanks are located. When fermentation is complete, many beers are now pasteurised, in the same manner as milk. The beer is then cooled to a temperature just above the freezing point, filtered and left to 'age'. Before final bottling, kegging or canning it will undergo a fine filtration to improve the clarity.

Refrigeration is required for the cold storage rooms and to provide chilled water for the plate heat exchangers. The 'cellars' are very wet areas, and the cooling plant should be designed to maintain as low a humidity as possible, to help preserve the building structure.

Beers at the point of sale are traditionally stored in cellars to keep them cool. Beers are in kegs or piped into bulk tanks. Artificial cooling of these areas is usual, using packaged beer cellar coolers (see Fig. 16.8), and these are somewhat similar to split-system air conditioners. Fig. 16.8a shows the outdoor scroll condensing unit and the indoor evaporator is in Fig. 16.8b. Bulk-storage tanks may have inbuilt refrigeration plant. Drinks such as lager beer, which are normally drunk colder than other beers, are passed through a chilled water bath or double-pipe heat exchanger for final cooling.

Bottled beers and other drinks are kept on refrigerated trays or bottle cooler cabinets, comprising a cooled base tray and an inbuilt refrigeration system.

(a)

(b)

Figure 16.8 ***Cellar cooling split system (Climate Center).***

16.10 WINES AND SPIRITS

The optimum temperature of fermentation of wine depends on the type, red wines working best at about 29°C whereas the white wines require a cooler condition of around 16°C. Heat is given off by the chemical process of fermentation. They are then traditionally matured and stored in caves or cellars at about 10°C. Much of the manufacture and most of the storage is now carried out in rooms controlled by mechanical refrigeration. Spirits do not need low-temperature storage.

The clarity of the final beverage is affected by small particles of tartrates and other substances which precipitate during storage. To obtain a product which will remain clear in storage, many wines and spirits are cooled by refrigeration to a temperature just above their freezing points and then fine-filtered.

16.11 SOFT DRINKS

The feature of most soft drinks is that they are 'carbonated', that is they have a proportion of dissolved carbon dioxide, which causes the bubbles and typical effervescent taste. The quantity of gas dissolved in the water will be 3.5–5 volumes, that is, each litre of water will have dissolved 3.5–5 L of carbon dioxide. The manufacturing technique is to dissolve the required amount of gas into the beverage, and then get it into its can or bottle.

The solubility of carbon dioxide in water depends on the pressure and temperature. The relationship between temperature and pressure for 3.5 and 5 volumes is shown in Fig. 16.9. It will also be affected by the amount of air already dissolved in the water. Therefore, the raw water is carefully filtered and de-oxygenated under vacuum before the sugars and flavourings are added.

Since the gas will dissolve at a much lower pressure at a low temperature, the beverage will be cooled to near 0°C, either before or during the introduction of the gas.

The liquid may be precooled in plate heat exchangers, using chilled water or ethyl alcohol–water although now more usually propylene glycol–water. One carbonisation method is to carry out the final cooling stage over a Baudelot cooler which is fitted within a pressure vessel. The gas is introduced at the pressure needed to dissolve the required proportion, and the gas meets the liquid as it flows in a thin film down the surface of the cooler.

It is then bottled as quickly as possible, before the gas has time to bubble out again. Once it is sealed in the bottle, cooling is not needed for storage.

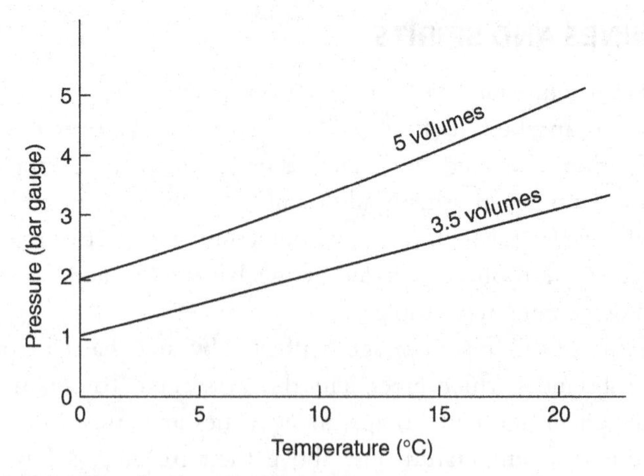

Figure 16.9 ***Solubility of carbon dioxide in water.***

Chilling of brines for pre-cooling will generally be in shell-and-tube evaporators. The Baudelot cooler within the pressure vessel may be cooled by flooded or dry expansion refrigerant, or by brine.

16.12 FRUITS

Fruits are seasonal in temperate climates, and a good harvest may be followed by a shortage if there is no method of preservation. The hard fruits, apples and pears, have traditionally been stored in cool places and may then last for several months, depending on the variety. Refrigeration has extended the storage life, and made this more reliable.

Artificial cooling has made it possible for fruit grown anywhere in the world to be brought to any market. Climacteric fruit such as bananas are picked whilst still green and undergo a controlled ripening on the ship. The conditions for refrigerated shipping depend on many factors, and the temperatures and humidities given in Table 16.1 are a general indication of the ranges. More precise information must be used for the operation for a particular product. A large amount of perishable food now travels by air and temporary protection against low temperatures may be needed if the cargo hold is not pressurised.

Storage of fruit requires careful control of the atmosphere in the store as well as temperature. Stores constructed to maintain such a controlled atmosphere, in addition to temperature control, are generally termed *gas stores.* They have a gas-tight structure to prevent diffusion. The fruits are loaded and the store is sealed. Within a few days they consume a proportion of the available oxygen, and the atmosphere is monitored to keep the right proportions by chemical removal or controlled ventilation. Climacteric fruits also require control of ethylene since this gas effects the ripening process. Considerable research over the past 60 years, mainly in the United Kingdom, has determined the correct balance of gases to prolong the storage life of the different varieties of apples and pears, both home grown and imported.

16.13 VEGETABLES

Most vegetables contain a very high proportion of water, and wilt rapidly as they dry out. Storage conditions demand a high humidity level of 90–98% saturation and temperatures as close to their general freezing point of 0°C as possible. Some leaf vegetables are sprinkled with ice chips, to maintain this damp, cold condition. Cold stores for vegetables have very large evaporators, to provide these high humidities. Apart from the preservation of the vegetable substance itself, mould growths and insect pests are also controlled by low temperature.

A few products, such as cucumbers and some crops of potato, are better kept at higher temperatures. These conditions vary with the variety, state of ripeness when picked and required time of storage.

Onions and garlic are susceptible to moist conditions, which encourage mould growth, and are stored at humidities of 65–70%. It is not possible to store these together with other vegetables for more than a very short time.

The convenience of having high-quality produce, graded and ready for cooking, has resulted in a high demand for frozen vegetables in the United Kingdom. Peas, carrot slices, beans and some leaf vegetables are frozen in air blast. There are slight changes in the texture, but the texture is further changed by cooking.

A few items, strawberries, other soft fruits and pieces of cauliflower, are quick frozen with liquid nitrogen. Frozen fruit and vegetables are sealed in plastic bags and stored at −18°C or lower. The humidity at this temperature is not important.

16.14 BAKERY PRODUCTS

Bread doughs become heated by the mixing process, and the yeast may begin to work too soon. The water content of the mix may be chilled, or the larger machines may have water-cooled jackets to take away this heat.

Doughs are prepared some time before the final baking process and will be left to 'prove', that is allow the yeast to commence working. The action can be retarded by cooling the dough at this stage, and this process permits the workload to be spread through the day. Typically, bread for the following morning can now be prepared on the previous day, up to the proving stage, and then kept under cold, humid storage until a few hours before baking is to commence.

Dough-retarding cabinets are now used in most bakeries. Bread doughs may be made up at any time and put into storage at a temperature between −4 and −3°C, depending on the required retard time, which may be up to 3 days. An automatic timer will terminate the cooling cycle and bring the doughs up to proving temperature when required. In this way, doughs can be ready for the oven when the bakery staff commence in the early morning. Also, stocks can be held ready for unexpected extra demands.

A high proportion of bread is sold sliced, but it will be too hot for this on leaving the oven. Large-scale bakeries have cooling tunnels to reduce the bread temperature so that it can be sliced. A high degree of hygiene is necessary, or the slicer will introduce airborne spores and the bread will grow moulds.

16.15 READY MEALS

There is an increasing demand for ready-prepared foods for final re-heating or cooking in microwave ovens. Applications are for retail sale of take-away meals and factory/office and institution catering. Such foods may be frozen and will then have a long storage life, but will require frozen storage.

It is possible to precook the product to a pasteurisation temperature and then cool, for short-term storage at a temperature above the freezing point. The required standards of temperature control and hygiene are very strict and the subsequent shelf life restricted. The process is cheaper than freezing. Product leaves an oven at 100°C may be allowed to cool in the ambient air to 70°C, if conditions of hygiene are satisfactory. During this time it may be split into meal size or other portions. Generally, it should then have an individual product thickness not more than 50 mm, or it will not cool in the specified time. Trays of the product are loaded into a chilling cabinet, and all parts must be reduced to storage temperature within the time allowed in the Department of Health guidelines. Since it is not required to freeze any part, the air to cool the product cannot be much below 0°C, and cabinets for this purpose have a built-in refrigeration plant which will provide air at −2°C, and with a speed over the product of some 6.5 m/s. The chilled product must be stored at 3°C or thereabouts. Shelf life may be up to a maximum of 5 days, but is usually only a day or so.

16.16 CHOCOLATE

Many confections are coated in a thin layer of chocolate. The latter is a mixture of chocolate, cocoa butter and other fats, blended to form a suitable coating material. This layer melts at a temperature generally in the range 27–34°C. The manufacturer wishes to coat the confection in a thin, continuous layer, and then harden this layer so that the product can be wrapped and packed with the least delay on the production line.

Chocolate enrobing starts with the item passing through the coating process, and then through a refrigerated air blast tunnel to harden the layer. The colder the air the quicker this will take place, but if the product leaves the tunnel too cold, atmospheric moisture may condense on the surface and spoil the glossy finish expected by the consumer. The average air temperature in the tunnel may be between 2 and 7°C, and the air is usually cooled with refrigerated or brine coils within the tunnel. It is sometimes necessary to air condition the entire working area so as to keep the dew point temperature lower than the temperature of the surface of the confection as it leaves the tunnel.

CHAPTER 17

The Cold Chain – Transport, Storage, Retail

17.1 INTRODUCTION

The *cold chain* is a term applied to food handling and distribution where the product is maintained at suitable temperature conditions all the way from harvesting, through the cooling or freezing process to the point of sale. This requires transport, various kinds of storage and display. The nature of the cold chain varies from product to product. For a relatively simple example, for frozen peas, the chain may look like this:

> Harvest → road transport → pre-storage treatment → freezing → packing → refrigerated road transport → supermarket cold room/store → display case → road transport → home freezer

The product must reach the freezing process within a short time after harvesting, and held below zero subsequently if quality is to be maintained.

The transport of cooled produce was one of the first major uses of mechanical refrigeration, dating back to 1880, only 20 years after the first static cold storage. World reefer trade has subsequently grown to 92.4 million tonnes in 2013 (Drewry, 2013). Logistics developments have enabled worldwide distribution of food under temperature-controlled conditions. This has opened up markets both in the major developed countries and also in the developing countries. Export of seasonal produce can represent a major portion of the production, and the ability to sell is dependent on its quality and safety at the consumer end of the chain. The temperature of the commodity must be maintained within specified limits (Fig. 17.1). Air, sea and land transport have each developed their own specialised segment, and where intermodal containers are used to transfer from one mode of transport to another, potentially damaging temperature excursions arising from trans-shipment can be avoided.

Although not a topic for this chapter, there is also a cold chain for the distribution of pharmaceuticals. Vaccines need to be stored at temperatures within the range 0–8°C. They are more or less stable at a given temperature, but will progressively lose their effectiveness if stored outside these limits, and this is a function of the temperature deviation and the length of time of exposure.

Refrigeration, Air Conditioning and Heat Pumps
http://dx.doi.org/10.1016/B978-0-08-100647-4.00017-6

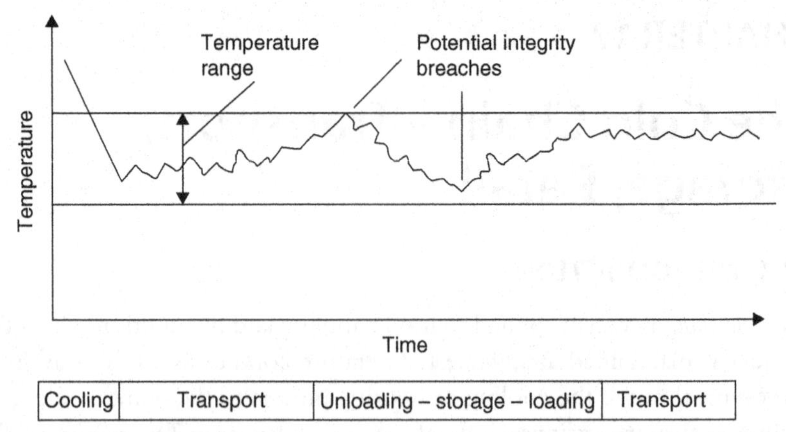

Figure 17.1 ***Illustrating possible product temperature changes along the cold chain.***

17.2 AIR TRANSPORT

Air transport enables highly perishable and valuable products to be moved fast over long distances, but it lacks the environment control that is possible for other modes. In-flight storage will be at hold temperature and whilst it may be quite low over most of the distance, the quality of the product will be highly dependent on prompt and speedy handling at the airports. Exposure to local weather conditions whilst waiting to be loaded onto a plane or being moved to and from the airport can constitute a major part of the total travelling time. Coldrooms are provided at some airports to store produce immediately before and after transit. Solid carbon dioxide (dry ice) is used for short-term cooling of airline passenger meals.

17.3 SEA AND INTERMODAL TRANSPORT

Sea transport was originally in insulated holds built into the ships. Few of these remain, owing to the high-handling costs, and most maritime trade now uses containers, either with their individual cooling plants or connected to a central refrigeration system on the vessel. Reefer is the generic name normally applied to a standard temperature controlled ISO container (see Fig. 17.2). The integrated refrigeration plant (Fig. 17.3) within these insulated containers has a control system that allows the set temperature to be maintained over a wide range of exterior temperature conditions. Monitoring and alarm devices ensure safety of the produce.

The standard width of ISO containers is 8 ft., the standard heights are 8 ft. 6 in., and 9 ft. 6 in., and the most common lengths are 20 ft. and

Figure 17.2 *Reefer container moving by road(Cambridge Refrigeration Technology).*

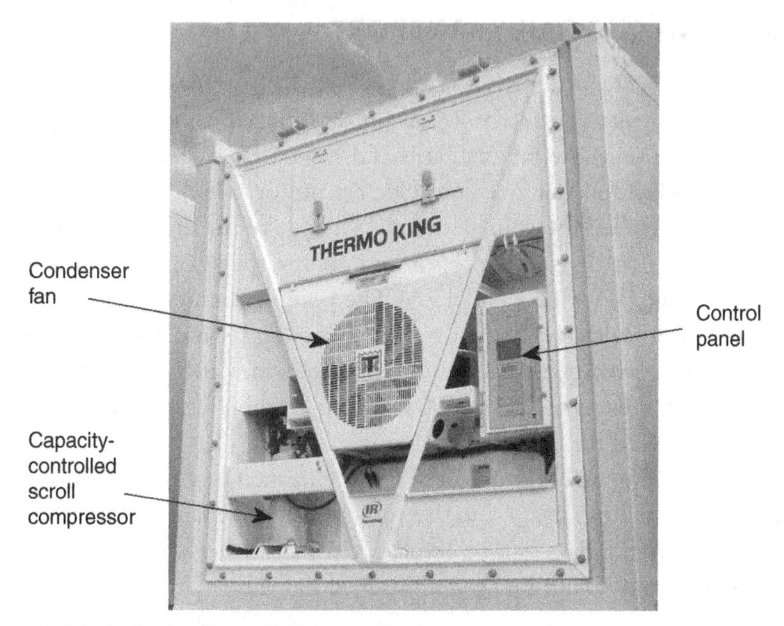

Figure 17.3 *Reefer cooling unit (Thermo King).*

40 ft. Cooling is normally required, but under some conditions heating may be necessary. It will normally be possible to set the box temperature to the appropriate condition for the specific consignment, and the in-built control system then maintains conditions. Typical temperature settings are 13.5°C for bananas, 0.5°C for some chilled and fresh produce, −18°C for frozen meat, −29°C for frozen fish or ice cream. For chilled and warmer temperature goods that must not be frozen, the temperature of air supplied is controlled, whereas for frozen goods control is by the temperature of air returning to the machinery. It is vital to stay within the pre-set temperature range to preserve the integrity of a shipment, otherwise irrevocable and expensive damage resulting in loss of market value may occur.

Container ships have slots with power supplies for reefers, and there are also specialised reefer container ships that are generally smaller in size. On-board the container ships the reefers are connected to the ship's power and can be monitored remotely. Monitoring can be extended via satellite to the shipper who in turn can supervise his expensive cargo of chilled or frozen goods. Slots are also provided on the dock from which the reefers are either dispersed to cold stores for trans-shipment or unloaded into road vehicles.

17.4 ROAD AND RAIL TRANSPORT

Reefer trailers are articulated semi-trailers with a maximum length of 15.5 m, an internal volume of 73 m^3 but holding up to 40 t. The majority of the cooling units are factory built and have their own diesel engine for use on the road and may have an electric motor which can be run from mains supplies when the vehicle is static. Change of the drive is by magnetic clutches. An open-drive compressor and engine speed options provide capacity flexibility (Figs 17.4 and 17.5). The complete unit will be of rugged construction to withstand vibration from poor roads and will be adaptable in being able to maintain any required temperature automatically. Heaters are also fitted, since vehicles may be working at ambient temperatures lower than that required for the produce being carried.

Direct cooling by evaporation of liquid nitrogen is also used. This is carried in metal vacuum flasks, and the vehicle will be reliant on depots where the liquid nitrogen flask can be refilled. The only mechanical equipment is a thermostatically controlled solenoid injection valve.

Figure 17.4 ***Self-contained transport refrigeration unit with diesel engine drive (Thermo King).***

Some vehicles have electrically powered units supplied from an alternator on the main engine; these can be mounted either over the cab, or beneath the vehicle.

Vehicles used only in the daytime can have cooling systems run from a mains electricity supply providing they can hold a sufficiently low temperature whilst on the road. Cooling the vehicle body overnight when in the garage, and relying on the cold mass of produce and good insulation to maintain conditions during delivery is sufficient. Additional daytime cooling capacity can be provided by eutectic plates in which beams or plates containing a phase change material are cooled during the night, and they absorb heat whilst the vehicle is on the road (Fig. 17.6). A conventional condensing unit mounted on the vehicle provides the refrigeration, but only runs when mains power supply is connected. This approach

Figure 17.5 *Self-powered diesel truck unit with electric standby option (Thermo King).*

Figure 17.6 *Eutectic plates within a local delivery vehicle interior – shelf supports for cargo baskets also visible (Cambridge Refrigeration Technology).*

also has the advantage of zero-noise emission from the cooling unit during deliveries. Some local delivery vehicles use liquid nitrogen.

Rail traffic is mainly in purpose-built, insulated wagons, many of these having self-contained refrigeration systems. Some produce is pre-cooled and/or iced. Re-icing stations are available on the longer routes in Europe.

17.5 LOGISTICS

During movement of goods between static cold stores and vehicles, every effort must be made to avoid any warming. The principle is to close the vehicle right up to the cold store wall.

The ideal arrangement is to back the vehicle up to a door with a sealing collar, so that the contents may move directly into the store without exposure to ambient temperatures. If the height differs from that in the store, adjustable platforms are fitted at the door. To avoid ingress of warm air (and loss of cold air) it is useful to have an airlock. These need to be at least the length of a loaded fork-lift truck, and require extra space together with double doors and extra movement time.

The major supermarkets have their own regional distribution centres which are supplied from the factory or container terminal on a daily basis. From there, distribution to individual stores takes place via a fleet of reefer trailers. Smaller retailers make use of independent distribution stores, where goods are delivered in bulk, stored for a short time, 'order-picked' and then sent out (Fig. 17.7).

Figure 17.7 ***Distribution centre (Star Refrigeration).***

Figure 17.8 ***Refrigerated compartment within a warehouse for a range of frozen and chilled products (Star Refrigeration).***

Distribution stores require adjacent or in-built refrigerated storage and order-picking areas, and may operate on a 24-h basis. For full access, the storage is on pallet racking. This occupies approximately two-thirds of the store, leaving the remainder for sorting the goods into the individual outgoing batches. The latter may be on pallets or wheeled racking. Operatives carry out the order-picking operation within the store and wear suitable protective clothing.

Stores within a warehouse are usually 5–8 m high, so that there is less air movement from the coolers at working level, with condensing units external to the building, imposing no additional cooling load on the warehouse (Fig. 17.8). Consignments of refrigerated produce are shipped from the distribution centres to the local cold storage areas in retail outlets.

17.6 REFRIGERATED DISPLAY

Refrigerated display, the final link in the commercial cold chain, is where the consumer purchases the product. The aim is to show the produce to the best advantage whilst maintaining it within prescribed temperature limits.

The first refrigerator for frozen foods was an integral type chest-freezer cabinet ice-cream conservator, a chest-freezer type of cabinet. The refrigeration system is in-built and the evaporator consisted of a coil of pipe in contact with the inner wall. These freezers are still commonly used for sale of ice-cream.

The two major types of display cabinet are those that operate from a remotely sited refrigeration plant (remote cabinets) and those that contain their own small refrigeration system, like a domestic refrigerator. The latter are sometimes termed 'integrals' or 'plug-in units'. Within each of these categories different types of cabinets exist. The most commonly used of these are; multi-deck, glass door, well and delicatessen. Additionally there are specialist designs for specific products and applications.

Multi-deck chill cabinets (Fig. 17.9) rely on an air curtain to maintain the temperature of products on the shelves. This type of cabinet offers excellent product display, and it is a well-established principle that goods that can be seen are more likely to be purchased than those hidden from sight. However even with careful air management multi-deck cabinets tend to have a proportionately larger heat gain load than other types. Store managers will favour their use where the turnover of goods is rapid.

Glass door displays and well types (Fig. 17.10) are more economical to run and are widely adopted for frozen produce. Users accept that there is a need to reduce wastage of energy and are happy to shop for frozen goods from this type of display. Good lighting and door seal heaters are necessary.

Open-top display can gain considerable heat from air infiltration and radiant heat from lighting. Temporary covers are frequently used when the building is closed, to reduce these gains and help preserve the foodstuffs. This is of considerable importance where cut meats are displayed, since the radiant heat from lights and the loss of the cold air blanket lead to surface moisture loss with severe darkening of the appearance. It helps to have glass walls at the sides to reduce draughts, which would disturb the layer of cold air in the cabinet. The evaporator may be pipe coils on the outside of the inner wall, or a finned coil at the back or sides. It is important that produce is kept below the design level of the cold air blanket. Integral reach-in displays are used for a variety of produce, and specialist designs are custom made for small retailers, hotels and restaurants (Fig. 17.11).

Evaporators need to be defrosted at regular intervals, usually every 6–8 hours. Build-up of frost on the evaporators can be limited by air-conditioning the shop area and so reducing the amount of moisture in the surrounding air.

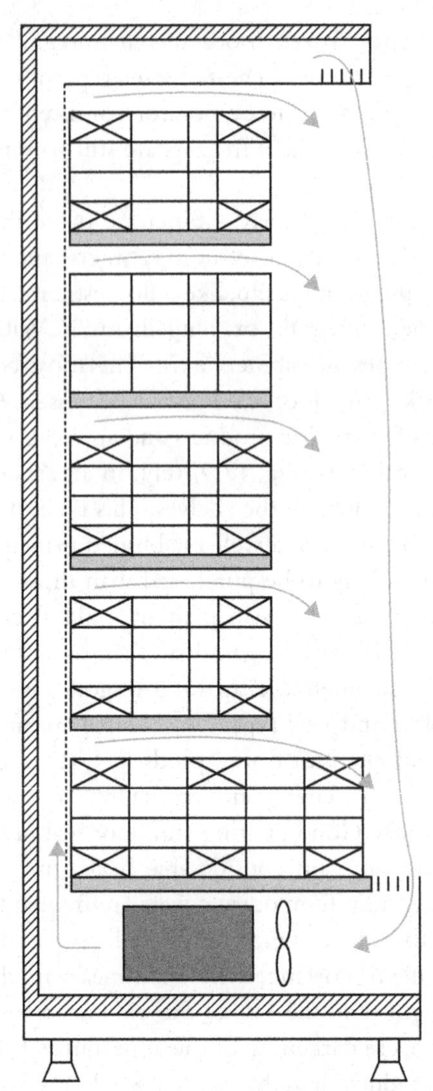

Figure 17.9 ***Sectional view of multi-deck cabinet for chilled food display (RD&T).***

Energy consumption of cabinets is determined according to standard methods that account for both the direct power input to the cabinet for lighting, fans, heaters and defrost heaters and the power input required for heat extraction. Testing has revealed a wide variation in consumption, and 15–20% energy saving could be achieved by many users by choice of efficient products (Evans et al., 2007). An analysis of heat loads for various cabinet types is shown in Fig. 17.12.

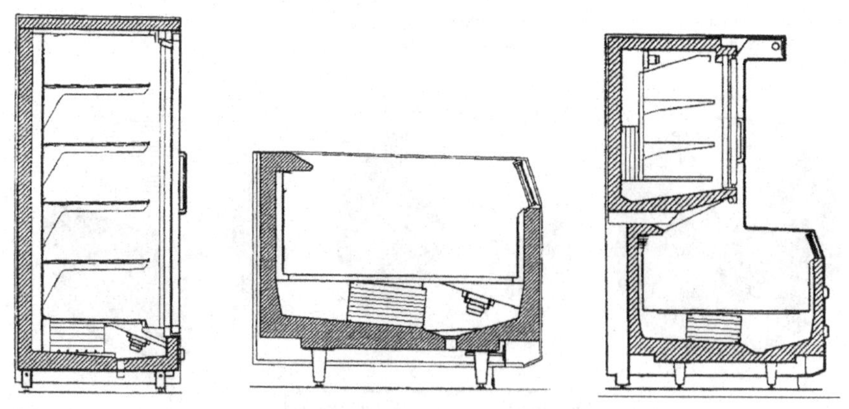

Figure 17.10 ***Sectional views of typical glass door and well-type cabinets for frozen food display (RD&T).***

17.7 RETAIL REFRIGERATION

A supermarket has a large number separate food storage areas – coldrooms and display cabinets, all of which require refrigeration. The original method was, as with the domestic food freezer, to have a condensing unit as part of each cabinet. Using 'plug-ins' in this way can cause excessive temperatures in

Figure 17.11 ***Refrigerated fish display counter (XL Refrigerators).***

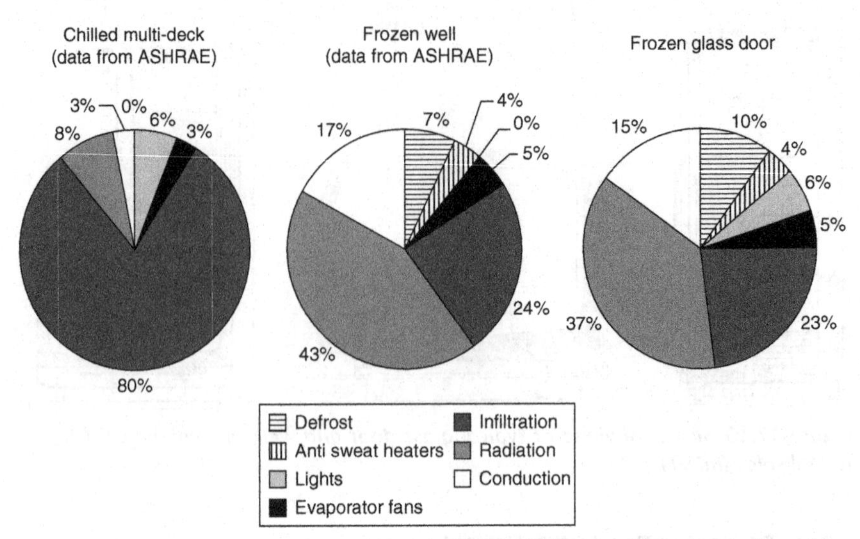

Figure 17.12 ***Percentage analysis of heat loads for different cabinet types.*** *(Data from RD&T and ASHRAE).*

the shopping areas where condenser heat is being injected. This places extra demands on the air-conditioning system. To avoid this, remote-condensing units, usually sited in a central plantroom, became a widely accepted solution.

With a continuing rise in cooling loads, and difficulty of providing sufficient space for condensing unit air circulation, the concept of a compressor pack (see Fig. 13.4) working in conjunction with remote condensers was developed. A bank of compressors provides for each evaporating temperature, with a common condensing pressure. These centralised systems are very flexible, with the compressors switched by a logic controller to maintain correct conditions, regardless of the number of coolers working at any one time. The grouped condensers give the opportunity to recover heat from the discharge gas for water heating, and in winter for heating the building.

Several defrost methods are possible. Off-cycle defrost does not use any form of artificial defrosting. The ice build-up that may occur on the evaporator heat transfer surface is removed during the off cycle of the equipment. This method is mainly used for chill cabinets. Electric defrost is simple to operate and control, and does not rely on any external service for the supply of heat. This method is used for frozen displays. Hot-gas defrost is a method in which the discharge gas from the compressors is taken to the evaporator to provide defrost heat. Many hot gas defrost systems have been installed, but reliability problems have led to the adoption of electric defrost in most installations today.

There is no 'one size fits all' for supermarket applications. There are situations where integral plug-ins or condensing units are a good solution. This is particularly true for smaller stores and filling station forecourt stores. With the increasing emphasis on refrigerant containment there has been a move towards use of small, localised compressor packs in preference to a single remote plant room (see Fig. 13.5). This type of installation uses less refrigerant charge and reduces so called *parasitic losses*, that is pressure drops and heat gains in long suction lines connecting remote plant rooms. It is best suited to a building where roof or wall space is available for the units, but this may not always be the case.

The requirement to provide adequate and unobtrusive refrigeration on a densely packed sales floor space remains a challenge for engineers, and the transport of HFC refrigerant through long pipe runs with many joints remains susceptible to leakage losses. Recognition of this problem over a period of time, and more recently with the introduction of the European F-Gas regulation, has driven innovations. These have included indirect systems, where brine rather than refrigerant is circulated to the display cabinets. This is an effective but a costly solution and potentially more energy-intensive when the pump power and extra heat transfer losses are accounted.

17.8 TRANSITION TO LOW GWP REFRIGERANTS

Cold chain applications mostly fall under the heading of commercial refrigeration. This sector is moving away from high GWP fluids, particularly R404A, in addition to implementing leakage reduction programmes and energy efficiency improvements. In the period following the phase out of CFCs and HCFCs, R404A dominated these applications, its low discharge temperature enabling relatively high-pressure ratios in a single stage without the need for additional compressor cooling. R404A (GWP 3922) now falls within the 2020 EU ban (see chapter: Refrigerants), and some applications have seen a transition to HFC blends such as R407A (GWP 2017) and R407F (GWP1825). Transition can take the form of retrofitting using existing equipment, or installation of new equipment. These blends, and others, some including HFO components, escape the 2020 ban, but may fall within a 2022 ban. They tend to provide slightly less capacity than R404A, for the same compressor displacement, similar efficiency, but with a higher discharge temperature. For long-term HFO/HFC blends it may be necessary to accept and work with mildly flammable (A2L) materials. For a blend to be non-flammable the GWP limit is approximately 400 – any lower is likely to be mildly flammable.

The alternative is a so-called natural refrigerant that contains no HFC or HFO component. Carbon dioxide, CO_2, is being applied in innovative ways in many commercial (and industrial) applications. High equipment costs may be justified by the knowledge that CO_2 is outside F-Gas regulations and represents a long-term solution. The key to its success is in taking advantage of its positive points, using the sub-critical cycle when conditions permit, cascade or booster cycle and heat recovery. Its high latent heat, heat transfer coefficient combined with high pressure and density result in the ability to produce large amounts of cooling with very small displacement compressors and small diameter pipelines. Its main disadvantage is its low critical temperature (see chapter: The Refrigeration Cycle). Fig. 17.13 shows a two-stage transcritical CO_2 unit serving both medium and low temperature display cases.

Figure 17.13 ***Two-stage transcritical R744 unit with gas cooler mounted above the compressors (Star Refrigeration).***

Ammonia (R717), traditionally little used in commercial systems, represents a possible alternative for the high-stage heat rejection if a self-contained unit can be roof mounted. It cannot be circulated to cabinets in occupied areas, but pump circulation, using R744 as an effective secondary is possible.

Flammables such as Propane (R290) and Propylene (R1270) are also possible within the charge limits for occupied areas. Quite widely used in domestic units, they can be suitable for integral cabinets in stores.

In short, all options are being re-considered and many examples of innovative applications may be found.

CHAPTER 18

Industrial Applications

18.1 INTRODUCTION

Industrial refrigeration can be defined as custom-built plant to provide cooling for large-scale processes. In this book the definition will be confined to processes other than air conditioning or food and drink applications. These would fall within a more general definition of industrial refrigeration but are covered elsewhere.

Applications of refrigeration for chemical plants, for example separation of gases, solidification of substances, removal of reaction heat, together with refrigeration in manufacturing and construction, ice and snow sports and environmental test chambers are included within the definition. Some environmental test chambers are quite small and could perhaps be better described as packaged equipment.

A further aspect of industrial refrigeration is the wide range of temperatures involved. Some processes require cooling at above 0°C which can be accommodated with chilled water. The next category would be the applications requiring temperatures below 0 and down to −40°C. These temperatures can be generated by conventional single-stage vapour compression cycle refrigeration, with two-stage systems in some cases. There is a variety of applications that call for temperatures below −40°C and down to approximately −70°C, and they are what could be called low temperature. The dividing line of −40°C marks the approximate boundary where commonly used refrigerants such as R717, HFCs and hydrocarbons retain a positive pressure. Below −40°C is a region where pressures below atmospheric can be encountered unless more specialised refrigerants are used. This coupled with a requirement for a large temperature lift often makes a cascade system a good solution when working below −40°C. Below −90°C is the realm of cryogenic refrigeration, which produces and uses liquefied gases such as liquid nitrogen or liquid oxygen and is outside the scope of this book.

Refrigeration, Air Conditioning and Heat Pumps
http://dx.doi.org/10.1016/B978-0-08-100647-4.00018-8

18.2 R717 AND R744 INDUSTRIAL SYSTEMS

Many industrial systems use ammonia, R717. The engineering and servicing requirements are well established. Its properties are summarized in Section 3.9.4. Ammonia can be detected by its characteristic odour at very low concentrations and this acts as an early warning signal. The safety aspects of ammonia plants are well documented and there is reason to expect a sustained increase in the use of ammonia as a refrigerant.

R744(CO_2) is now being applied in low stage for cascade systems. Fig. 18.1 shows a CO_2 condenser cascade. CO_2 is used as a phase-change refrigerant in a cascade system with ammonia on the high-pressure side. In this industrial warehouse system the R717 suction separator (surge drum) and economiser are located on the roof of the plant room next to the evaporative condensers. The cascade heat exchanger which acts as both the R717 evaporator and the CO_2 condenser is directly below the R717 drum. The condensed CO_2 runs down into a high-pressure receiver in the plant room below. The CO_2 is circulated to the evaporators and the vapour is compressed in a CO_2 compressor in the plant room below.

Figure 18.1 ***Condenser cascade (Star Refrigeration).***

18.3 CHILLED LIQUIDS FOR COOLING

The use of chilled water or a non-freeze solution for heat transfer is now replacing many applications where direct expansion of refrigerant has been used in the past. This is due to increasing concern about safety and about the quantity of refrigerant within systems. A secondary coolant enables the refrigerant to be confined to a dedicated part of the building and also enables packaged liquid chillers to be applied. Examples of applications include the following:

- Chemical/pharmaceutical processes
- Plastic processing: injection moulding, blow moulding, extruders, PVC pipes, woven sacks, film lamination etc
- Lubricants/oil cooling and de-waxing of oils
- Plating, anodising and induction heat treatment
- Pulp and paper processing
- Printing ink/paper waxing cooling
- Bleaching/dyeing industry
- EDM machine cooling
- Concrete curing

Fig. 18.2 is an ethylene glycol chiller used for a wax moulder. The chiller uses R134a and has a capacity of 2320 kW at 0°C and 990 kW at −15°C. It is designed for use with a remote air cooled condenser (not shown). The lower vessel in the foreground is the direct expansion glycol chiller, above it is the liquid receiver with pressure relief valve. Behind these vessels are the

Figure 18.2 *Air-cooled glycol chiller (J & E Hall).*

two open-drive single-screw compressors mounted on one horizontal oil separator. The compressors are enclosed to reduce noise levels.

18.4 SOLVENT RECOVERY

It is necessary to control the emission of volatile organic compounds (VOCs) that originate from many organic-chemical-manufacturing processes as well as from paint spraying, coating, printing and cleaning activities. Any loss of these creates an environmental hazard, apart from the cost of the material itself. Techniques include adsorption and also condensation by cooling. Here, refrigerated condensation units working in the range from −50 to −70°C are used, and also cryogenic techniques which can deliver more rapid condensation.

All these solvents are volatile liquids and each has a pressure–temperature characteristic (see Section 1.4), so can be condensed if cooled to its saturation temperature. Finned-tube evaporators are generally used, but the condensation may be at a high pressure, requiring heat exchangers of the shell-and-tube or shell-and-plate type.

The size of equipment can vary from a 200W unit for a commercial dry-cleaning machine to systems of megawatt size for synthetic fibre processes.

18.5 LOW-TEMPERATURE LIQUID STORAGE AND TRANSPORT

Many volatile liquids can only be stored or transported at reduced temperatures, or excessive pressures will build up in the vessel. The important application is in the storage and transport of liquid methane, at temperatures of around −250°C.

Liquid carbon dioxide has many industrial uses and is stored at power stations for purging boiler furnaces and in oil tankers to purge petrol tanks. The vapour pressure of carbon dioxide is high, and storage vessels might possibly reach the critical temperature of 31°C. Storage temperatures of −20 to −4°C corresponding to vessel pressures of 19–30 bar can be maintained by single-stage refrigeration systems, with the evaporator coil inside the insulated storage vessel. For safety, most cooling systems are in duplicate.

The bulk transport of volatile liquids such as ammonia can be in insulated, non-refrigerated tanks, providing the liquid is cold on entry and the journey time is limited.

18.6 DE-WAXING OF OILS

Wax has the same boiling-point range as lubricating oil fractions, but has a much higher freezing point. Therefore, cooling crystallisation is a very effective way to separate the two materials.

Many of the lube-oil processing plants are quite large and require many scraped surface continuous crystallisers, often with a number of units in a series. Larger plants usually require several parallel trains of crystallisers.

Impurities may be removed from lubricating oils in the same way that wines and spirits are cooled and filtered. The base liquid is cooled down to a temperature at which the impurity will solidify, and then passed through a filter to take out the solids. The general principle is applied to many manufacturing and refining processes. The pre-cooling of the base liquid and its subsequent reheating can be achieved by counterflow heat exchangers, as in the pasteurisation and cooling of milk. Most waxes have a by-product value, and it may be necessary to chill them in a warm climate, to set the wax into blocks for packaging.

18.7 ICE AND SNOW SPORTS

Ice skating rinks are ice sheets, formed and maintained using a brine in tubes buried under the surface. Tubes may be steel or plastic for a permanent rink or plastic for a temporary installation. The temperature within the pipes is approximately −11°C and must be lower for rinks in the open air, owing to high solar radiation loads. Packaged liquid chillers are generally used, and for temporary installations they can be transportable, complete with brine pumps and other apparatus. Fig. 18.3 shows an indoor ice rink cooled by an ammonia system. R744 has been used in ice rinks in Europe as an alternative to ammonia. Applied as a phase-change secondary refrigerant, its high capacity and low pressure-drop characteristics result in smaller diameter floor tubing. This is lower cost and enables longer circuits, also suitable for ski circuits.

Outdoor artificial snow can be made by mimicking the natural formation of snowflakes. In the snow machine pressurised water is forced through a nozzle, breaking it into a mist of tiny droplets, and compressed air is then used to blast them into the air. The molecules align and crystallise into ice particles as the droplets cool. The art to snow-making is adjusting the water and air to ensure that the water drops are small enough and sent far enough that they will freeze before they hit the ground. To make the snow for indoor snow domes very rapid cooling is an additional ingredient, and this is typically provided by liquid nitrogen.

Figure 18.3 *Ice rink (Star Refrigeration).*

18.8 COOLING CONCRETE

The setting of concrete is an exothermic reaction, and large masses of concrete in building foundations, bridges and dams will heat up, causing expansion cracks if not checked. To counteract this heating, the materials are cooled before and as they are mixed, so that the concrete is laid some 15 K colder than ambient, and warms to ambient on setting. In practice, the final mix temperature can be held down to 10°C.

Methods are to pre-cool the aggregate with cold air, to chill the mix water and to provide part of the mix in the form of flake ice. Chilled water pipes may be buried in the concrete mass.

18.9 GROUND FREEZING

Ground freezing is a highly effective way of stabilising soil during construction projects, and it has the following advantages:

- *Strength*: The strength of the soil and resistance to vibration is increased and this enables the necessary excavations.
- *Non-polluting*: No materials are added or removed from the ground, and it provides a safe working environment.
- *Removal of water ingress*: Flow of water into the excavation is halted, thus avoiding the need for pumping and depletion of aquifer reserves.

- *Versatility*: Can be applied in many situations.
- *Speed and cost*: Very effective when compared with alternative methods.

The freezing is accomplished by driving a ring of vertical freeze tubes into the ground and passing chilled brine down through an inner pipe so that it flows up the annulus, to cool and eventually freeze the surrounding wet soil. This process is continued until the frozen soil builds up a continuous wall around the proposed excavation. Depths of over 650 m have been excavated in this way. Calcium chloride brine, cooled by skid-mounted chillers, is used as the secondary coolant. Typically ammonia is the refrigerant, cooling the brine to −25°C in a chiller having a capacity of 100 to 500 kW; lower temperatures are possible if necessary.

Cryogenic methods employing liquid nitrogen are also used, and this is particularly effective for rapid operations and in rescue situations.

18.10 LOW-TEMPERATURE TESTING

Environment test chamber systems are used in a diverse number of industries including automotive, consumer, pharmaceutical and aerospace. They are used throughout industry to test everything from semiconductors to reefers.

Sizes range from bench-top units to reach-in chambers and full walk-in and drive-in chambers and rooms. In addition to temperature control they may provide the following:

- Temperature and/or humidity simulation
- Solar simulation
- Vibration simulation
- Altitude simulation
- Wind simulation
- Rain simulation
- Sound attenuation

An environmental chamber suitable for testing equipment up to the size of reefer units is shown in Fig. 18.4.

Mechanisms and electronics for the aerospace industry are tested at temperatures which may prevail under working conditions. A typical specification might be to test at −70°C. BioStore freezer rooms are designed for cold storage at −75°C up to −40°C. Where the component is large, it must be contained within a cold chamber which is capable of reaching this condition. Smaller items are tested in self-contained cabinets with a chamber the size of a large domestic refrigerator.

Sometimes expendable refrigerants are used. They are liquids or gases that can be injected directly into the space being cooled or into heat

Figure 18.4 ***Test chamber capable of taking the largest permissible road vehicles (Cambridge Refrigeration Technology).***

exchangers, similar to mechanical systems. As the liquid enters the chamber (directly or through a fin coil) it absorbs heat and flashes to a gas. The gas is then vented out of the chamber and should be ducted outdoors. The two most popular refrigerants are liquid nitrogen and liquid carbon dioxide (CO_2). Cryogenic temperatures down to −184°C can be

(a)

(b)

Figure 18.5 ***Industrial R1270 water-cooled chiller installed in a petro-chemical plant (J&E Hall).***

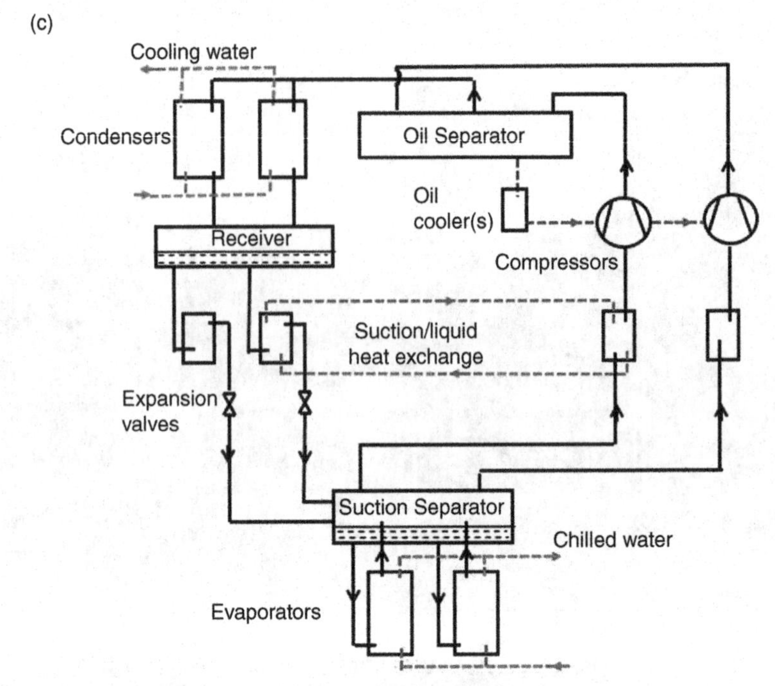

Figure 18.5 ***(cont.)***

achieved with liquid nitrogen. CO_2 on the other hand is limited by its triple point temperature and can only achieve temperature down to −68°C without risk of solidification.

Two-stage and three-stage systems are used traditionally with R23 in cascade at the lower end and R404A or similar for the high stage(s).

18.11 CHEMICAL INDUSTRY

Processes in the chemical industry require the control of temperatures of reactions when heat is liberated. Where direct expansion refrigerant coils could constitute a hazard, heat exchangers using chilled water or brine are employed. Coolers of this sort will be found in every branch of the chemical industry.

An example is the water chiller in Fig. 18.5. The compressors and oil separator can be seen in view 18.5a, and view 18.5b shows the suction separator and plate heat exchanger. A simplified circuit diagram is provided in Fig. 18.5c. It has a cooling capacity of 4,435 kW and is designed for good turndown capability. Located on a petrochemical site, this chiller has two single-screw compressors mounted above one oil separator. They are enclosed in stainless steel hoods to reduce the noise level. The plant has plate heat exchangers for both evaporators and condensers. The refrigerant circulates through the evaporators from the suction separator or *surge drum* by natural convection (thermosyphon). Suction/liquid heat exchange between the liquid and the suction gas is effected with the aid of water circulation (pumps not shown) and separate plate heat exchangers. One such circuit is shown.

CHAPTER 19

Air and Water Vapour Mixtures

19.1 INTRODUCTION

The atmosphere consists of a mixture of dry air and water vapour. Air is itself a mixture of several elemental gases, mainly oxygen and nitrogen, but the proportions of these are consistent throughout the atmosphere and it is convenient to consider dry air as one gas. This has a *molecular mass* of 28.97 and the standard atmospheric pressure is 1013.25 mbar or 101 325 Pa.

At ambient and indoor conditions water present in the air is in vapour form, that is as superheated low-pressure steam. Air conditioning and many cooling and heating systems ultimately change the air temperature and the proportion of water vapour in the air *(humidity)* must be considered for system calculations. The study of humid air is called *psychrometry*. Water may also be present in air in the liquid form, as rain or mist, or as a solid (snow, hail).

19.2 CALCULATION OF PROPERTIES

If air and water are present together in a confined space, a balance condition will be reached where the air becomes saturated with water vapour. If the temperature of the mixture is known, then the pressure of the water vapour will be the pressure of steam at this temperature, determined by experiment (Table 19.1). Dalton's law of partial pressures (see also chapter: Fundamentals) states that the total pressure of a mixture of gases is equal to the sum of the individual pressures of the constituent gases, taken at the same temperature and occupying the same volume. Since the water-saturation vapour pressure is dependent only on temperature, this pressure can be obtained from steam tables. Therefore, the partial pressure exerted by the dry air must be the remainder.

Thus, for an air–water vapour mixture at 25°C:

$$\begin{aligned} \text{Total (standard) pressure} &= 1013.25\ \text{mbar} \\ \text{Partial pressure of saturated vapour} &= \underline{\ \ 31.66}\ \text{mbar} \\ \text{Partial pressure of dry air} &= \ \ 971.59\ \text{mbar} \end{aligned}$$

Refrigeration, Air Conditioning and Heat Pumps
http://dx.doi.org/10.1016/B978-0-08-100647-4.00019-X

Table 19.1 Water vapour pressure at various temperatures

Temperature (°C)	Vapour pressure (mbar)
0	6.10
10	12.27
15	17.04
20	23.37
25	31.66

This calculation of the proportions by partial pressure can be converted to proportions by mass, by multiplying each pressure by the molecular mass of the constituent (Avogadro's hypothesis). The molecular mass of water is 18.016, and for dry air 28.97, treating it as one gas. This gives:

$$\text{Proportion by mass of water} = 31.66 \times 18.016 = 570.4$$
$$\text{Proportion by mass of dry air} = 971.59 \times 28.97 = 28{,}146$$

$$\text{Proportion by weight of } \frac{\text{water}}{\text{dry air}} = \frac{570.4}{28{,}146} = 0.0203\,\text{kg/kg}$$

Since neither dry air nor water vapour is a perfect gas, there is a slight difference between published value, typically 0.02016 and this simplified calculation.

The specific enthalpy (or total heat) of the mixture can be taken from 0 K (−273.15°C) or from any convenient arbitrary zero. Since most air-conditioning processes take place above the freezing point of water, and we are concerned mostly with differences rather than absolute values, the reference is commonly taken as 0°C, dry air. For conditions of 25°C, saturated, the specific enthalpy of the mixture, per kilogram of dry air, is

$$\text{Sensible heat of dry air} = 1.006 \times 25 = 25.15\,\text{kJ/kg}$$
$$\text{Sensible heat of water} = 0.020\,16 \times 25 \times 4.187 = 2.11$$
$$\text{Latent heat of water} = 0.02016 \times 2440 = 49.19$$
$$\text{Total} \quad \overline{76.45}\,\text{kJ/kg}$$

(Taking specific heat of dry air as 1.006. Again, some slight variations in published values may be found.)

The specific volume of the mixture can be obtained, taking either of the two gases at their respective partial pressures, and using the General Gas

Law. Absolute pressures must be expressed in pascals and temperatures in Kelvin:

$$pV = mRT$$

$$\text{or} \quad V = mRT/p$$

$$\text{For the dry air } V_a = \frac{1 \times 287 \times (25 + 273.15)}{97159}$$

$$= 0.8807 \text{m}^3$$

$$\text{For the water vapour } V_v = \frac{0.02016 \times 461 \times (25 + 273.15)}{3166}$$

$$= 0.8752 \text{ m}^3$$

(The published figure is 0.871 5 m³/kg.)

19.3 MOISTURE CONTENT, PERCENTAGE SATURATION, AND RELATIVE HUMIDITY

The moisture content of 0.02016 kg/kg dry air in the example at 25°C, saturated, is also termed its specific humidity. Air will not always be saturated with water vapour in this way, but may contain a lower proportion of this figure, possibly 50%:

$$\frac{0.02016}{2} = 0.01008 \text{ kg/kg dry air}$$

This lower figure can be expressed as a percentage of the saturation quantity:

$$\text{Percentage saturation} = 100 \times \frac{g}{g_{ss}}$$

$$= 100 \times \frac{0.01008}{0.02016}$$

$$= 50\% \text{ sat.}$$

Properties for this new mixture can be calculated as done previously to obtain the specific enthalpy and specific volume. The proportion of moisture can also be expressed as the ratio of the vapour pressures, and is then termed *relative humidity:*

$$\text{Relative humidity} = 100 \times \frac{P_s}{P_{ss}}$$

$$= 50.8\% \text{ relative humidity}$$
(for the example taken)

Since most air conditioning calculations are based on weights of air and moisture, *percentage saturation* is usually employed, and moisture content is expressed as kilograms per kilogram of dry air. Published data may be found where it is still expressed in the original quantities of Willis H. Carrier, that is grains per pound, where 1 lb = 7000 grains.

19.4 DEW POINT

Saturated air at 25°C, having a water vapour content of 0.020,16 kg/kg, can be shown as a point *A* on a graph of moisture content against temperature (Fig. 19.1). Air which is 50% saturated at this temperature will contain 0.010 08 kg/kg and will appear on this graph as point *B*. If this 50% saturation mixture is slowly cooled, the change of condition will be along

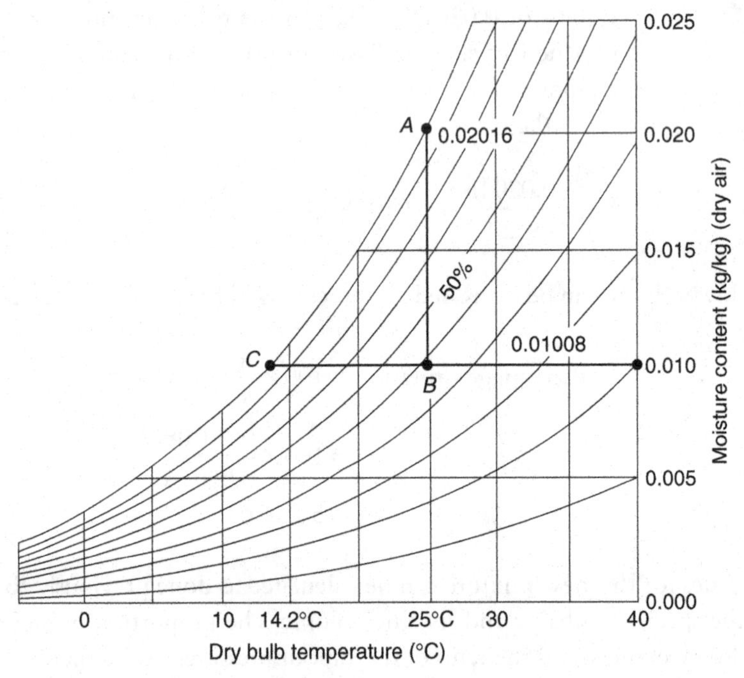

Figure 19.1 ***Temperature–moisture content graph.***

the line BC, with constant moisture content but decreasing temperature. It will eventually reach point C on the saturation line, where the maximum moisture it can hold is 0.010 08 kg/kg (approximately 14.2°C). It cannot be cooled below this temperature and still hold this proportion of water vapour, so moisture will be precipitated as dew. The point C for the mixture originally at B is termed the *dew point temperature.*

19.5 WET BULB TEMPERATURE

If the percentage saturation of an air sample is less than 100, that is it is less than saturated, and it comes into contact with water at the same temperature, there will be a difference in vapour pressures. As a result, some of the water will evaporate. The latent heat required for this change of state will be drawn from the sensible heat of the water, which will be slightly cooled. This drop in the water temperature provides a temperature difference, and a thermal balance will be reached where the flow of sensible heat from the air to the water (Fig. 19.2) provides the latent heat to evaporate a part of it.

The effect can be observed and measured by using two similar thermometers (Fig. 19.3), one of which has its bulb enclosed in a wet wick. The drier the air passing over them, the greater will be the rate of evaporation from the wick and the greater the difference between the two readings. In the case of air at 25°C, 50% saturation, the difference will be approximately 6.5 K. The measurements are termed the *dry bulb* and *wet bulb* temperatures, and the difference the *wet bulb depression.*

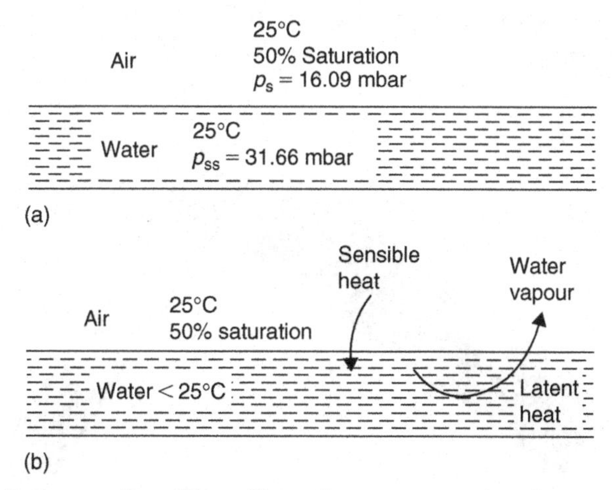

Figure 19.2 ***Exchange of sensible and latent heat at water–air surface.***

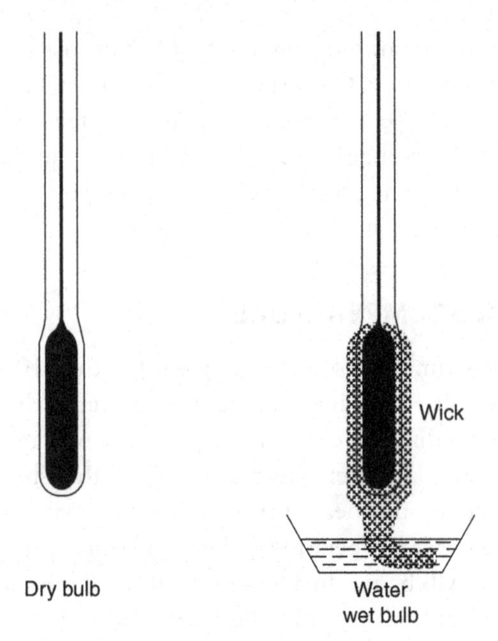

Figure 19.3 ***Thermometers, dry bulb and wet bulb.***

In order that consistent conditions can be obtained, the air speed over the thermometers should be not less than 1 m/s. This can be done with a mechanical aspiration fan (the Assmann psychrometer) or by rotating the thermometers manually on a radius arm (the sling psychrometer, Fig. 19.4). If the thermometers cannot be in a moving airstream, they are shielded from draughts by a perforated screen and rely only on natural convection. In this case the wet bulb depression will be less and the reading is termed the *screen wet bulb*.

Figure 19.4 ***Sling psychrometer (Business Edge).***

It follows that the drier the air, the greater will be the difference between the dry bulb, wet bulb and dew point temperatures and, conversely, at 100% saturation these three will coincide.

19.6 THE PSYCHROMETRIC CHART

All the aforementioned properties may be tabulated, but can be displayed more effectively in graphical form. The basic properties to be shown are dry bulb temperature, moisture content and specific enthalpy. Within the limits of the graph required for ordinary air-conditioning processes, the grid lines can be assumed as parallel and form the basis of the psychrometric chart (Fig. 19.5). It will be seen from the full chart (Fig. 19.6) that the dry-bulb lines are slightly divergent. The moisture content and enthalpy grids are parallel.

On this chart, the wet bulb temperatures appear as diagonal lines, coinciding with the dry bulb at the saturation line. If measurements are taken with the two thermometers of the sling psychrometer, the condition can be plotted on the psychrometric chart by taking the intersection of the dry

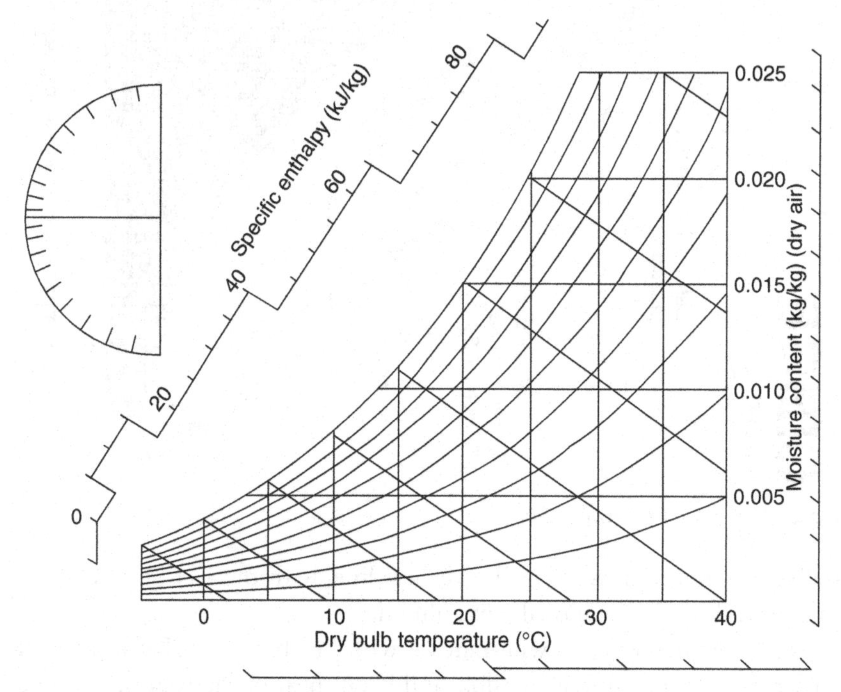

Figure 19.5 ***Basic CIBSE psychrometric chart (CIBSE).***

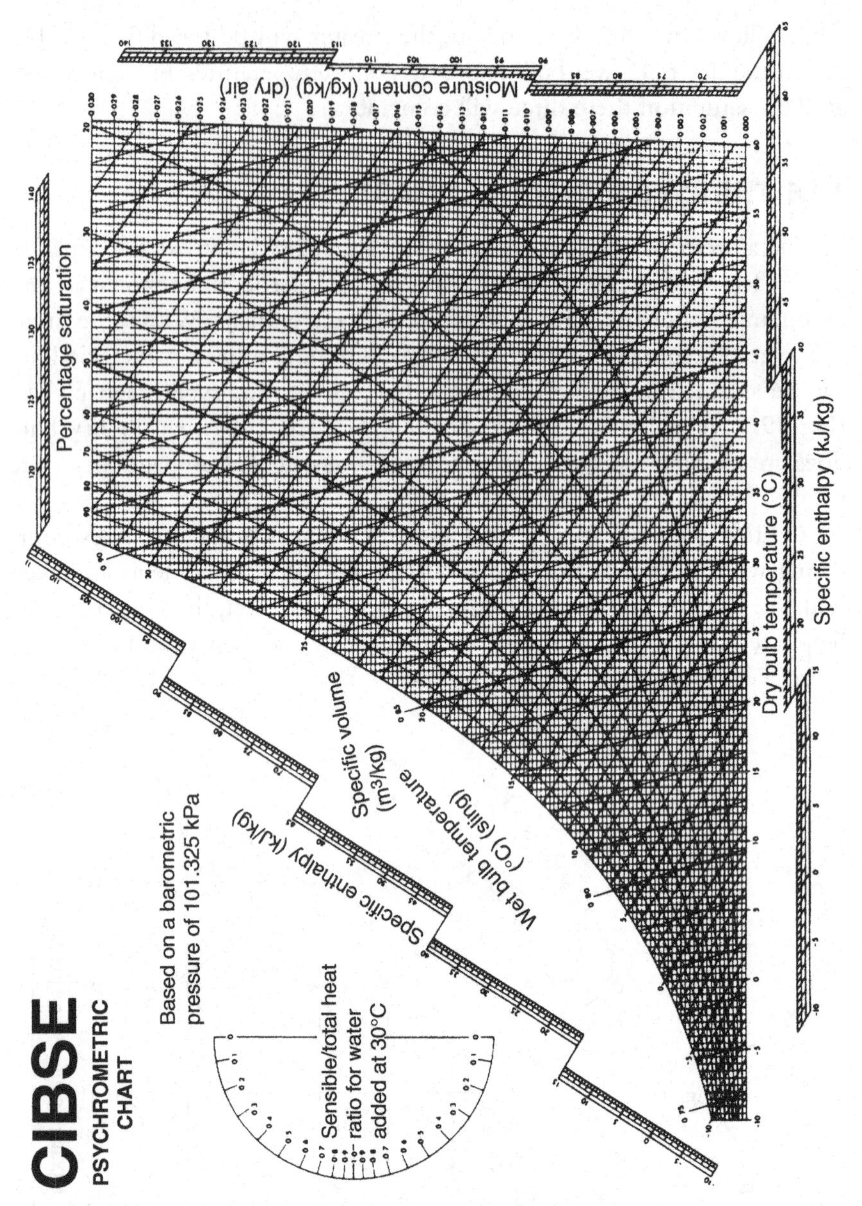

Figure 19.6 ***Psychrometric chart.***

bulb temperature, as read on the vertical line, with the wet bulb temperature, read down the diagonal wet bulb line.

The specific enthalpy will increase with dry bulb (sensible heat of the air) and moisture content (sensible and latent heat of the water). The adiabatic (isoenthalpic) lines for an air–water vapour mixture are almost parallel

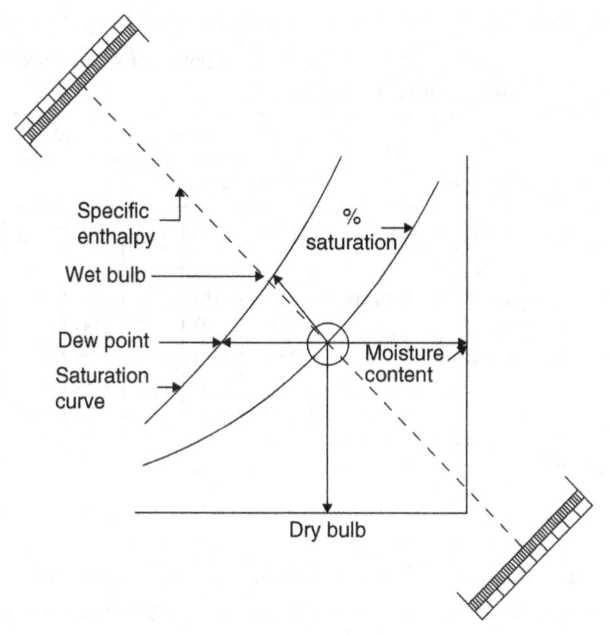

Figure 19.7 ***Reading the CIBSE psychrometric chart.***

with the wet bulb lines so, to avoid any confusion, the enthalpy scale is placed outside the body of the chart, and readings must be taken using a straight edge (see Fig. 19.7).

Another property which is shown on the psychrometric chart is the specific volume of the mixture, measured in cubic metres per kilogram. This appears as a series of diagonal lines, at intervals of 0.01 m^3.

19.7 EFFECTS ON HUMAN COMFORT

The human body takes in chemical energy as food, drink and oxygen, and consumes these to provide the energy of the metabolism. Some mechanical work may be done, but the greater proportion is liberated as heat, at a rate between 90 W when resting and 440 W when doing heavy work.

A little of this is lost by radiation if the surrounding surfaces are cold and some as sensible heat, by convection from the skin. The remainder is taken up as latent heat of moisture from the respiratory tissues and perspiration from the skin (see Table 19.2). Radiant loss will be very small if the person is clothed, and is ignored in this table.

Convective heat loss will depend on the area of skin exposed, the air speed, and the temperature difference between the skin and the ambient. As

Table 19.2 Heat emission from the human body (adult male, body surface area 2 m^2)

Application		Sensible (s) and latent (l) heat emissions, W, at the stated dry-bulb temperature (°C)						
			20		22		24	
Degree of activity	**Typical**	**Total**	**s**	**l**	**s**	**l**	**s**	**l**
Seated at rest	Theatre, hotel lounge	115	90	25	80	35	75	40
Light work	Office, restaurant	140	100	40	90	50	80	60
Walking slowly	Store, bank	160	110	50	100	60	85	75
Light bench work	Factory	235	130	105	115	120	100	135
Medium work	Factory, dance hall	265	140	125	125	140	105	160
Heavy work	Factory	440	190	250	165	275	135	305

(From CIBSE Guide A).

the dry bulb approaches body temperature (36.9°C) the possible convective loss will diminish to zero. At the same time, loss by latent heat must increase to keep the body cooled. This, too, must diminish to zero when the wet bulb reaches 36.9°C.

In practice, the human body can exist in dry bulb temperatures well above blood temperature, provided the wet bulb is low enough to permit evaporation. Therefore, the limiting factor is one of wet-bulb rather than dry bulb temperature, and the closer the upper limits are approached, the less heat can be rejected and so the less work can be done.

19.8 CLIMATIC CONDITIONS

Fig. 19.8 shows the maximum climatic conditions in different areas of the world. The humid tropical zones have high humidities but the dry bulb rarely exceeds 35°C. The deserts have an arid climate, with higher dry bulb temperatures. Approximate limits for human activities are related to the enthalpy lines and indicate the ability of the ambient air to carry away the 90–440 W of body heat.

The opposite effect will take place at the colder end of the scale. Evaporative and convective loss will take place much more easily and the loss by radiation may become significant, removing heat faster than the body can generate it. The rate of heat production can be increased by greater bodily activity, but this cannot be sustained, so losses must be prevented by thicker insulation against convective loss and reduced skin exposure in the form of

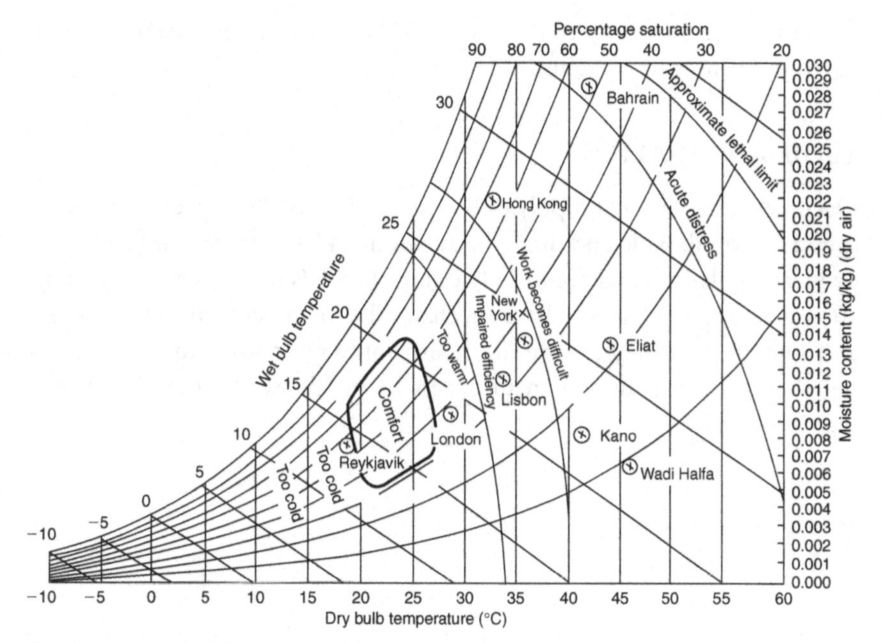

Figure 19.8 *Typical climate conditions.*

more clothing. The body itself can compensate by closing sweat pores and reducing the skin temperature.

19.9 OTHER COMFORT FACTORS

A total assessment of bodily comfort must take into account changes in convective heat transfer arising from air velocity, and the effects of radiant heat gain or loss. These effects have been quantified in several objective formulas, to give equivalent, corrected effective, globe, dry resultant and environmental temperatures, all of which give fairly close agreement. This more complex approach is required where air speeds may be high, there is exposure to hot or cold surfaces, or other special conditions call for particular care.

For comfort in normal office or residential occupation, with percentage saturations between 35 and 70%, control of the dry bulb will result in comfortable conditions for most persons. Feelings of personal comfort are as variable as human nature and at any one time 10% of the occupants of a space may feel too hot and 10% too cold, whereas the 80% majority are comfortable. Such variations frequently arise from lack or excess of local air movement, or proximity to cold windows, rather than an extreme of temperature or moisture content.

19.10 AIR QUALITY

Occupied spaces need a supply of outside air to provide oxygen, remove respired carbon dioxide and dilute body odours. The quantities are laid down by local regulations and commonly call for 6–8 L/s per occupant. Buildings are usually required also to have mechanical extract ventilation from toilets and some service areas, so the fresh air supply must make up for this loss, together with providing a small excess to pressurise the building against ingress of dirt.

CHAPTER 20

Air Treatment Fundamentals

20.1 INTRODUCTION

This chapter briefly describes the commonly used air treatment processes with the aid of calculation examples. Manufacturers' software with the ability to perform these calculations is available but the theory is best understood by study of examples.

20.2 HEATING

Buildings lose heat in winter by conduction out through the fabric, convection of cold air and some radiation. The air from the conditioning system must enter the spaces warmer than the required internal condition, to provide the heat to counteract this loss.

Heating methods are the following:

1. Hot water or steam coils
2. Direct-fired – gas and sometimes oil
3. Electric resistance elements
4. Refrigerant condenser coils of heat pump or heat reclaim systems.

Fig. 20.1 shows the sensible heating of air at the conditions given in Example 20.2.

Example 20.1

Air circulates at the rate of 68 kg/s and is to be heated from 16 to 34°C. Calculate the heat input and the water mass flow for an air heater coil having hot water entering at 85°C and leaving at 74°C.

$$Q = 68 \times 1.02 \times (34 - 16) = 1248 \text{ kW}$$

$$m_w = \frac{1248}{4.187 \times (85 - 74)} = 27 \text{ kg/s}$$

Note: the 1.02 here is a general figure for the specific heat capacity of indoor air which contains some moisture, and is used in preference to 1.006, which is for dry air.

Refrigeration, Air Conditioning and Heat Pumps
http://dx.doi.org/10.1016/B978-0-08-100647-4.00020-6

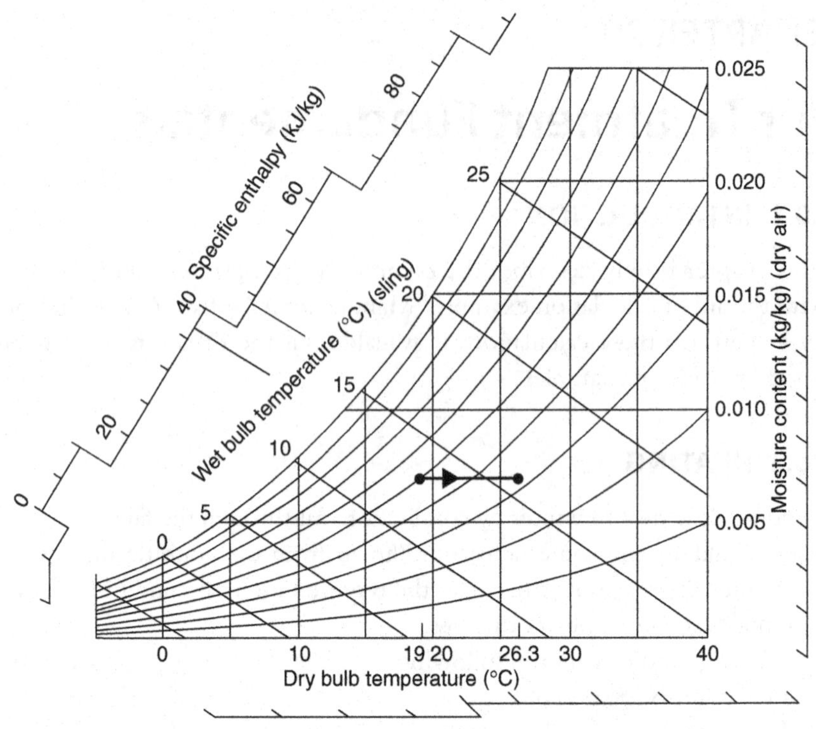

Figure 20.1 ***Sensible heating of air.***

Example 20.2

A building requires 500 kW of heating. Air enters the heater coil at 19°C at the rate of 68 kg/s. What is the air supply temperature?

$$t = 19 + \frac{500}{68 \times 1.02} = 19 + 7.2$$
$$= 26.2°\text{C}$$

If the cycle is being traced out on a psychrometric chart, the enthalpy can be read off for the coil inlet and outlet conditions. In Example 20.1, the enthalpy increase as measured on the chart is 7.35 kJ/kg dry air (taken at any value of humidity), giving

$$68 \times 7.35 \sim 500\,\text{kW}.$$

20.3 MIXING OF AIRSTREAMS

Air entering the conditioning plant will probably be a mixture of return air from the conditioned space and outside air. Since no heat or moisture is gained or lost in mixing,

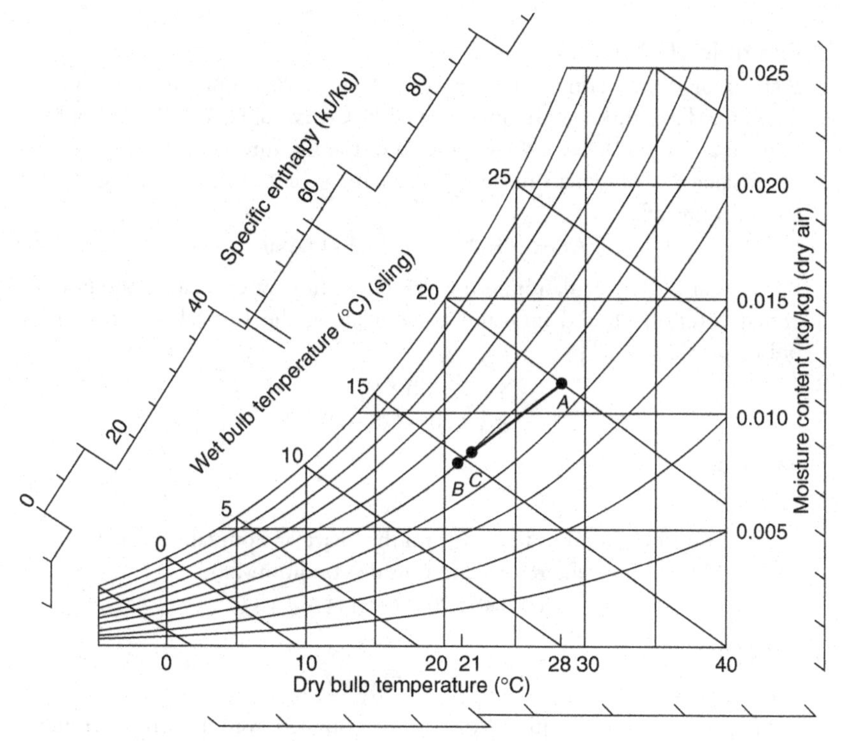

Figure 20.2 ***Mixing of two airstreams.***

$$\text{Sensible heat before} = \text{sensible heat after}$$

and

$$\text{Latent heat before} = \text{latent heat after}$$

The conditions after mixing can be calculated, but can also be shown graphically by a mix line joining the condition A and B (see Fig. 20.2). The position C along the line will be such that

$$AC \times m_a = CB \times m_b$$

This straight-line proportioning holds good to close limits of accuracy. The horizontal divisions of dry bulb temperature are almost evenly spaced, so indicating sensible heat. The vertical intervals of moisture content indicate latent heat.

Example 20.3

Return air from a conditioned space at 21°C, 50% saturation, and a mass flow of 20 kg/s, mixes with outside air at 28°C dry bulb and 20°C wet bulb, flowing at 3 kg/s. What is the condition of the mixture?

Method 1. Construct on the psychrometric chart as shown in Fig. 20.2 and measure off:

$$\text{Answer} = 22°\text{C dry bulb}, 49\% \text{ sat.}$$

Method 2. By calculation, using dry bulb temperatures along the horizontal component, and moisture content along the vertical. For the dry bulb, using

$$AC \times m_a = CB \times m_b$$
$$(t_c - 21) \times 20 = (28 - t_c) \times 3$$

giving

$$t_c = 21.9°\text{C}$$

The moisture content figures, from the chart or from tables, are 0.0079 and 0.0111 kg/kg at the return and outside conditions, so

$$(g_c - 0.0079) \times 20 = (0.0111 - g_c) \times 3$$

giving

$$g_c = 0.0083 \text{ kg/kg}$$

If only enthalpy is required, this can be obtained from the same formula in a single equation:

$$(h_c - h_a) \times m_a = (h_b - h_c) \times m_b$$
$$(h_c - 41.8) \times 20 = (56.6 - h_c) \times 3$$

giving

$$h_c = 43.7 \text{ kJ/kg dry air.}$$

20.4 SENSIBLE COOLING

If air at 21°C dry bulb, 50% saturation, is brought into contact with a surface at 12°C, it will give up some of its heat by convection. The cold surface is warmer than the dew point, so no condensation will take place, and cooling will be sensible only (Fig. 20.3).

This process is shown as a horizontal line on the chart, since there is no change in the moisture content. The loss of sensible heat can be read off the chart in terms of enthalpy, or calculated from the dry bulb reduction, considering the drop in the sensible heat of both the dry air and the water vapour in it.

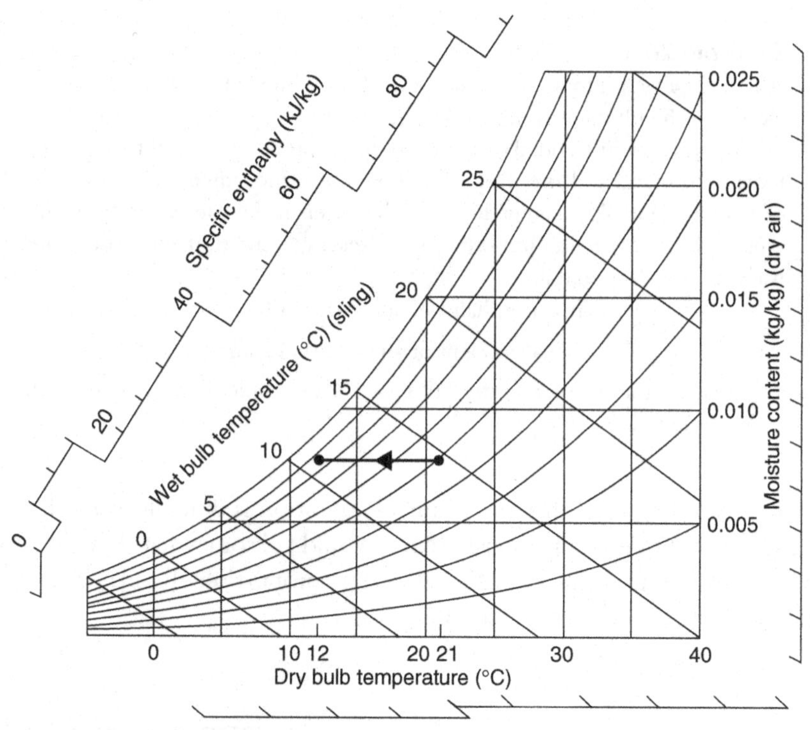

Figure 20.3 ***Sensible cooling of air.***

20.5 WATER SPRAY (ADIABATIC SATURATION)

The effect of spraying water into an airstream is shown in Fig. 20.2, assuming that the air is not already saturated. Evaporation takes place and the water draws its latent heat from the air, reducing the sensible heat and therefore the dry bulb temperature of the air (Fig. 20.4).

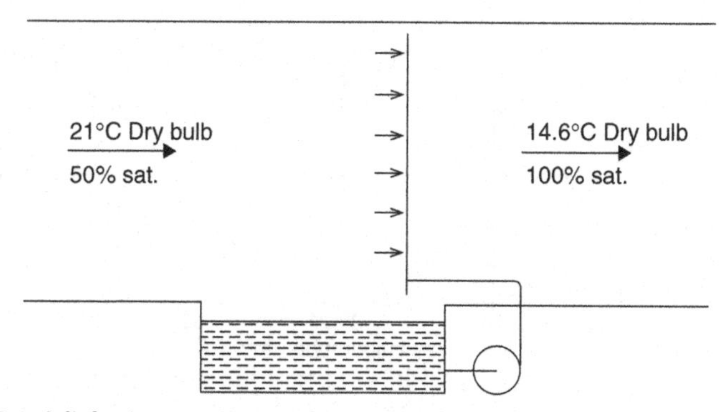

Figure 20.4 ***Adiabatic saturation to ultimate condition.***

Example 20.4

Water is sprayed into an airstream at 21°C dry bulb, 50% saturation. What would be the ultimate condition of the mixture?

No heat is being added or removed in this process, so the enthalpy must remain constant, and the process is shown as a movement along the line of constant enthalpy. Latent heat will be taken in by the water, from the sensible heat of the air, until the mixture reaches saturation, when no more water can be evaporated.

$$\text{Initial enthalpy of air} = 41.08 \text{ kJ/kg}$$

$$\text{Final enthalpy of air} = 41.08 \text{ kJ/kg}$$

Final condition, 14.6°C dry bulb, 14.6°C wet bulb, 14.6°C dew point, 100% saturated.

It should be noted that this ultimate condition is difficult to reach, and the final condition in a practical process would fall somewhat short of saturation, possibly to point C in Fig. 20.5. The proportion AC/AB is termed the *effectiveness* of the spray system.

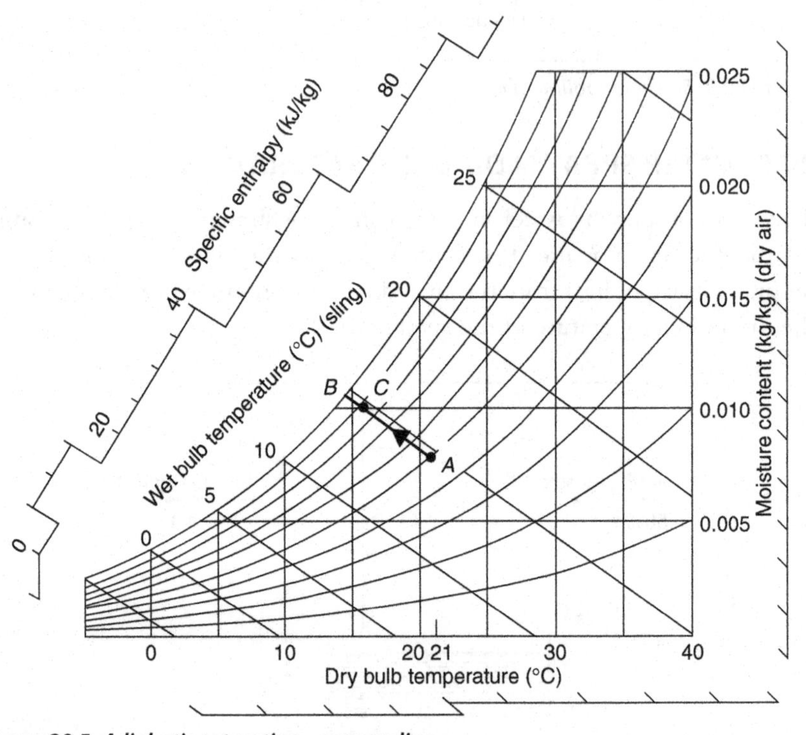

Figure 20.5 ***Adiabatic saturation – process line.***

The adiabatic (constant enthalpy) line *AC* is almost parallel to the line of constant wet bulb. Had the latter been used, the final error would have been about 0.2 K, and it is sometimes convenient and quicker to calculate on the basis of constant wet bulb. (This correlation applies only to the mixture of dry air and water vapour, and not to other gas mixtures.)

20.6 STEAM INJECTION

Moisture can be added to air by injecting steam, that is water which is already in vapour form and does not require the addition of latent heat (Fig. 20.6). Under these conditions, the air will not be cooled and will stay at about the same dry bulb temperature. The steam is at 100°C when released to the atmosphere (or may be slightly superheated), and so raises the final temperature of the mixture.

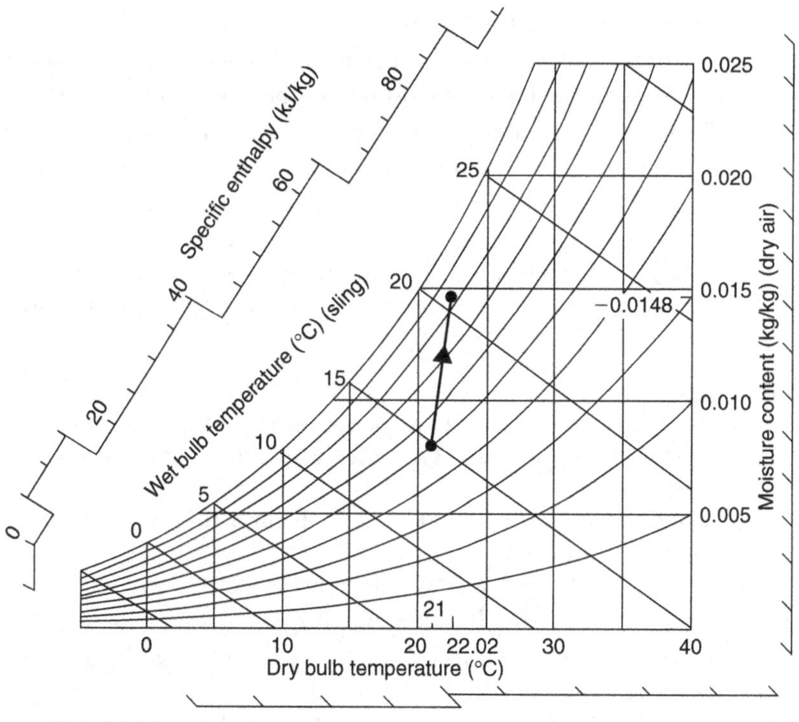

Figure 20.6 ***Addition of steam to air.***

Example 20.5

Steam at 100°C is injected into an airstream at 21°C dry bulb, 50% saturation, at the rate of 1 kg steam/150 kg dry air. What is the final condition?

$$\text{Moisture content of air before} = 0.0079 \text{ kg/kg}$$

$$\text{Moisture added, 1 kg/150 kg} = \underline{0.0067} \text{ kg/kg}$$

$$\text{Final moisture content} = 0.0148 \text{ kg/kg}$$

An approximate figure for the final dry bulb temperature can be obtained, using the specific heat capacity of the steam through the range 20–100°C, which is about 1.972 kJ/kg. This gives

$$\text{Heat lost by steam} = \text{heat gained by air}$$

$$0.0067 \times 1.972(100 - t) = 1.006(t - 21)$$

giving

$$t = 22.02°\text{C}$$

Where steam is used to raise the humidity slightly, the increase in dry-bulb temperature is small.

20.7 AIR WASHER WITH CHILLED WATER

The process of adiabatic saturation in Section 20.4 assumed that the spray water temperature had no effect on the final air condition. If, however, a large mass of water is used in comparison with the mass of air, the final condition will approach the water temperature. If this water is chilled below the dew point of the entering air, moisture will condense out of the air, and it will leave the washer with a lower moisture content (see Fig. 20.7).

The ultimate condition will be at the initial water temperature *B*. Practical saturation efficiencies (the ratio AC/AB) are about 50–80% for air washers having a single bank of sprays and 80–95% for double spray banks (see Fig. 20.8).

Example 20.6

Air at 23°C dry bulb, 50% saturation, enters a single-bank air washer having a saturation efficiency of 70% and is sprayed with water at 5°C. What is the final condition?

1. By construction on the chart (Fig. 20.7), the final condition is 10.4°C dry bulb, 82% saturation.
2. By proportion: Dry bulb is 70% of the way from 23°C down to 5°C

$$23 - [0.7(23 - 5)] = 10.4°\text{C}$$

Moisture content is 70% down from 0.008 9 to 0.005 4 kg/kg (ie saturated air at 5°C)

$$0.008\,9 - [0.7(0.008\,9 - 0.005\,4)] = 0.006\,45 \text{ kg/kg}$$

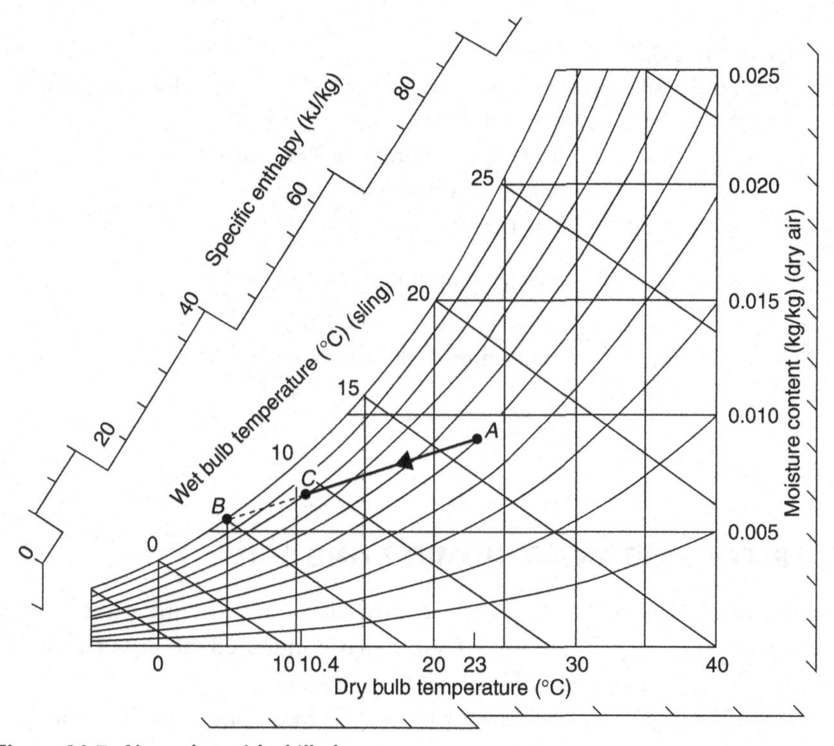

Figure 20.7 ***Air washer with chilled water.***

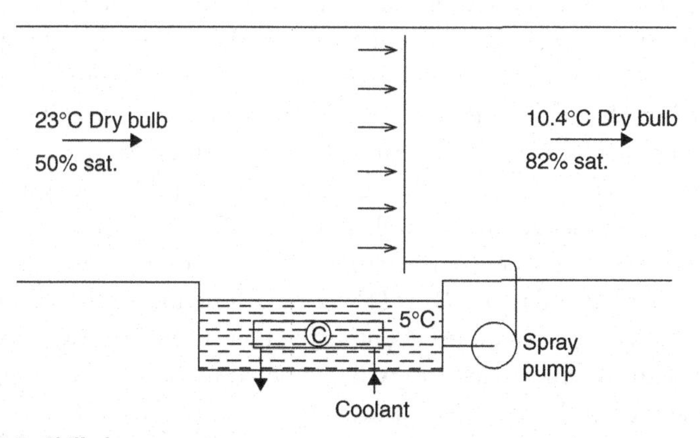

Figure 20.8 ***Chilled water spray.***

Example 20.7

In Example 20.6, water is sprayed at the rate of 4 kg water for every 1 kg air. What is the water temperature rise?

$$\begin{aligned}
\text{Enthalpy of air before} &= 45.79 \text{ kJ/kg} \\
\text{Enthalpy of air after} &= \underline{26.7\ \ \text{kJ/kg}} \\
\text{Heat lost per kilogram air} &= 19.09 \text{ kJ}
\end{aligned}$$

$$\begin{aligned}
\text{Heat gain per kilogram water} &= 19.09/4 \\
&= 4.77 \text{ kJ}
\end{aligned}$$

$$\begin{aligned}
\text{Temperature rise of water} &= \frac{4.77}{4.187} \\
&= 1.1 \text{ K}
\end{aligned}$$

20.8 COOLING AND DEHUMIDIFYING COIL

In the previous process, air was cooled by close contact with a water spray. No water was evaporated, in fact some was condensed, because the water was colder than the dew point of the entering air.

A similar effect occurs if the air is brought into contact with a solid surface, maintained at a temperature below its dew point. Sensible heat will be transferred to the surface by convection and condensation of water vapour will take place at the same time. Both the sensible and latent heats must be conducted through the solid and removed. The simplest form is a metal tube, and the heat is carried away by refrigerant or a chilled fluid within the pipes. This coolant must be colder than the tube surface to transfer the heat inwards through the metal.

The process is indicated on the chart in Fig. 20.9, taking point B as the tube temperature. Since this would be the ultimate dew point temperature of the air for an infinitely sized coil, the point B is termed the apparatus dew point (ADP). In practice, the cooling element will be made of tubes, probably with extended outer surface in the form of fins. Heat transfer from the air to the coolant will vary with the fin height from the tube wall, the materials, and any changes in the coolant temperature which may not be constant. The average coolant temperature will be at some lower point D, and the temperature difference $B - D$ will be a function of the conductivity of the coil. As air at condition A enters the coil, a thin layer will come into contact with the fin surface and will be cooled to B. It will then mix with the remainder of the air between the fins, so that the line AB is a mix line.

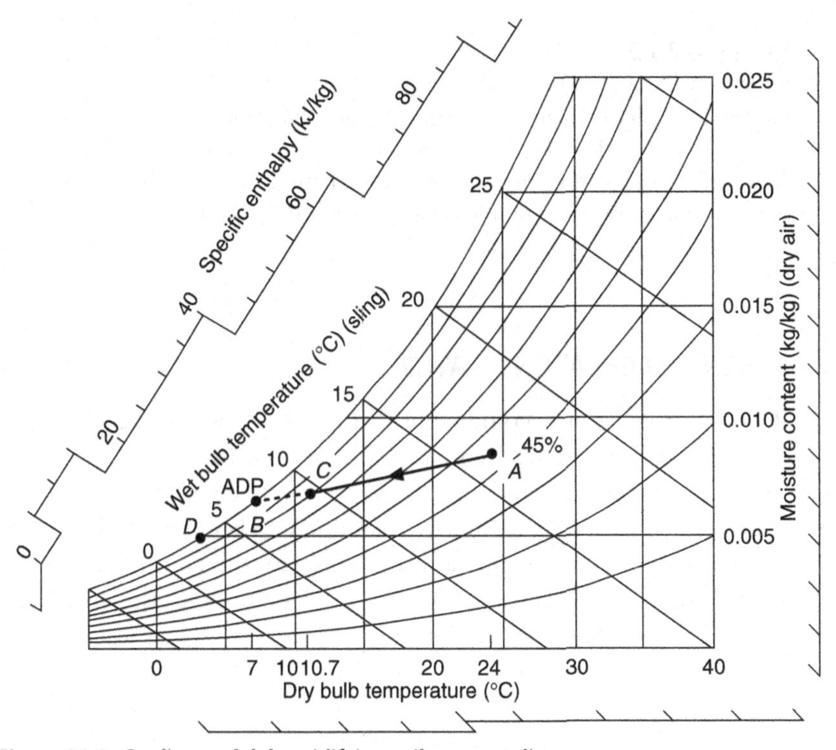

Figure 20.9 ***Cooling and dehumidifying coil – process line.***

The process line AB is shown here as a straight line for convenience of working. Analysis of the air as it passes through a cooling coil shows the line to be a slight curve.

The proportion AC/AB is termed the coil contact factor. The proportion CB/AB is sometimes used, and is termed the bypass factor.

Example 20.8

Air at 24°C dry bulb, 45% saturation, passes through a coil having an ADP of 7°C and a contact factor of 78%. What is the off-coil condition?

1. By construction on the chart (Fig. 20.9), 10.7°C dry bulb, 85% saturation.
2. By calculation, the dry bulb will drop 78% of 24 to 7°C:

$$24 - [0.78 \times (24 - 7)] = 10.7°\text{C}$$

and the enthalpy will drop 78% of 45.85 to 22.72 kJ/kg:

$$45.85 - [0.78 \times (45.85 - 22.72)] = 27.81 \text{ kJ/kg}$$

The two results obtained here can be compared with tabulated figures for saturation and give about 84% saturation.

Example 20.9

Air is to be cooled by a chilled water coil from 27°C dry bulb, 52% saturation, to 15°C dry bulb, 80% saturation. What is the ADP?

This must be done by construction on the chart, and gives an ADP of 9°C. The intersection of the process and saturation lines can also be computed. Again, it has been assumed that the process line is straight.

20.9 SENSIBLE–LATENT RATIO

In all cases the horizontal component of the process line is the change of sensible heat, and the vertical component gives the latent heat. It follows that the slope of the line shows the ratio between them, and the angle, if measured, can be used to give the ratio of sensible to latent to total heat. On the psychrometric chart in general use (Fig. 20.5), the ratio of sensible to total heat is indicated as angles in a segment to one side of the chart. This can be used as a guide to coil and plant selection.

Example 20.10

Air enters a coil at 23°C dry bulb, 40% saturation. The sensible heat to be removed is 36 kW and the latent 14 kW. What are the ADP and the coil contact factor if air is to leave the coil at 5°C?

Plotting on the chart (Fig. 20.10) from 23°C/40% and using the ratio

$$\frac{\text{Sensible heat}}{\text{Total heat}} = \frac{36}{36+14} = \frac{36}{50} = 0.72$$

The process line meets the saturation curve at −1°C, giving the ADP (which means that condensate will collect on the fins as frost).

Taking the 'off' condition at 5°C dry bulb and measuring the proportion along the process line gives a coil contact factor of 75%.

20.10 MULTI-STEP PROCESSES

Some air treatment processes cannot be made in a single operation, and the air must pass through two or more consecutive steps to obtain the required leaving condition.

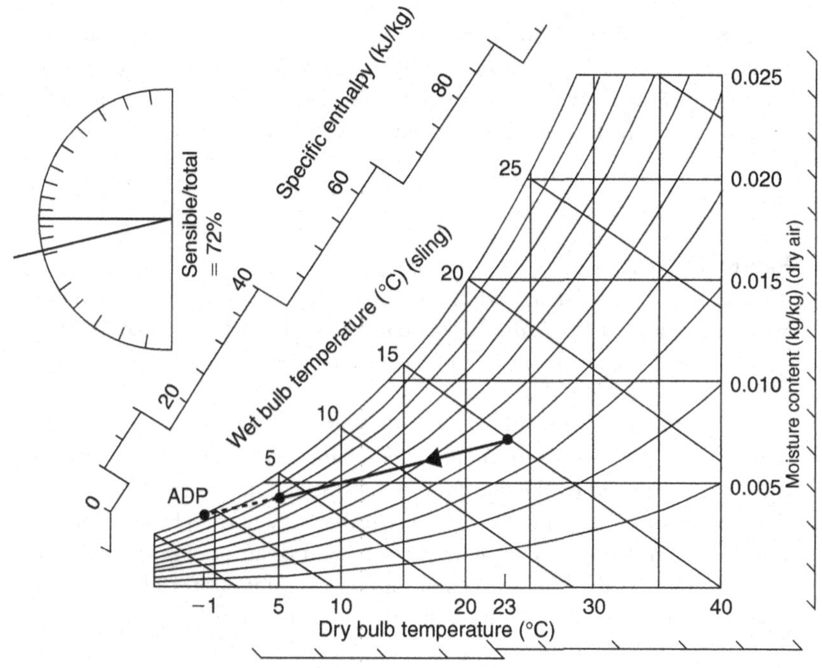

Figure 20.10 ***Cooling and dehumidifying coil.***

Example 20.11

If air is to be cooled and dehumidified, it may be found that the process line joining the inlet and outlet conditions does not meet the saturation line, for example in cooling air from 24°C dry bulb, 45% saturation, to 19°C dry bulb, 50% saturation, the process line shows this to be impossible in one step (Fig. 20.11). The air must first be cooled and dehumidified to reach the right moisture level of 0.006 9 kg/kg and then re-heated to get it back to 19°C. The first part is identical to that in Example 20.8, and the second step is the addition of sensible heat in a reheat coil.

Example 20.12

Winter outside air enters at 0°C dry bulb, 90% saturation, and is to be heated to 30°C, with a moisture content of 0.012 kg/kg.

This can be done in several ways, depending on the method of adding the moisture and final dry bulb control (see Fig. 20.12). If by steam injection, the air can be preheated to just below 30°C and the steam injected (line *ABC*). To give better control of the final temperature, the steam may be injected at a lower condition, with final re-heat to get to the right point (line *ADEC*).

If by water spray or washer, the necessary heat must be put into the air first to provide the latent heat of evaporation. This can be done in two stages, A to F to C, or three stages A to H to J to C, if reheat is required to get the exact final temperature. The latter is easier to control.

Example 20.13

Air enters a packaged dehumidifier (see chapter: Air Conditioning Methods and Applications) at 25°C dry bulb and 60% saturation. It is cooled to 10°C dry bulb and 90% saturation, and then re-heated by its own condenser. What is the final condition?

All of the heat extracted from the air, both sensible and latent, passes to the refrigerant and is given up at the condenser to re-heat, together with the energy supplied to the compressor and the fan motor (since the latter is in the airstream). Figures for this electrical energy have to be determined and assessed in terms of kilojoules per kilogram of air passing through the apparatus. A typical cycle is shown in Fig. 20.13 and indicates a final condition of approximately 47°C dry bulb and 10% saturation.

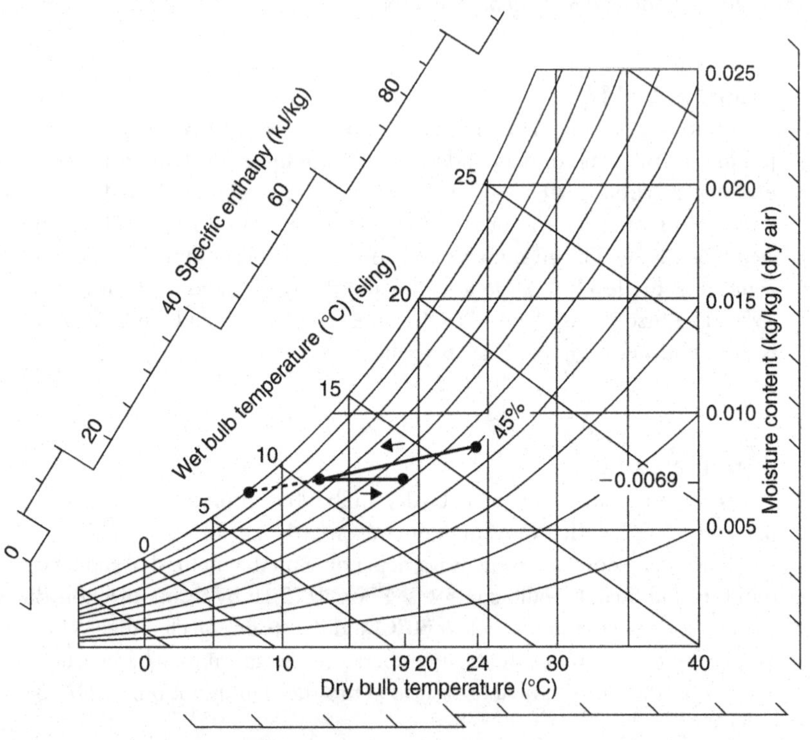

Figure 20.11 ***Cooling with dehumidifying, followed by re-heat – process lines.***

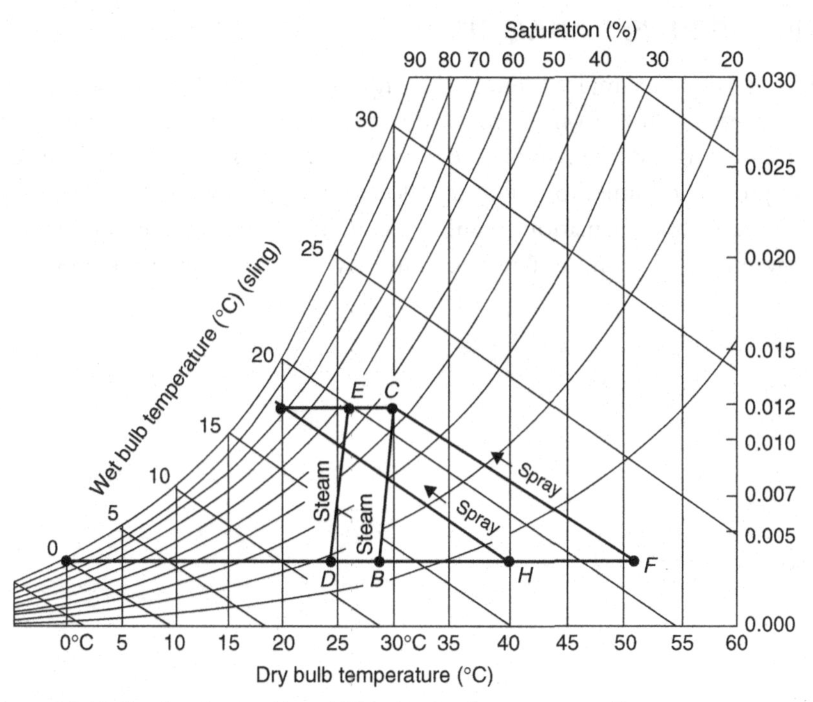

Figure 20.12 ***Pre-heating and humidification in winter – process lines.***

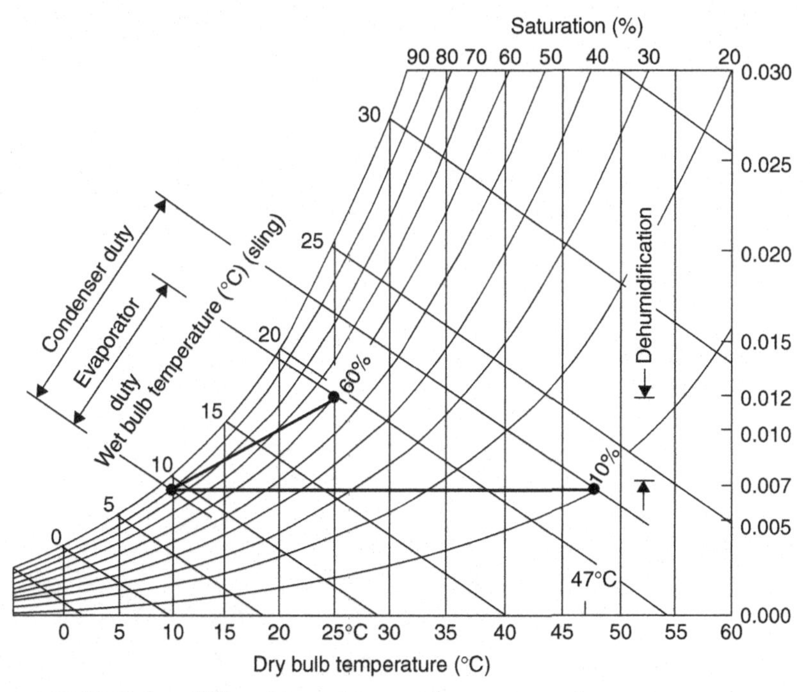

Figure 20.13 ***Dehumidifier with condenser re-heat – process lines.***

20.11 PROCESS ANALYSIS

The last three examples indicate the importance of analysis of the required air treatment process on the psychrometric chart as a guide to the methods which can be adopted and those which are not possible. This analysis can also provide optimisation of energy flows for a process.

Direct desk calculations would have indicated the overall energy flows between the inlet and outlet states, but may not have shown the process.

CHAPTER 21

Practical Air Treatment

21.1 INTRODUCTION

The air in a building may require heating, cooling, humidification adjustment or introduction of fresh air depending on the circumstances and in order to maintain comfort conditions. The needs and methods that may be adopted are outlined here, followed by some exemplar numerical calculations. The air conditions of refrigerated spaces such as cold stores are considered in the relevant refrigeration chapters.

21.2 HEATING

The majority of air-conditioned buildings are offices or are used for similar indoor activities, and are occupied intermittently. The heating system must bring them up to comfortable working conditions by the time work is due to start, so the heating must come into operation earlier to warm up the building.

A large part of the heating load when operating in daytime is for heating fresh incoming air, which is not needed before occupation, and the heat-up time can be reduced if the fresh air supply can remain inoperative for this time.

The required warm-up time varies with ambient conditions, being longer in cold weather and least in warm. Optimum-start controllers are programmed for the building warm-up characteristics and sense the inside and ambient conditions. They then determine the required start-up period and start the heating accordingly.

Air cooling systems commonly have a mass flow of approximately 0.065 kg/(s kW) of cooling load. The normal heating load will be less than the cooling load for most of the time and, if this full airflow is maintained, the air inlet temperature will be of the order of 30–32°C. This is below body temperature and may give the effect of a cold draught, although heat is being supplied to the room. Where possible, the heating airflow should be reduced to give warmer inlet air. This is particularly so with packaged air-conditioners of all sizes, which may have to be located for convenience rather than for the best airflow pattern.

Refrigeration, Air Conditioning and Heat Pumps
http://dx.doi.org/10.1016/B978-0-08-100647-4.00021-8

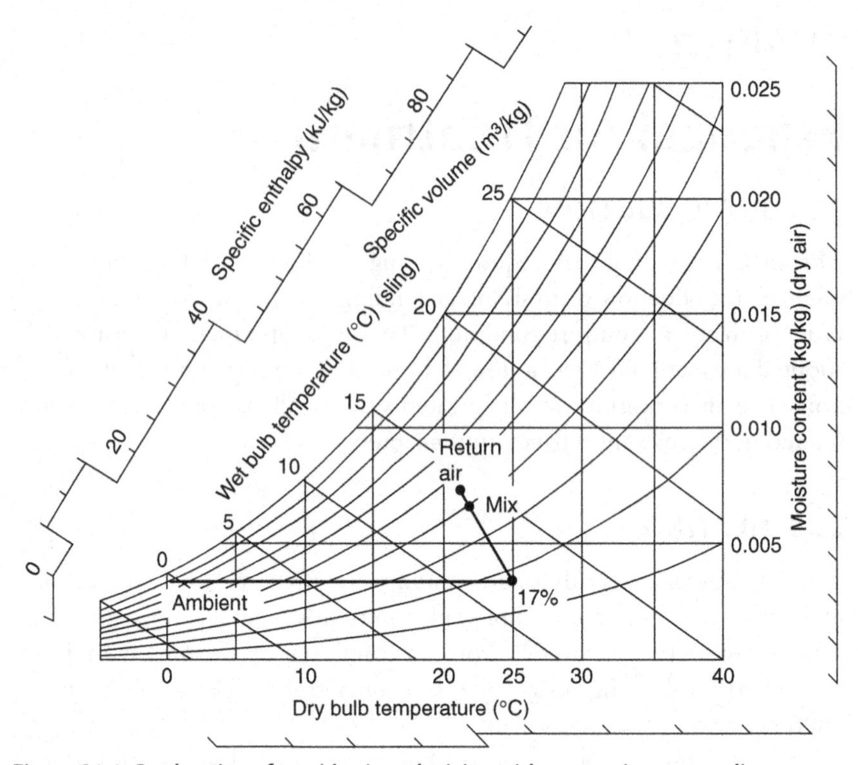

Figure 21.1 ***Pre-heating of outside air and mixing with return air – process lines.***

The addition of moisture to the winter air in the United Kingdom is not usually necessary, except for systems using all outside air, or where persons with severe respiratory trouble are accommodated. With a winter ambient of 0°C dry bulb, 90% saturation, outside air pre-heated to 25°C will then be 17% saturation, which could cause discomfort. However, if this is diluted with the return air, it is unlikely that indoor humidity will fall below 35% saturation. Humidification of this to 50% saturation would permit a slightly lower dry bulb (0.5 K less) to give a similar degree of comfort, thus slightly reducing the conduction losses from the building fabric. However, this is at the cost of the latent heat to evaporate this moisture and a higher dew point (10.4°C instead of 5°C) with increased condensation on cold building surfaces and greater deterioration (see Fig. 21.1).

21.2.1 Addition of Moisture

Addition of moisture to the airstream (see Sections 20.4 and 20.5) is difficult to control, since water remains in the apparatus at the moment of

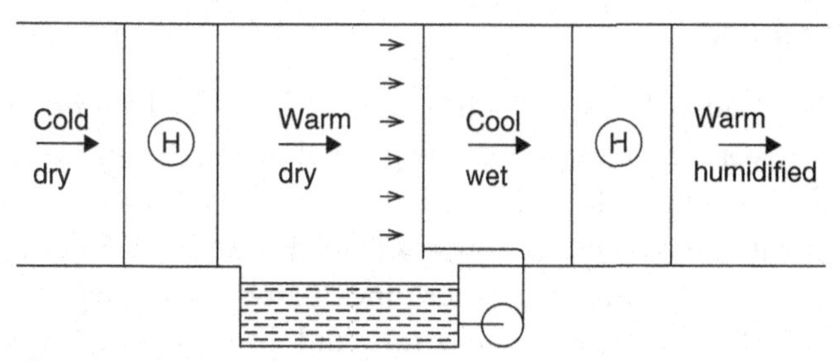

Figure 21.2 ***Pre-heat, humidify, re-heat cycle – apparatus.***

switching off humidification. For this reason, the heat–humidify–re-heat cycle as shown in Fig. 21.2 is to be preferred, as the final heater control can compensate for overshoot.

Air washers require water treatment and bleed-off, since they concentrate salts in the tank. Steam will be free from such impurities, but the boiler will need attention to remove accumulations of hardness.

Mist and spray humidifiers, unless the water is pure, leave a powder deposit of these salts in the conditioned space.

The use of standard factory-packaged air-conditioners to hold close humidity, together with a separate humidifier to correct for over-drying, is a common source of energy wastage, since both may operate at the same time. Packaged units, unless specifically built for the duty, will pull down to 45% saturation or lower under UK conditions. Humidity tolerances for process conditioning such as computer and standards rooms can often be 45–55% saturation, and this differential gap should be wide enough to prevent simultaneous operation of both humidifying and dehumidifying plant.

21.2.2 Outside Air Proportion

The high internal heat load of many buildings means that comfort cooling may be needed even when the ambient is down to 10°C or lower. Under these conditions, a high proportion of outside air can remove building heat and save refrigeration energy. This presupposes the following:

1. The fresh air ducting and fan can provide more air.
2. This outside air can be filtered.
3. There are adequate automatic controls to admit this extra air only when wanted.
4. Surplus air in the building can be extracted.

See also chapter: Efficiency, Running Cost and Carbon Footprint.

21.3 COOLING

The cooling load will normally be greatest in the early afternoon, so no extra start-up capacity is required. The general practice of using a single coil for cooling and dehumidification without reheat, for comfort cooling, will give design balance conditions only at full load conditions. Slightly different conditions must be accepted at other times. Closer control can be obtained by variation of the coolant temperature and air mass flow over the coil, but such systems can be thrown out of calibration, and measures should be taken to avoid unauthorised persons changing the control settings or energy will be wasted with no benefit in the final conditions.

21.3.1 Evaporative Coolers

Many of the warmer climates have a dry atmosphere (see Fig. 19.8). In such areas, considerable dry bulb temperature reduction can be gained by the adiabatic saturation cycle (Section 20.5). The apparatus draws air over a wetted pad and discharges it into the conditioned space. It is termed an *evaporative cooler* or *adiabatic cooler*. The apparatus is illustrated in Fig. 21.3a and the process is plotted on the chart in Fig. 21.3b. The lowest temperature that can theoretically be achieved is the dew point temperature of the treated air. Water spray is also used and this type of cooling can be advantageously applied to reduce the air inlet to mechanical refrigeration condensers in high temperature conditions.

A two-stage evaporative cooler uses the cooled water from the first stage to pre-cool the air entering the second stage. The apparatus is illustrated in Fig. 21.4a and the process is plotted on the chart in Fig. 21.4b. The two air systems are separate. Outside air is drawn through the first stage, passing through the upper wetted pad, and so cools the water down to a temperature approaching the ambient wet bulb. This chilled water then circulates through a dry coil to cool another supply of outside air, thus reducing its wet bulb temperature. This second-stage air then passes through the lower wetted pad and into the cooled space. Water make-up is required to both circuits. In an indirect version the first stage air may be spray cooled and used to cool the primary air in an air-to-air heat exchanger. This is less efficient.

The evaporative cooler has no refrigeration system and only requires electric power for fans and water pumps plus, of course, an adequate supply of water. No moisture can be removed from the air. It is very widely used in arid climates. In practice, within the United Kingdom, the technology is used as a supplementary cooling measure only. It requires three to four times the airflow rate of a conventional air conditioning systems and larger

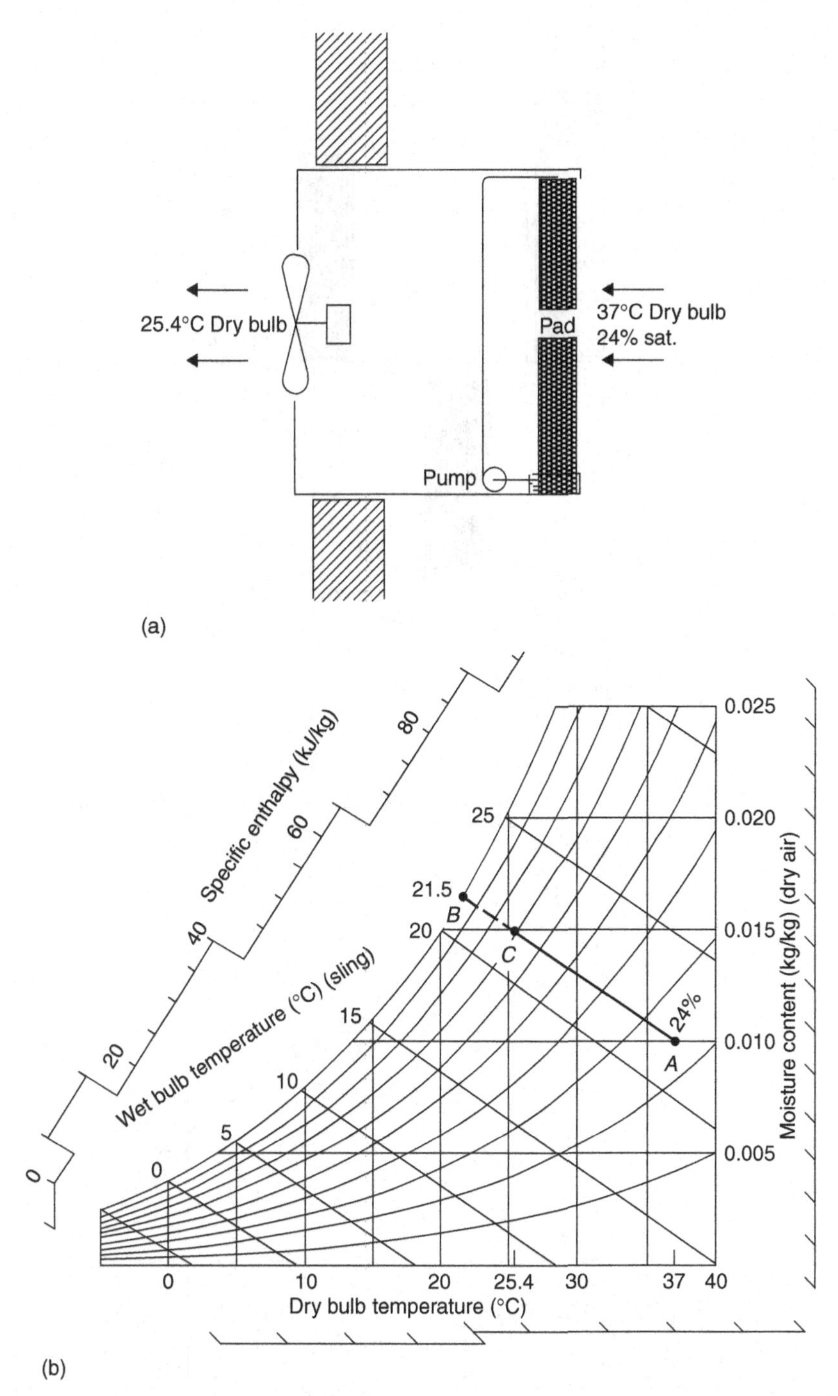

Figure 21.3 ***Evaporative cooler.***

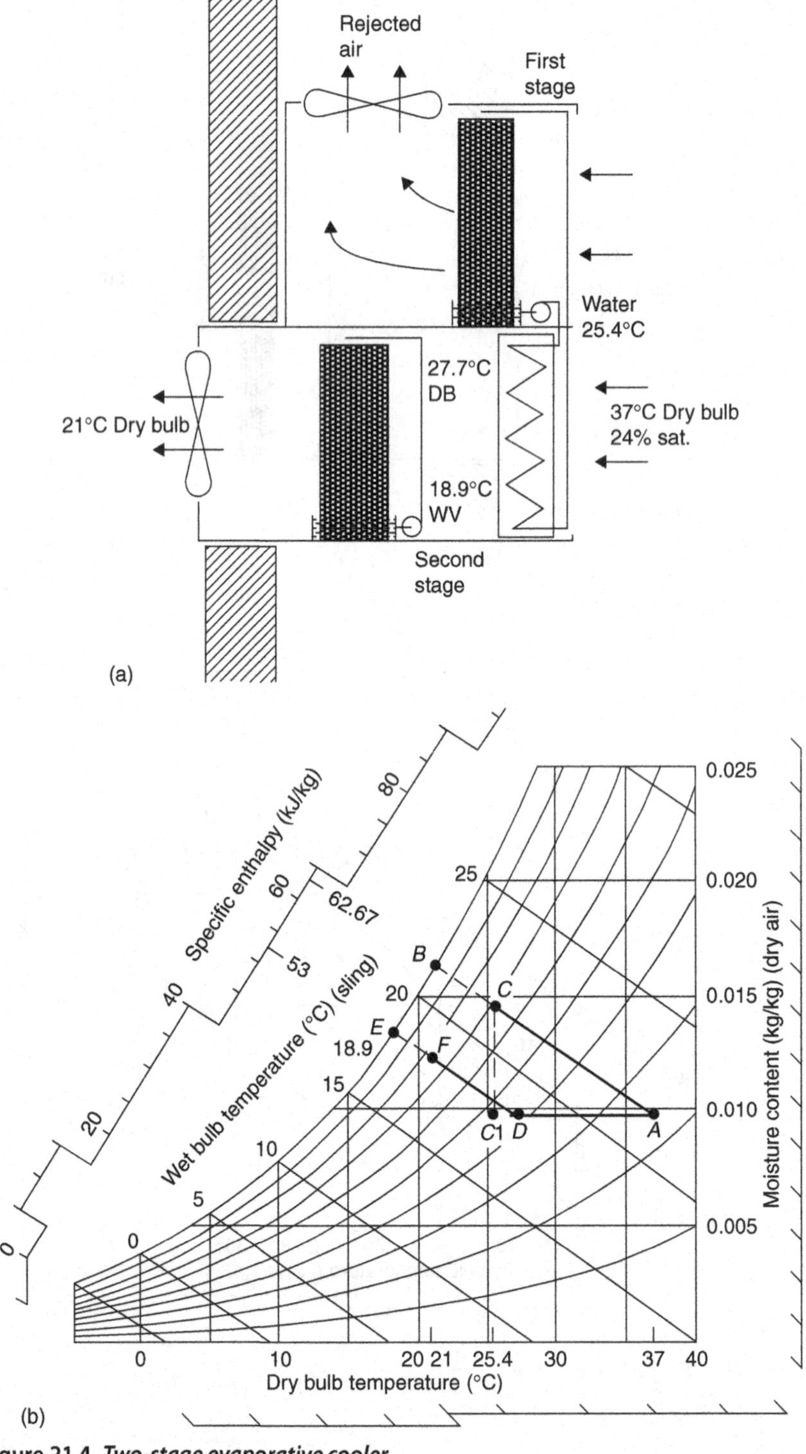

Figure 21.4 ***Two-stage evaporative cooler.***

ducts are required. The high airflow rates and the absence of re-circulation promotes good indoor air quality. In practice, because of the limited cooling capacity of an indirect evaporative cycle, the primary air is often cooled again by direct evaporation or by a mechanical cooling system. This is called an indirect–direct system.

21.3.2 Cooling Towers and Evaporative Condensers

The cooling tower is an evaporative cooler with the prime purpose of cooling the water rather than the air. The normal forms have the water and air in counter flow or cross-flow, although towers are built with parallel flow (see Fig. 21.5).

The process is complex and cannot be simplified into an ultimate balance condition, from which to work back to a supposed operating point, so factors are used for design and application which are based on similar apparatus.

Heat from the water is transferred to the air, so the available heat gain by the air will depend on its initial enthalpy. This is usually expressed in

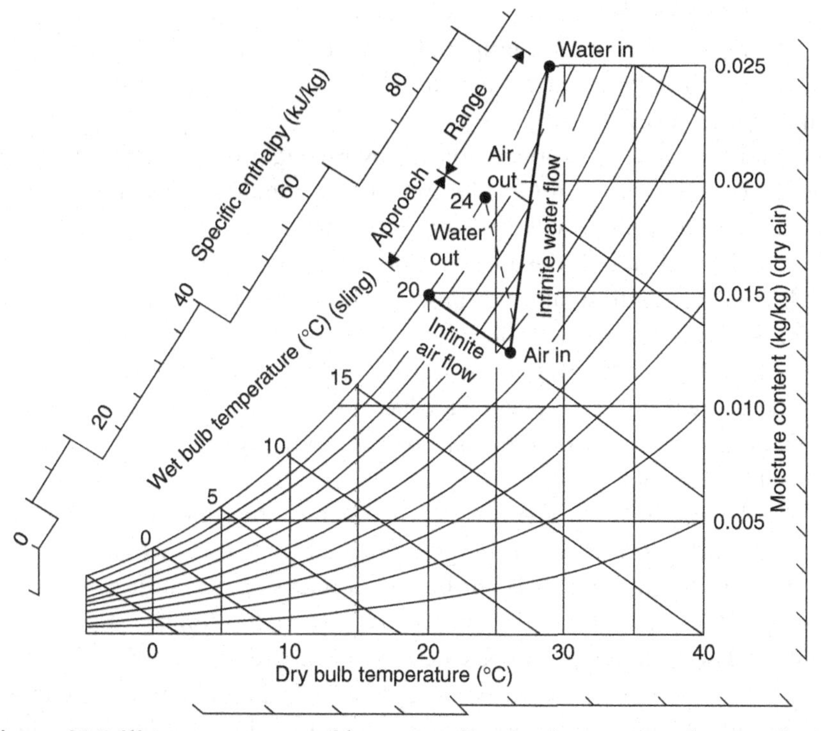

Figure 21.5 ***Water tower – possible process lines under one set of atmospheric conditions.***

terms of ambient wet bulb temperature, since the two are almost synonymous and the wet bulb is more easily recognised. This is used as a yardstick to describe performance in terms of the approach of the leaving water temperature to the ambient wet bulb. The two could only meet ultimately in a tower having an airflow infinitely larger than the water flow, so the term is descriptive rather than a clear indication of tower efficiency.

Assuming the mass flows of air and water to be equal, an approximate balance can be found.

In the case of the evaporative condenser, the heat is input to the condenser coils, which are kept wet by the spray. The water acts both as a heat-transfer medium and an evaporative coolant, and its temperature will vary through the stack of tubes. The overall process is complex and ratings are determined from practical tests on a complete condenser.

21.4 DEHUMIDIFICATION

Moisture can be removed from any material which is to be dried by passing air over it which has a lower water-vapour pressure. Also, in removing this moisture, the latent heat of evaporation must be supplied, either directly by heating, or by taking sensible heat from the airstream which is carrying out the drying process.

Moisture may be removed from air by passing it over a surface which is colder than its dew point (see Fig. 20.9). In air-conditioning systems this is a continuous process, providing that the moisture condenses out as water and can be drained away. If the apparatus dew point is below 0°C, the moisture will condense as frost, and the process must be interrupted from time to time to defrost the evaporator.

Air will leave the evaporator section with a reduced moisture content, but at a low-temperature and high-percentage saturation. In this state, it may not be effective in removing moisture from any subsequent process, as it will be too cold.

In the unit dehumidifier process, all or part of the condenser heat is used to re-heat the air leaving the evaporator (see Fig. 21.6). Since the moisture in the air has given up its latent heat in condensing, this heat is reclaimed and put back into the outlet air. In a typical application, air at 25°C dry bulb and 60% saturation can be dried and re-heated to a condition of 46°C dry bulb and 10% saturation (see Fig. 20.13). In this state, it is hot enough to provide the necessary latent heat to dry out the load product. The entire system is in one unit, requiring only an electrical supply and a water drain, so there is no constraint on location.

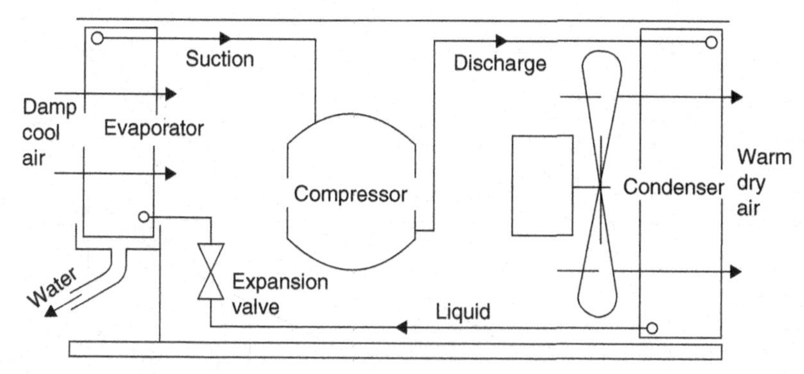

Figure 21.6 ***Dehumidifier.***

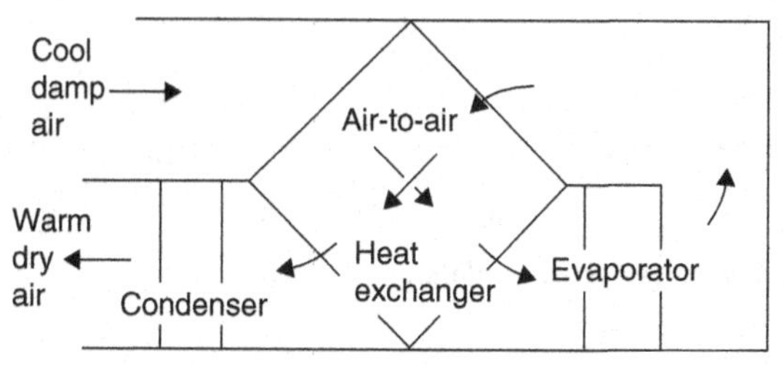

Figure 21.7 ***Dehumidifier with heat recovery.***

The efficiency of a unit dehumidifier can be improved by a heat exchanger which pre-cools the incoming air by using the cold air leaving the evaporator (see Fig. 21.7).

The performance of a dehumidifier in terms of moisture removal will vary considerably with the condition of the incoming air. Typical capacity figures are shown in Fig. 21.8.

The refrigeration method of drying air is the most energy efficient, down to a lower limit of about 0.005 kg/kg moisture content at atmospheric pressure. Equipment to work at frosting conditions can be duplicated, one evaporator defrosting whist the other is operating. Below this limit, chemical or adsorption drying must be used.

21.5 NUMERICAL EXAMPLES

Examples of different methods of controlling air conditions are given in the following examples. These include evaporative cooling (Ex. 21.1, 21.2), cooling tower (Ex. 21.3), air cooling with air flow control, chilled water, refrigerant and both air and refrigerant (Ex 21.4 to 21.8).

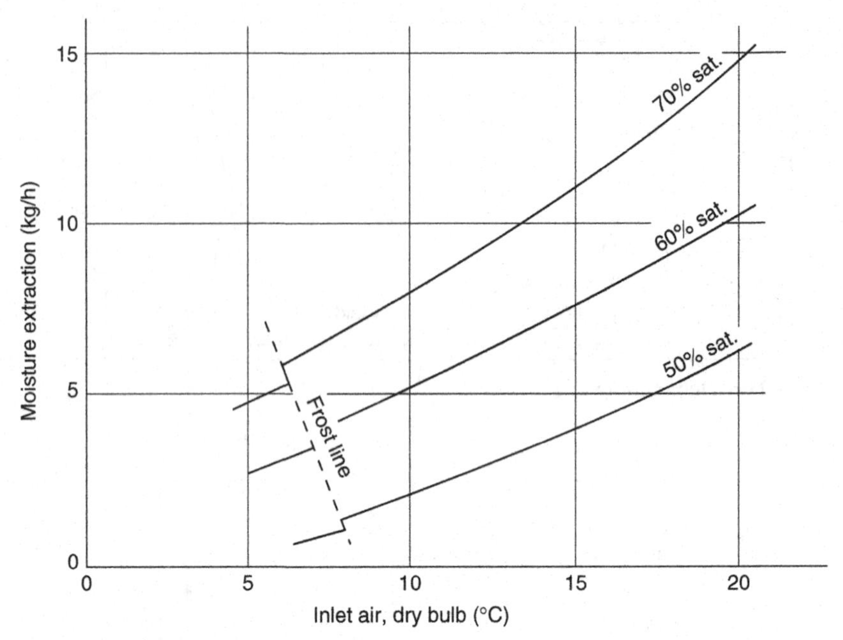

Figure 21.8 ***Performance of dehumidifier.***

Example 21.1

Air at 37°C dry bulb, 24% saturation, is drawn through an evaporative cooler having an adiabatic saturation efficiency of 75%. What is the final dry bulb, and how much water is required?

The entering enthalpy is 62.67 kJ/kg, and this remains constant through the process.

By construction on the chart, or from tables, the ultimate saturation condition would be 21.5°C, and 75% of the drop from 37 to 21.5°C gives a final dry bulb of 25.4°C.

The water requirement can be calculated from the average latent heat of water over the working range, which is 2425 kJ/kg. The amount of water to be evaporated is $1/2425 = 0.4 \times 10^{-3}$ kg/(s kW).

Example 21.2

Taking the first stage as Example 21.1, the water would be cooled to 25.4°C and could be used in a coil of 80% contact factor to pre-cool outside air to

$37 - 0.8\,(37 - 25.4) = 27.7$°C (point D, Fig. 21.4)

The wet bulb is now 18.9°C and the enthalpy is 53 kJ/kg. A second-stage evaporative cooler with an efficiency of 75% will bring this down to 21°C dry bulb (point F).

Example 21.3

Air enters a cooling tower at 26°C dry bulb and 20°C wet bulb. Water at the same mass flow enters at 29°C and leaves at 24°C. If the air leaves the tower at 98% saturation, what is its final condition?

From the chart, the air leaves at approximately 25.7°C dry bulb.

Calculations of this sort are only of importance to the tower designer. Manufacturers' application data gives the cooling range or capacity in terms of wet bulb, inlet water temperature and mass flow.

Example 21.4

A space is to be held at 21°C dry bulb and 50% saturation and has an internal load of 14 kW sensible and 1.5 kW latent heat gain. The inlet air temperature is 12°C. What are the required inlet air conditions and the mass air flow?

$$\text{Inlet air temperature} = 12.0°\text{C}$$

$$\text{Air temperature rise through room, } 21 - 12 = 9.0\,\text{K}$$

$$\text{Air flow for sensible heat, } \frac{14}{9 \times 1.02} = 1.525\,\text{kg/s}$$

$$\text{Moisture content of room air, } 21°\text{C}, 50\% = 0.007\,857\,\text{kg/kg}$$

$$\text{Moisture to pick up, } \frac{1.5}{2440 \times 1.525} = 0.000\,403$$

$$\text{Moisture content of entering air} = 0.007\,454$$

From tables this gives approximately 85% saturation.

Note that the value 1.02 in the third line is a general figure for the specific heat capacity of moist air, commonly used in such calculations. (The true figure for this particular example is slightly higher.) The value of 2440 for the latent heat is, again, a quantity in common use, and is close enough for most purposes.

Example 21.5

For the same duty, a chilled water fan coil unit is fitted within the space. Water enters at 5°C and leaves at 10.5°C. The fan motor power is 0.9 kW. What water flow is required?

$$\text{Total cooling load, } 14.0 + 1.5 + 0.9 = 16.4\,\text{kW}$$

$$\text{Mass water flow, } \frac{16.4}{4.19 \times (10.5 - 5)} = 0.71\,\text{kg/s}$$

Example 21.6

For the same duty, liquid R134a enters the expansion valve at 33°C, evaporates at 5°C and leaves the cooler at 9°C. Fan power is 0.9 kW. What mass flow of refrigerant is required?

Total load, as Example 25.2 = 16.4 kW

Enthalpy of R134a, evaporated at 5°C

$$\text{superheated to } 9°\text{C} = 405.23\,\text{kJ/kg}$$

$$\text{Enthalpy of liquid R134a at } 33°\text{C} = 246.71\,\text{kJ/kg}$$

$$\text{Refrigerating effect} = 158.52\,\text{kJ/kg}$$

$$\text{Required refrigerant mass flow} = \frac{16.4}{158.52} = 0.103\,\text{kg/s}$$

Example 21.7

For the same application, primary air reaches induction units at the rate of 0.4 kg/s and at conditions of 13°C dry bulb and 72% saturation. Chilled water enters the coils at 12°C and leaves at 16°C. What will be the room condition and how much water will be used?

The chilled water enters higher than the room dew point temperature, so any latent heat must be removed by the primary air, and this may result in a higher indoor condition to remove the design latent load:

$$\text{Moisture in primary air, } 13°\text{C dry bulb, } 72\%\text{ sat.} = 0.006744\,\text{kg/kg}$$

$$\text{Moisture removed, } \frac{1.5}{2440 \times 0.4} = \underline{0.001537\,\text{kg/kg}}$$

$$\text{Moisture in room air will rise to} = 0.008\,281\,\text{kg/kg}$$

which corresponds to a room condition of 21°C dry bulb, 53% saturation.

$$\text{Sensible heat removed by primary air, } 0.4 \times 1.02 \times (21 - 13) = 3.26\,\text{kW}$$

$$\text{Heat to be removed by water, } 14.0 - 3.26 = 10.74\,\text{kW}$$

$$\text{Mass water flow, } \frac{10.74}{4.19 \times (16 - 12)} = 0.64\,\text{kg/s}$$

CHAPTER 22

Air Conditioning Load Estimation

22.1 INTRODUCTION

The air in a building gains heat from internal activities such as electronic devices and includes occupancy. Also the air quality will tend to deteriorate and air changes will be necessary. In addition, and depending on ambient conditions, there will be net heat gain from the outside. Smaller buildings in northern Europe may normally experience a net heat loss during most of the year, and sufficient ventilation is obtained through openings to the outside. Larger buildings have a smaller surface to volume ratio and consequently build-up of heat and contaminants becomes unacceptable without air conditioning. It is necessary to distinguish air conditioning and comfort cooling. Comfort cooling may be defined as the use of mechanical cooling to maintain control over the maximum air temperature. There may be some incidental dehumidification of the supply air as a consequence. Air conditioning involves control over both temperature and humidity.

22.2 COMPONENTS OF LOAD

The cooling load to maintain steady temperature and humidity in a conditioned space consists of following four components:

1. Heat leakage through the fabric by conduction from warmer surroundings.
2. Heat gain by radiation through transparent surfaces – usually solar but occasionally by other means (radiant heat from a process, such as furnaces).
3. Heat gain by forced or natural convection – air infiltration and fresh air supply – sensible and latent heat.
4. Internal heat sources – lights, people, machines, etc. – sensible and latent heat.

Refrigeration, Air Conditioning and Heat Pumps
http://dx.doi.org/10.1016/B978-0-08-100647-4.00022-X

22.3 CONDUCTION HEAT GAINS

Conduction of heat through plain surfaces under steady-state conditions is given by the product of the area, temperature difference and overall conductance of the surface (see Section 1.7):

$$Q = A \times \Delta T \times U$$

where

$$U = \frac{1}{R_{si} + R_1 + R_2 + R_3 + \cdots + R_{so}}$$

and R_{si} is the inside surface thermal resistance, R_{so} is the outside surface thermal resistance, and R_1, R_2, etc. are the thermal resistances of the composite layers of the fabric.

Example 22.1

A building wall is made up of pre-cast concrete panels 40 mm thick, lined with 50 mm insulation and 12 mm plasterboard. The inside resistance is 0.3 (m² K)/W and the outside resistance 0.07 (m² K)/W. What is the *U* factor?

$$U = \frac{1}{0.3 + 0.040/0.09 + 0.050/0.037 + 0.012/0.16 + 0.07}$$

$$= \frac{1}{2.24}$$

$$= 0.45\ \mathrm{W/(m^2\,K)}$$

Conductivity values for all building materials, of the surface coefficients, and many overall conductivities can be found in the CIBSE Guides.

The dominant factor in building surface conduction is the absence of steady-state conditions, since the ambient temperature, wind speed and solar radiation are not constant. It will be readily seen that the ambient will be cold in the morning, will rise during the day, and will fall again at night. As heat starts to pass inwards through the surface, some will be absorbed in warming the outer layers and there will be a time lag before the effect reaches the inner face, depending on the mass, conductivity and specific heat capacity of the materials. Some of the absorbed heat will be retained in the material and then lost to ambient at night. The effect of thermal time lag can be expressed mathematically.

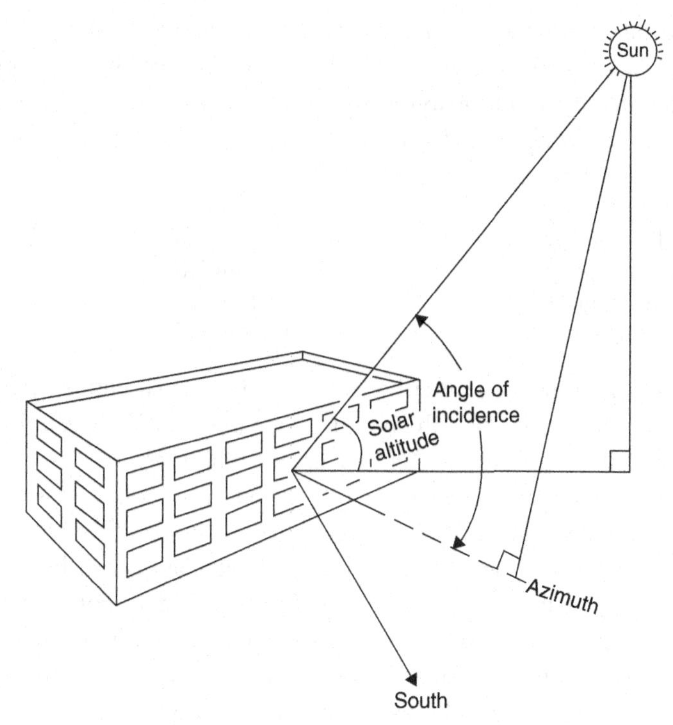

Figure 22.1 ***Angle of incidence of the Sun's rays on window.***

The rate of heat conduction is further complicated by the effect of solar radiation, which reaches the Earth's surface at a maximum intensity of about 0.9 kW/m^2. The amount of incident solar radiation absorbed by a plane surface will depend on the absorption coefficient and the angle at which the radiation strikes. The angle of the sun's rays to a surface (see Fig. 22.1) is always changing, so this must be estimated on an hour-to-hour basis. Various methods of reaching an estimate of heat flow are used, and the sol-air temperature provides a simplification of the factors involved. This, also, is subject to time lag as the heat passes through the surface.

22.4 SOLAR HEAT

Solar radiation through windows has no time lag and must be estimated by finite elements (ie on an hour-to-hour basis), using calculated or published data for angles of incidence and taking into account the type of window glass.

Since solar gain can be a large part of the building load, special glasses and window constructions have been developed, having two or more layers and with reflective and heat absorbing surfaces. These can reduce the energy passing into the conditioned space by as much as 75%. Typical transmission figures are as follows:

Plain single glass	0.75 transmitted
Heat-absorbing glass	0.45 transmitted
Coated glass, single	0.55 transmitted
Metallised reflecting glass	0.25 transmitted

Windows may be shaded by either internal or external blinds, or by overhangs or projections beyond the building face (see Fig. 22.2). Windows may also be shaded for part of the day by adjacent buildings.

All these factors need to be taken into account, and solar transmission estimates are usually calculated or computed for the hours of daylight through the hotter months, although the amount of calculation can be much reduced if the probable worst conditions can be guessed. For example, the greatest solar gain for a window facing west will obviously be after midday. Comprehensive data on solar radiation factors, absorption coefficients and methods of calculation can be found in reference books.

Figure 22.2 ***Solar shading (Business Edge).***

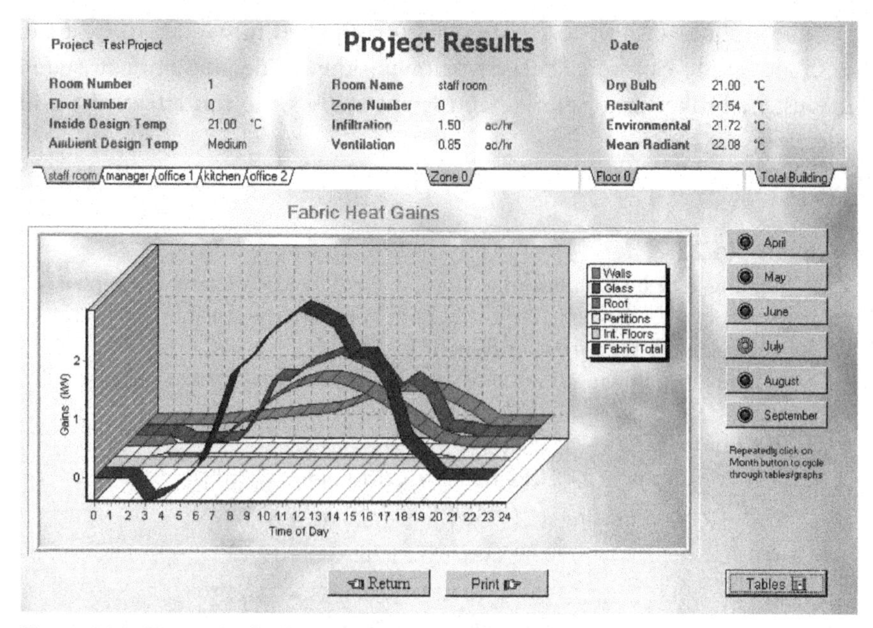

Figure 22.3 ***Heat gain display screen (Business Edge).***

Computer software is widely used for the task of load calculation and can emulate the energy flow of a room or building using mathematical models. Recognised design methods are applied to calculate thermal gains. A typical screen display of heat gains during a 24 h period is shown in Fig. 22.3.

It will be helpful to arrive at total loads for zones, floors and the complete installation where cooling loads are required for a large building of many separate rooms, and this will help to determine the best method of conditioning and the overall size of plant.

22.5 FRESH AIR

The movement of outside air into a conditioned building is balanced by the loss of an equal amount at the inside condition, whether by intent (positive fresh air supply or stale air extract) or by accident (infiltration through window and door gaps, and door openings). Since a building for human occupation must have some fresh air supply and some mechanical extract from toilets and service areas, it is usual to arrange an excess of supply over extract, to maintain an internal slight pressure and so reduce accidental air movement and ingress of dirt.

The amount of heat to be removed (or supplied in winter) to treat the fresh air supply can be calculated, knowing the inside and ambient states. It must be broken into sensible and latent loads, since this affects the coil selection.

Example 22.2

A building is to be maintained at 21°C dry bulb and 45% saturation in an ambient of 27°C dry bulb, 20°C wet bulb. What are the sensible and latent air-cooling loads for a fresh air flow of 1.35 kg/s?

There are following three possible calculations, which cross-check the same:

1. Total heat:

$$\text{Enthalpy at } 27\,^\circ\text{C dry bulb}, 20\,^\circ\text{C wt bulb} = 57.00\,\text{kJ/kg}$$

$$\text{Enthalpy at } 21\,^\circ\text{C dry bulb}, 45\%\,\text{sat.} = 39.08\,\text{kJ/kg}$$

$$\text{Heat to be removed} = \overline{17.92}$$

$$Q_t = 17.92 \times 1.35 = 24.2\,\text{kW}$$

2. Latent heat:

$$\text{Moisture at } 27\,^\circ\text{C DB}, 20\,^\circ\text{C WB} = 0.0117\,\text{kg/kg}$$

$$\text{Moisture at } 21\,^\circ\text{C DB}, 45\%\,\text{sat.} = 0.0070\,\text{kg/kg}$$

$$\text{Moisture to be removed} = \overline{0.0047}$$

$$Q_l = 0.0047 \times 1.35 \times 2440 = 15.5\,\text{kW}$$

3. Sensible heat:

$$Q_s = [1.006 + (4.187 \times 0.0117)]\,(27 - 21) \times 1.35 = 8.6\,\text{kW}$$

Where there is no mechanical supply or extract, factors are used to estimate possible natural infiltration rates. Empirical values may be found in several standard references.

Where positive extract is provided, and this duct system is close to the supply duct, heat exchange apparatus (see Fig. 22.4) can be used between them to pre-treat the incoming air. For the air flow in Example 23.2, and in Fig. 22.5, it would be possible to save 5.5 kW of energy by use of a heat exchanger. An alternative is the thermal wheel is illustrated in Fig. 25.1.

Figure 22.4 *Air-to-air heat exchanger (Business Edge).*

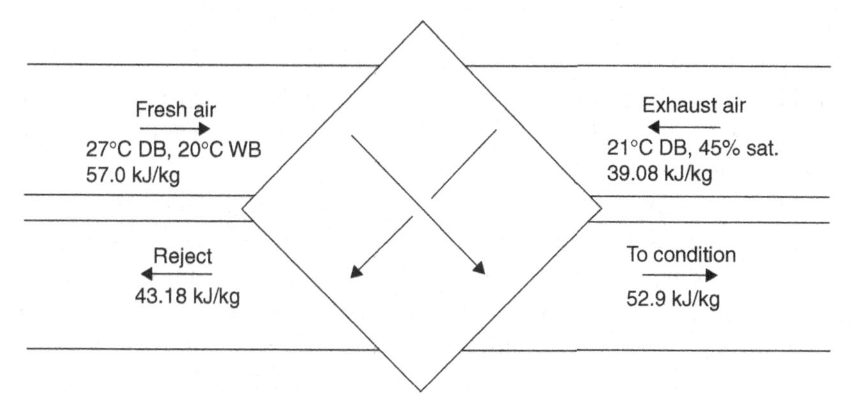

Figure 22.5 *Heat recovery to pre-cool summer fresh air.*

22.6 INTERNAL HEAT SOURCES

Electric lights, office machines and other items of a direct energy-consuming nature will liberate all their heat into the conditioned space, and this load may be taken as part of the total cooling load. Particular care should be taken to check the numbers of office electronic devices, and their probable proliferation within the life of the building.

Lighting, especially in offices, can consume a great deal of energy and justifies the expertise of specialist advice. Switching should be arranged so that a minimum of the lights can be used in daylight hours. It should always be borne in mind that in addition to running the lighting, energy is required to remove the heat generated, whilst cooling is in operation.

Ceiling extract systems are now commonly arranged to take air through the light fittings, so that a proportion of this load can be rejected with the exhausted air.

Example 22.3

Return air from an office picks up 90% of the input of 15 kW to the lighting fittings. Of this return air flow, 25% is rejected to ambient. What is the resulting heat gain from the lights?

$$\text{Total lighting load} = 15\ \text{kW}$$

$$\text{Picked up by return air}, 15 \times 0.9 = 13.5\ \text{kW}$$

$$\text{Rejected to ambient}, 13.5 \times 0.25 = 3.375\ \text{kW}$$

$$\text{Net room load}, 15.0 - 3.375 = 11.625\ \text{kW}$$

The heat input from human occupants depends on their number (or an estimate of the probable number) and intensity of activity. This must be split into sensible and latent loads.

The energy input of part of the plant itself must be included in the cooling load. All cases include fan heat, either net motor power or gross motor input, depending on whether the motors are in the conditioned space or not. Also, in the case of packaged units within the space, the compressors will dissipate some heat, and this should be accounted.

22.7 ASSESSMENT OF TOTAL LOAD ESTIMATES

Examination of the items which comprise the total cooling load may throw up peak loads which can be reduced by localised treatment such as shading, modification of lighting, removal of machines, etc. A detailed analysis of this sort can result in substantial savings in plant size and future running costs.

A careful site survey should be carried out if the building is already erected, to verify the given data and search for load factors which may not be apparent from the available information.

It will be seen that the total cooling load at any one time comprises a large number of elements, some of which may be known with a degree of certainty, but many of which are transient and which can only be estimated to a reasonable closeness. Even the most sophisticated and time-consuming of calculations will contain a number of approximations, so short-cuts and empirical methods are very much in use. Abbreviated tables are given by Gosling (1980) and Jones (2000). Full physical data will be found in the CIBSE Guide Books.

Since the estimation will be based on a desired indoor condition at all times, it may not be readily seen how the plant size can be reduced at the expense of some temporary relaxation of the standard specified. Some of the programs available can be used to indicate possible savings both in capital cost and running energy under such conditions. See also chapter: Efficiency, Running Cost and Carbon Footprint.

CHAPTER 23

Air Movement

23.1 INTRODUCTION

Conditioned air has to be taken and circulated in the required spaces, and a study of air movement essentials is given here.

23.2 STATIC AND TOTAL PRESSURE

Air at sea level exerts a static pressure, due to the weight of the atmosphere, of 1013.25 mbar. The density, or specific mass, at 20°C is 1.2 kg/m³. Densities at other conditions of pressure and temperature can be calculated from the Gas Laws:

$$\rho = 1.2\left(\frac{p}{1013.25}\right)\left(\frac{273.15+20}{273.15+t}\right)$$

where p is the new pressure, in mbar, and t is the new temperature in °C.

Example 23.1

What is the density of dry air at an altitude of 4500 m (575 mbar barometric pressure) and a temperature of −10°C?

$$\rho = 1.2\left(\frac{575}{1013.25}\right)\left(\frac{293.15}{263.15}\right)$$
$$= 0.76\ \text{kg/m}^3$$

Air passing through a closed duct will lose pressure due to friction and turbulence in the duct.

An air-moving device such as a fan will be required to increase the static pressure in order to overcome this resistance loss (see Fig. 23.1).

If air is in motion, it will have kinetic energy of

$$0.5 \times \text{mass} \times (\text{velocity})^2$$

Refrigeration, Air Conditioning and Heat Pumps
http://dx.doi.org/10.1016/B978-0-08-100647-4.00023-1

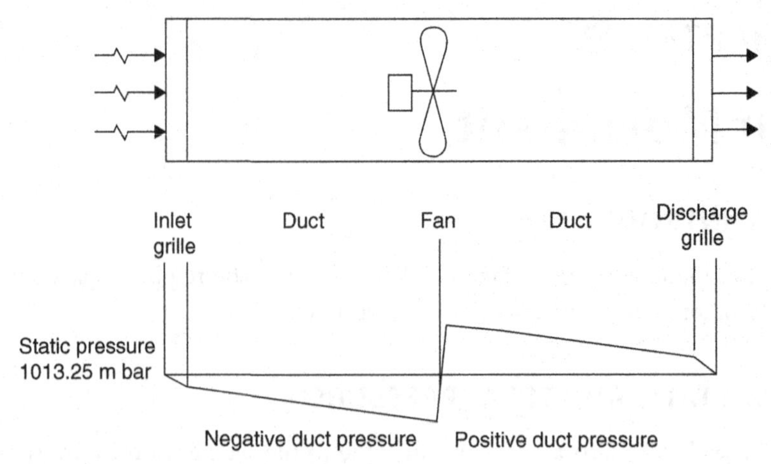

Figure 23.1 ***Static pressure in ducted system.***

Example 23.2

If 1 m^3 of air at 20°C dry bulb, 60% saturation, and a static pressure of 101.325 kPa is moving at 7 m/s, what is its kinetic energy?

Air at this condition, from psychrometric tables, has a specific volume of 0.8419, so 1 m^3 will weigh 1/0.8419 or 1.188 kg, giving:

$$\text{Kinetic energy} = 0.5 \times 1.188 \times (7)^2$$
$$= 29.1\ \text{kg}/(\text{m}\,\text{s}^2)$$

The dimensions of this kinetic energy are seen to be the dimensions of pascals. This kinetic energy can therefore be expressed as a pressure and is termed the velocity pressure.

The total pressure of the air at any point in a closed system will be the sum of the static and velocity pressures. Losses of pressure due to friction will occur throughout the system and will show as a loss of total pressure, and this energy must be supplied by the air-moving device, usually a fan.

23.3 MEASURING DEVICES

The static pressure within a duct is too small to be measured by a bourdon tube pressure gauge, and an inclined manometer is usually employed (Fig. 23.2). Also, there are electromechanical anemometers. The pressure tapping into the duct must be normal to the airflow.

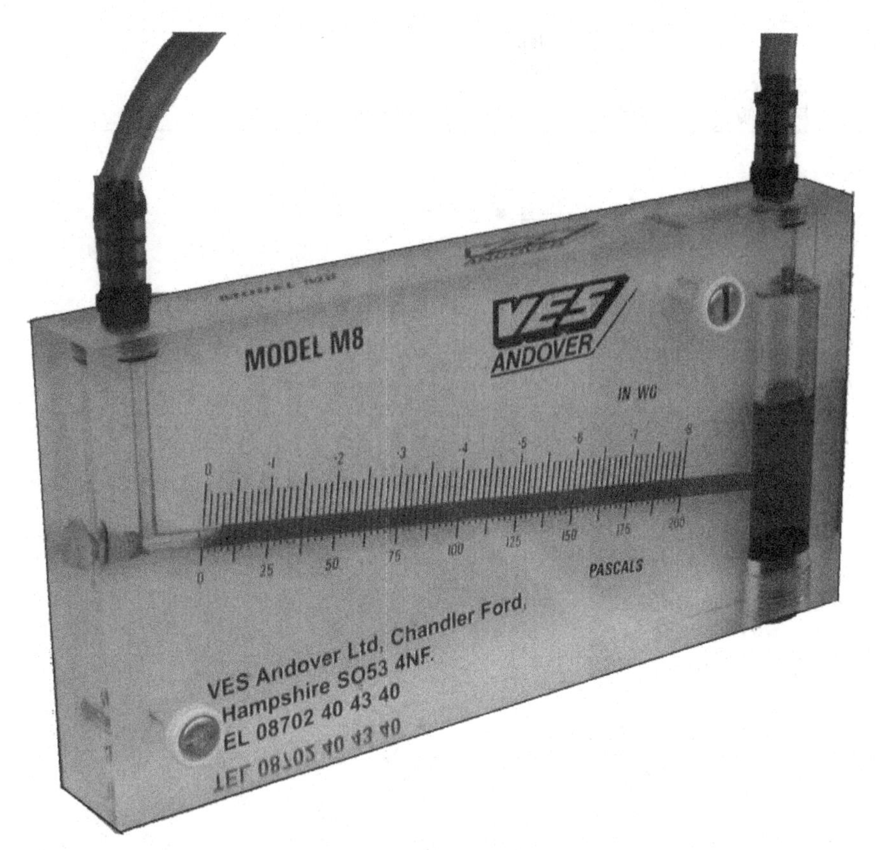

Figure 23.2 ***Inclined manometer (Business Edge).***

Instruments for measuring the velocity as a pressure effectively convert this energy into pressure. The transducer used is the Pitot tube (Fig. 23.3), which faces into the airstream and is connected to a manometer. The outer tube of a standard pitot tube has side tappings which will be normal to the airflow, giving static pressure. By connecting the inner and outer tappings to the ends of the manometer, the difference will be the velocity pressure.

Sensitive and accurate manometers are required to measure pressures below 15 Pa, equivalent to a duct velocity of 5 m/s, and accuracy of this method falls off below 3.5 m/s. The pitot head diameter should not be larger than 4% of the duct width, and heads down to 2.3 mm diameter can be obtained. The manometer must be carefully levelled.

Air speed can be measured with mechanical devices, the best known of which is the vane anemometer (Figs 23.4 and 23.5). The air turns the vanes of the meter, and the airflow is read from the hand held unit. These

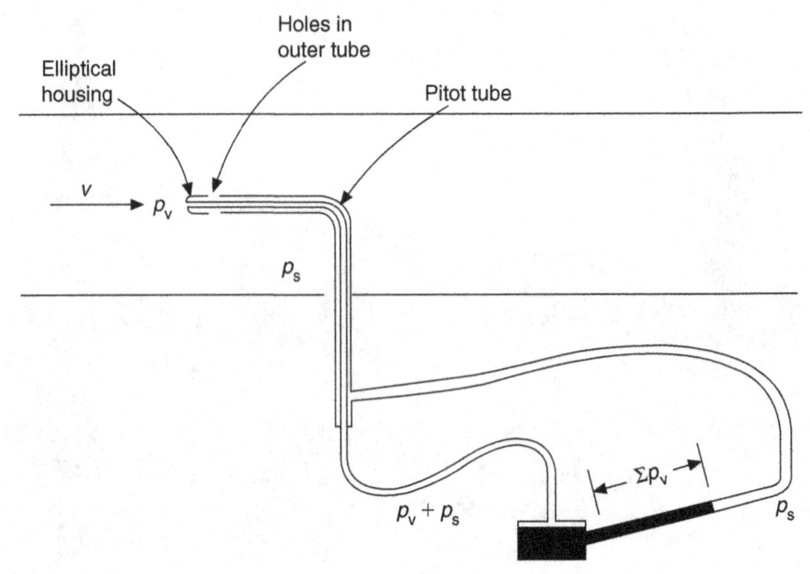

Figure 23.3 ***Pitot tube.***

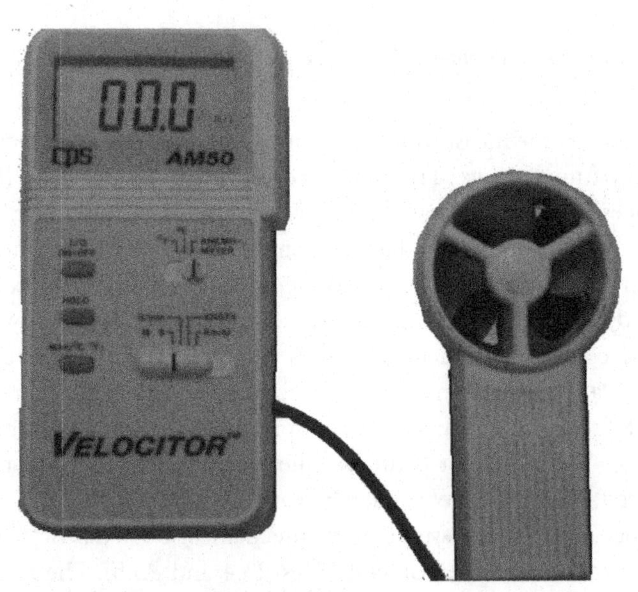

Figure 23.4 ***Vane anemometer (Business Edge).***

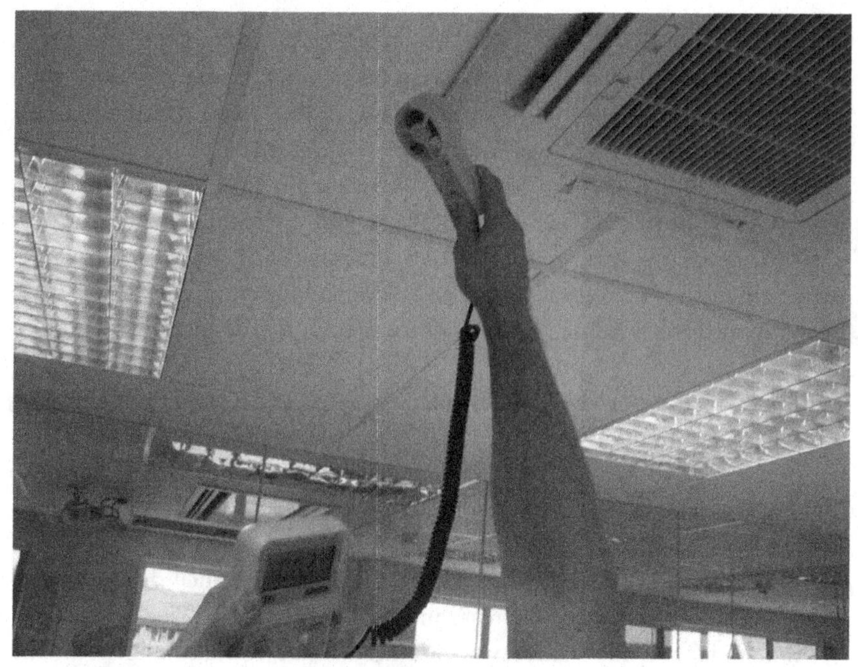

Figure 23.5 ***Vane anemometer in use (Business Edge).***

instruments need to be calibrated if close accuracy is required. Accuracies of 3% are claimed.

The hot wire anemometer detects the cooling effect of the moving air over a heated wire or thermistor, and converts the signal to velocity. Air velocities down to 1 m/s can be measured with claimed accuracies of 5%, and lower velocities can be indicated.

Airflow will not be uniform across the face of a duct, the velocity being highest in the middle and lower near the duct faces, where the flow is slowed by friction. Readings must be taken at a number of positions and an average calculated. Methods of testing and positions for measurements are covered in British Standard BS1042. In particular, airflow will be very uneven after bends or changes in shape, so measurements should be taken in a long, straight section of duct.

More accurate measurement of airflow can be achieved with nozzles or orifice plates. In such cases, the measuring device imposes a considerable resistance to the airflow, so that a compensating fan is required. This method is not applicable to an installed system and is used mainly as a development tool for factory-built packages, or for fan testing.

23.4 FANS

Fans are extensively used for circulating air through evaporators and condensers and in air conditioning systems. They can be divided into two types: *axial* and *centrifugal*. Axial fan blades generate aerodynamic lift that causes the blades to impart a force on the air, propelling it forward. Low-pressure, high-volume propeller fans (Fig. 23.6) are used in conjunction with heat exchangers such as evaporators and condensers. They move large quantities of air and generate low pressures. The efficiency is improved by a close fitting inlet shroud. Some blade types are specifically designed to minimise noise.

Other axial types used for circulating air in ducts include vane-axial, which has good downstream air distribution and tube-axial. A typical axial fan performance characteristic is shown in Fig. 23.7. Fan curves have the volume flow rate on the x-axis and the pressure development on the y-axis. The fan should be selected as near to the peak efficiency as possible. When the fan enters the stall region there is a dip in performance due to turbulence and this can cause an increase in noise level.

Vane-axial and tube-axial types are used in air conditioning. With the vane-axial type straightening vanes are used on the output side of the impeller, tube-axial types (Fig. 23.8) are used where air distribution downstream of the fan is not critical.

Centrifugal fans (Figs 23.9 and 23.10) rotate the air and use centrifugal force to generate the airflow. They can generate significantly more pressure

(a) (b)

Figure 23.6 ***Low-pressure, high-volume ducted propeller fan (a) fan blades, (b) in duct (CIBSE/FMA).***

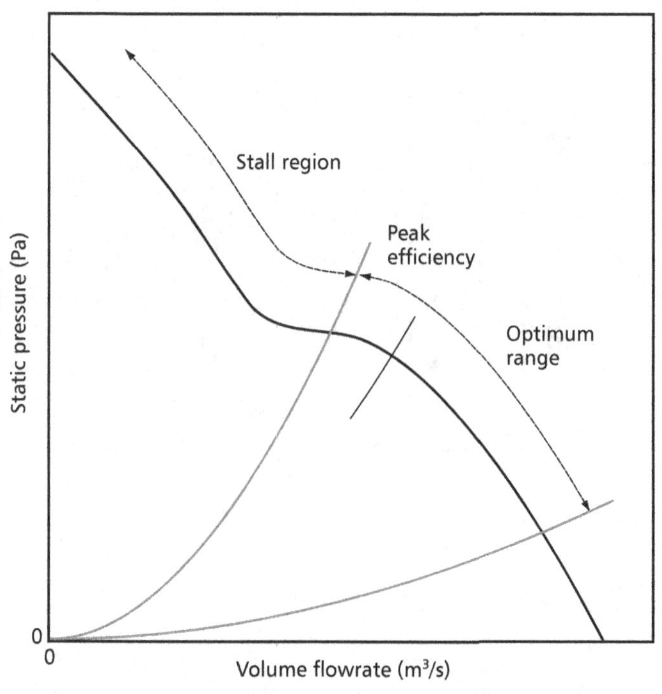

Figure 23.7 ***Typical fan performance characteristic (CIBSE/FMA).***

Figure 23.8 ***Vane- axial fan (CIBSE/FMA).***

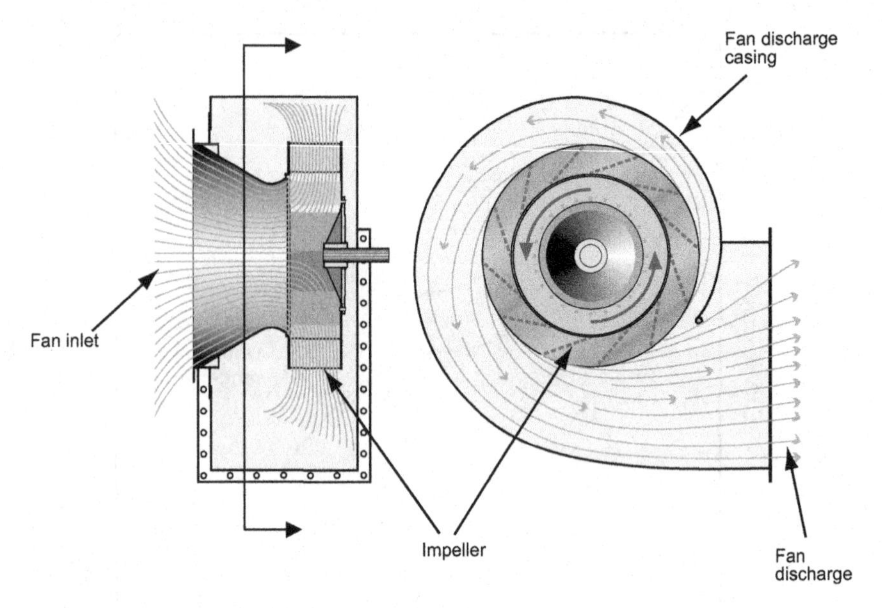

Figure 23.9 ***Cased centrifugal fan (CIBSE/FMA).***

than an axial fan. Most centrifugal fans operate within a casing and this is used to recover a significant portion of the kinetic energy in the flow leaving the impeller, converting it to static pressure. Air leaving the tips of the blades has both radial and tangential velocities, so the shape of the blade determines the fan characteristics. Forward-curved fan blades increase the

Figure 23.10 ***Double inlet centrifugal machine (CIBSE/FMA).***

tangential velocity considerably. As a result, the power required increases with mass flow, although the external resistance pressure is low, and oversize drive motors are required if the system resistance can change in operation. Backward-curved fans have a flatter power curve, since the air leaves the blade at less than the tip speed. This is the most efficient type.

The robust nature of the impeller makes the centrifugal fan much more tolerant of stall operation compared to the axial type, delivering almost the same pressure/volume even when stall is encountered. However, they should not be operated in stall mode because of pressure fluctuations.

The cross flow type has impellers similar to those of the multi-vane forward curved centrifugal, but the action is different. A vortex is formed by the blade forces and has its axis parallel to the shaft and near the impeller circumference. Efficiency is low, but these fans are relatively quiet. Its particular shape is very suitable for many kinds of air-handling devices such as fan coil units and fan convectors.

Fan efficiency varies across the operating range and so it is important to ensure that the fan is correctly selected. The overall efficiency of a fan is the electrical power input compared to the air power output. The specific fan power (SFP) is defined as the power in terms of watts per litre per second of air delivered. This is not usually quoted because the value is dependent on the position on the fan curve. The SFP is affected by the system design, reduced system resistance giving a lower SFP. It is used as a limit for the amount of fan power consumed in a building, and takes into account all the ancillary losses such as variable speed drives.

23.5 FAN LAWS AND FAN CONTROL

The performance of geometrically similar fans can be predicted quite accurately using the fan laws. The fan laws may be summarised as the following; for a fixed diameter and inlet air density:

- volume varies as speed
- pressure varies as $(\text{speed})^2$
- power varies as $(\text{speed})^3$

Similarly for a fixed speed:

- volume varies as $(\text{diameter})^3$
- pressure varies as $(\text{diameter})^2$
- power varies as $(\text{diameter})^5$

The laws apply only to the same point of operation on the fan characteristic, so they cannot be used to predict other points on the fan curve.

They are used most often to calculate changes in flow rate, pressure and power when size or rotational speed is changed.

Airflow control is an essential element of most applications and fans can be wasteful of energy if their operation is not regulated effectively. There are several possible methods of control and the choice of method will depend on its effectiveness and the size of the equipment.

Speed control using a fixed frequency is possible for small induction motors by using a variable voltage speed controller. This is frequently used for single-phase motors. Variable frequency control (inverter) drives can be applied for motors of 150 W upwards, and can reduce the speed down to approximately 20%. These devices will incur their own electrical losses, both in the device and in the motor itself where the electrical waveform deviates from the true sinusoidal form. The effect on the fan itself is summarised in Fig. 23.11.

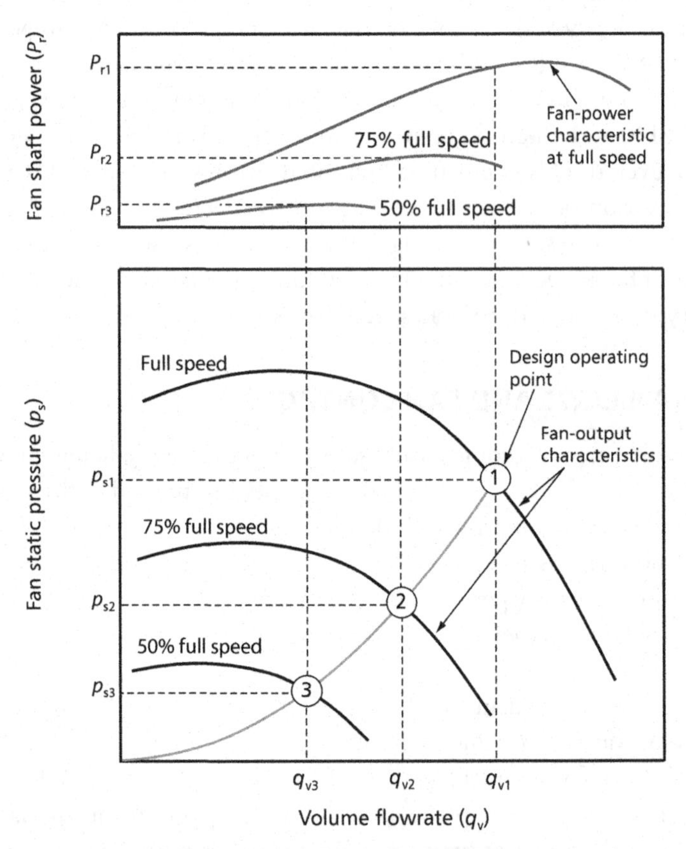

Figure 23.11 ***Flow control by speed regulation (CIBSE/FMA).***

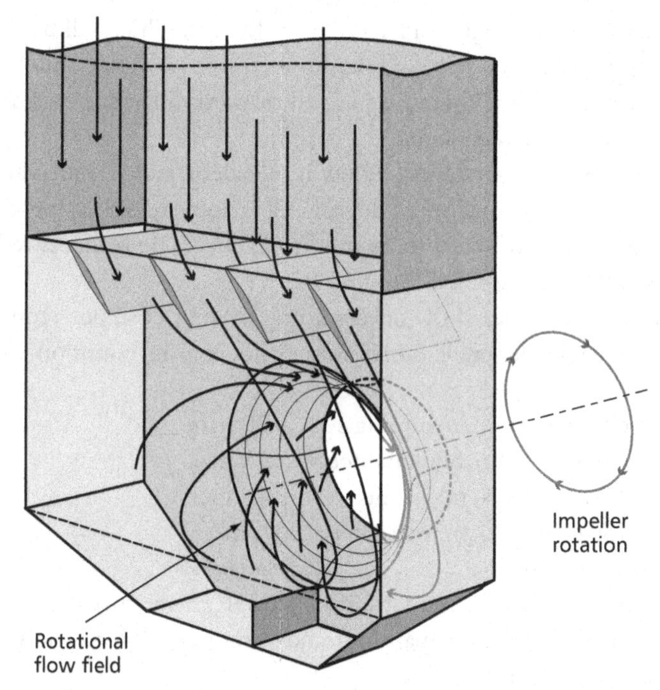

Figure 23.12 ***Pre-rotation of inlet air by louver dampers in a fan inlet box prior to entering the impeller for part-load duty (CIBSE/FMA).***

Alternatives to speed control are alteration of the fan or system geometry. Volume flow control for a fixed speed fan can be achieved with inlet vane angle changes. By spinning the inlet air in the same direction as the impeller high efficiencies can be maintained even at low duties (see Fig. 23.12).

The interaction of the fan blades with the air may be altered by changing the direction of the airflow as it enters or leaves the impeller. Alteration of the blade angle may be a manual process. Fans for which the blade pitch may be varied whilst the impeller is in motion are more sophisticated and can be controlled right down to zero flow. An alternative method is to change the system geometry by using volume control dampers to increase the system resistance. This approach may not achieve worthwhile energy savings.

23.6 FLOW OF AIR IN DUCTS

General laws for the flow of fluids were determined by Reynolds, who recognised two flow patterns, laminar and turbulent. In laminar flow the fluid can be considered as a series of parallel strata, each moving at its own speed, and not mixing. Strata adjacent to walls of the duct will be slowed by friction and will move slowest, while those remote from the walls will move fastest. In turbulent flow there is a general forward movement together with irregular transfer between strata.

In air conditioning systems, all flow is turbulent, and formulas and charts show the resistance to airflow of ducting of various materials, together with fittings and changes of shape to be met in practice. The reader is referred to the tables and charts in CIBSE Guide B2.

High duct velocities show an economy in duct cost, but require more power which will generate more noise. Velocities in common use are as follows:

High-velocity system, main ducts	20 m/s
High-velocity system, branch ducts	15 m/s
Low-velocity system, main ducts	10 m/s
Low-velocity system, branch ducts	6 m/s
Ducts in quiet areas	3–4 m/s

Ducting construction must be stiff enough to retain its shape, be free from air-induced vibration (panting) and strong enough to allow air-tight joints along its length.

The frictional resistance to airflow within a duct system follows the general law

$$H = a\frac{v^2}{d}$$

where a is a coefficient based on the roughness of the duct surface and the density of the air. Duct-sizing charts are based on this law. Since such charts cannot cater for all shapes, they give resistances for circular ducts, and a subsidiary chart shows how to convert rectangular shapes to an equivalent resistance round duct.

It should be noted that the energy for a pressure drop must come from static pressure, since the velocity, and hence the velocity pressure, remains constant.

Frictional resistance to airflow of fittings such as bends, branches and other changes of shape or direction will depend on the shape of the fitting and the velocity, and such figures are tabulated with factors to be multiplied by the velocity pressure. Tables of such factors can be found in standard works of reference.

The sizing of ductwork for a system commences with an assumption of an average pressure-loss figure, based on a working compromise between small ducts with a high-pressure drop and large ducts with a small pressure drop. An initial figure for a commercial air conditioning plant could be 0.8–1.0 Pa/m. This will permit higher velocities in the larger ducts with lower velocity in the branches within the conditioned spaces, where noise may be more noticeable.

Pressure drops for proprietary items such as grilles and filters can be obtained from manufacturers.

An approximate total system resistance can be estimated from the design average duct loss and the maximum duct length, adding the major fittings. However, it is safer to calculate each item and tabulate as shown in Table 23.1 for the system shown in Fig. 23.13. Only the longest branch need be taken for fan pressure.

It will be seen that where there are a number of branches from a main duct, there will be an excess of available pressure in these branches. In order to adjust the airflows on commissioning, dampers will be required in the branch ducts or, as is more usually provided, in the necks of the outlet grilles. The latter arrangement may be noisy if some of these dampers have to be closed very far to balance the airflow, with a resulting high velocity over the grille blades.

23.7 FLOW OF AIR UNDER KINETIC ENERGY

Any static pressure at the outlet of a duct will be lost as the air expands to atmospheric pressure. This expansion, which is very small, will be in all directions, with no perceptible gain of forward velocity. Static pressure can be converted to velocity at the outlet by means of a converging nozzle or by a grille. In both cases the air outlet area is less than the duct area, and extra forward velocity is generated from the static pressure. The leaving air will form a jet, the centre of which will continue to move at its original velocity, the edges being slowed by friction and by entrainment of the surrounding air (see Fig. 23.14). The effect is to form a cone, the edges of which will form an included angle of 20–25°, depending on the initial velocity and the shape of the outlet. Since the total energy of the moving air cannot increase, the velocity will fall as the mass is increased by entrained air, and the jet will lose all appreciable forward velocity when this has fallen to 0.25–0.5 m/s.

Table 23.1 System pressure loss for system shown in Fig. 23.13

Item	Type	Size (mm)	Length (m)	Airflow (m^3/s)	Velocity (m/s)	p_v (Pa)	Resistance factor	Pressure loss (Pa)	p_t (Pa)	p_s (Pa)
1	Inlet louvers	900 × 600	–	1.3	2.41	3.5	0.40	2.1	−2.1	−5.6
2	Duct	900 × 600	2	1.3	2.41	3.5	0.1	0.2	−2.3	−5.8
3	Filter	900 × 600	–	1.3	2.41	3.5	60 Pa	60[a]	−62.3	−65.8
4	Cooling coil	900 × 600	–	1.3	2.41	3.5	97 Pa	97[a]	−159.3	−162.8
5	Reduce	900 × 600 to 500 diameter	–	1.3	6.62	26.3	0.04	1.1	−160.4	−186.7
6	Fan	500 diameter	–	1.3						
7	Enlarge	500 diameter to 600 × 600	–	1.3	3.61	7.8	0.4	3.1	34.1	26.3
8	Duct	600 × 600	8	1.3	3.61	7.8	0.2	1.6	31.0	23.2
9	Branch, straight		–	1.3	3.61	7.8	0.04	0.3	29.4	21.6
10	Duct	600 × 400	6	0.65	2.7	4.4	0.18	1.1	29.1	24.7
11	Outlet grille	600 × 400	–	0.65		4.4		28[a]	28.0	23.6

Required fan pressure = 186.7 + 26.3 = 213 Pa.
[a] Typical catalogue figures.

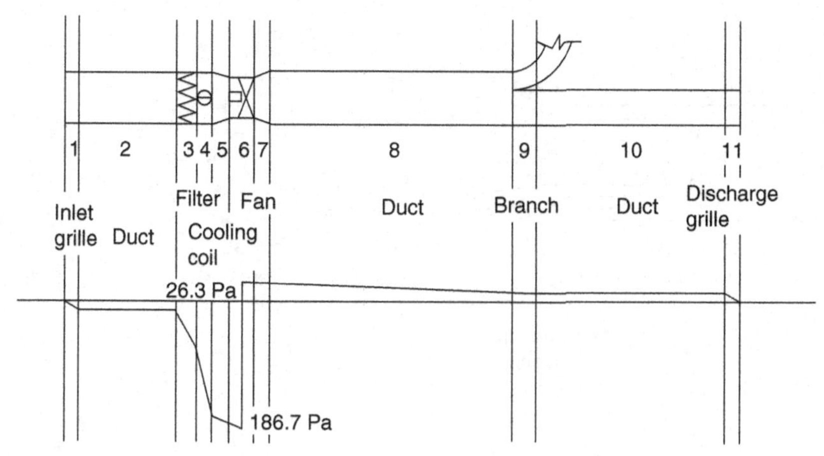

Figure 23.13 ***Ducted system with fittings and fans, showing static pressure.***

If the air in a horizontal jet is warmer or cooler than the surrounding air, it will tend to rise or fall. This effect will lessen as the jet entrains air, but may be important if wide temperature differences have to be used or in large rooms.

If an air jet is released close to a plane surface (ceiling or wall usually), the layer of air closest to the surface will be retarded by friction and the jet will tend to cling to the surface. Use of this effect is made to distribute air across a ceiling from ceiling slots or from grilles high on the walls (see Fig. 23.15). Air is entrained on one side only and the cone angle is about half of that with a free jet. This produces a more coherent flow of input air with a longer throw.

If the air jet is held within a duct expansion having an included angle less than 20°, only duct friction losses will occur. Since there is no entrained air to take up some of the kinetic energy of the jet, a large proportion of the drop in kinetic energy will be regained as static pressure, that is the static pressure within the duct after the expansion will be greater than it was before the expansion (see Fig. 23.16).

The optimum angle for such a duct expansion will depend on the air velocity, since the air must flow smoothly through the transition and not 'break away' from the duct side with consequent turbulence and loss of energy. This included angle is about 14 degrees. With such an expansion, between 50 and 90% of the loss of velocity pressure will be regained as static pressure.

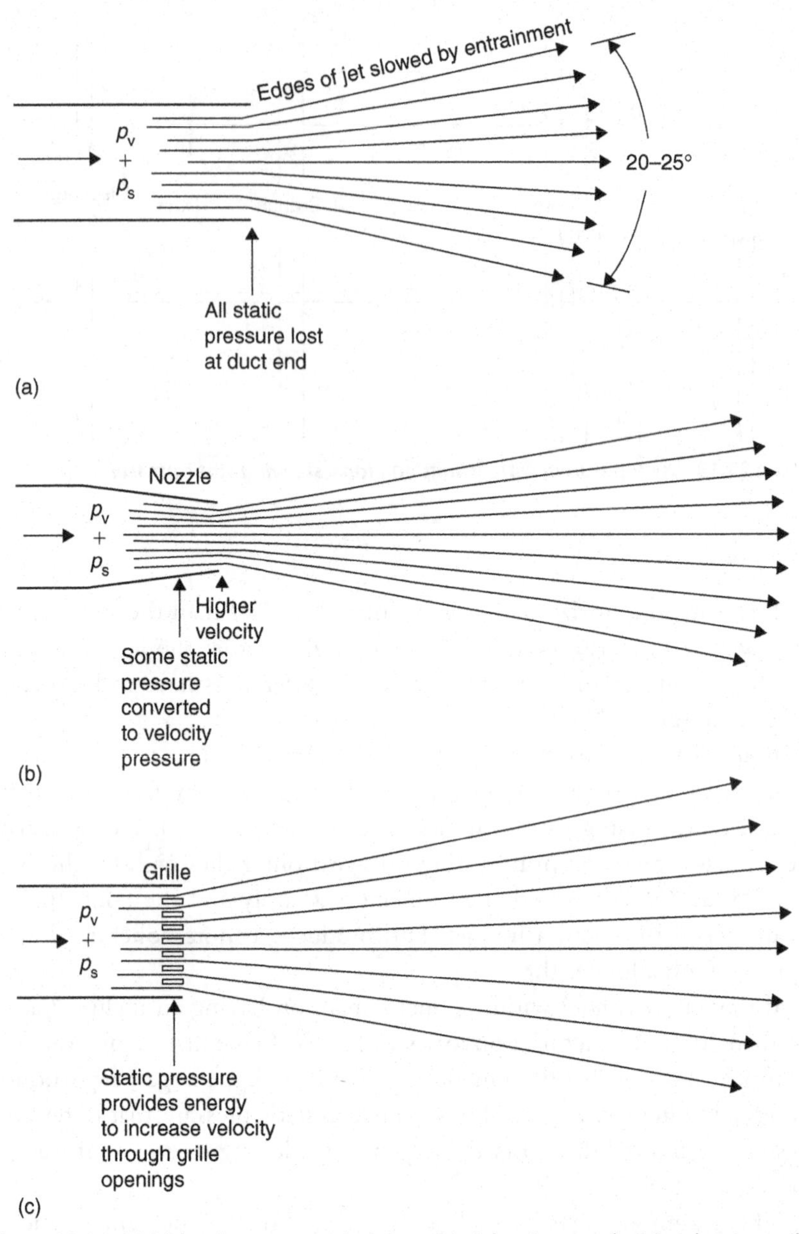

Figure 23.14 (a) Air leaving open-ended duct, (b) air leaving nozzle, (c) air leaving grille.

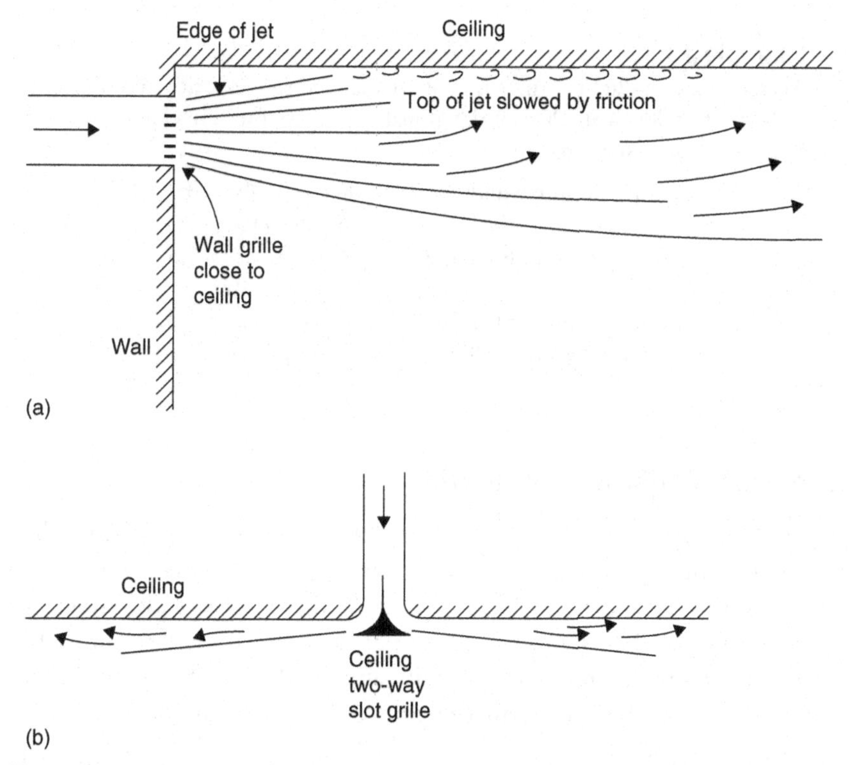

Figure 23.15 ***Restriction of jet angle by adjacent surface.*** (a) Wall grille close to ceiling. (b) Ceiling slots.

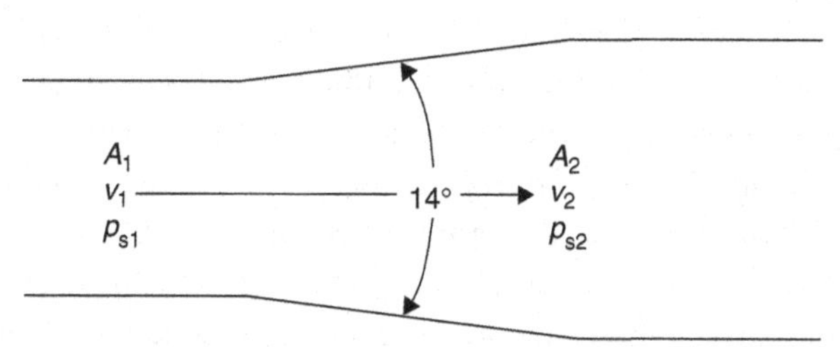

Figure 23.16 ***Duct expansion with static pressure regain.***

Example 23.3

Air moving in a duct at 8 m/s is gently expanded to a velocity of 5.5 m/s. If the friction losses are 20% of the available velocity pressure change, what is the amount of static regain?

$$\text{Velocity presure extering expansion} = 0.5 \times 1.2 \times 8^2 = 38.4\,\text{Pa}$$

$$\text{Velocity presure leaving expansion} = 0.5 \times 1.2 \times 5.5^2 = 18.15\,\text{Pa}$$

$$\text{Friction losses} = 0.2(38.4 - 18.15) = 4.05\,\text{Pa}$$

$$\text{Static regain} = 0.8(38.4 - 18.15) = 16.2\,\text{Pa}$$

23.8 FLOW OF AIR IN A ROOM

Since incoming air may be as much as 11 K colder or 25 K warmer than the conditioned space, the object of the duct and grille system must be to distribute this air and mix it with the room air with the least discomfort to the occupants. The subjective feeling of discomfort will depend on the final temperature difference, the velocity, and the degree of activity, cold air being less acceptable than warm. Velocities at head level should be between 0.1 and 0.45 m/s.

Fig. 23.17a shows a typical office or hotel room with supply duct in the central corridor ceiling space and a wall grille directing air towards the window, which will usually be the greatest source of heat gain or loss. High-level discharges of this type work best when cooling, since the incoming air jet will fall as it crosses the room. On heating it will tend to rise, and so must have enough velocity to set up a forced circulation in the pattern shown.

Fig. 23.17b shows perimeter units under the window and discharging upwards to absorb the heat flow through the window. The angle and velocity of discharge should be enough for the air to set up a circulation within the room to reach the far wall or, in an open-plan room, to the area covered by an adjacent grille. In such units, air returns to the face or underside of the unit.

Fig. 23.17c shows floor slots, setting up a pattern similar to the perimeter units. This arrangement has been adopted in some buildings having all glass walls. The position of the return grille varies with room layout and stagnant zones can occur.

Fig. 23.17d shows ceiling grilles or slots, requiring all ducting to be within the ceiling void. This system is generally adopted for open plan rooms, since the area can be divided into strips with alternate supply and extract slots, or into squares (or near-squares) for supply and extract by grilles.

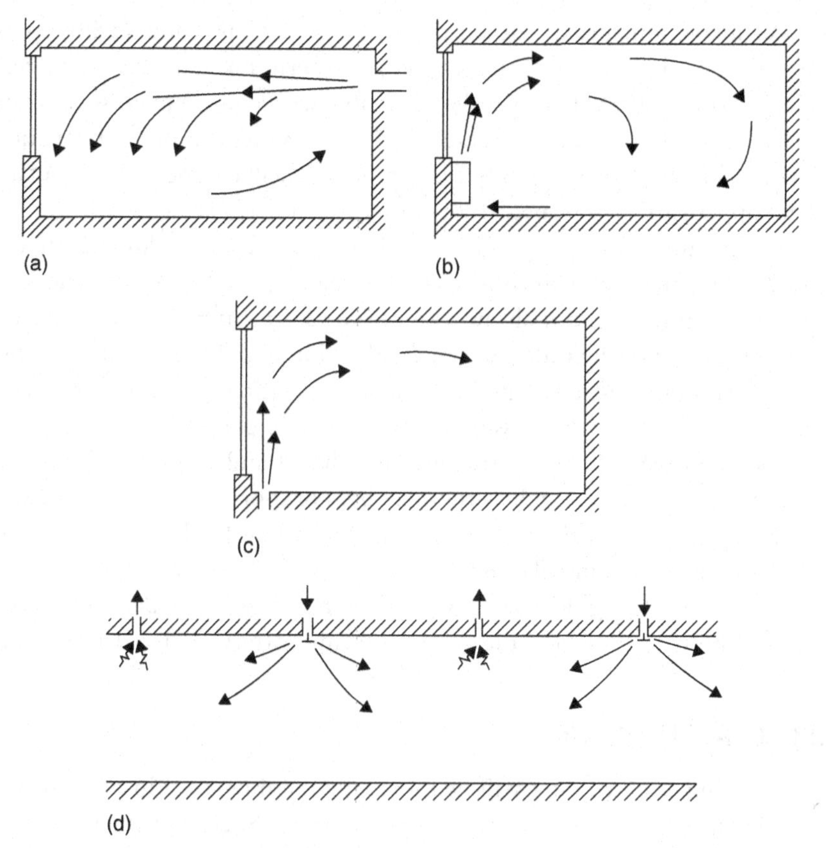

Figure 23.17 ***Room air circulation patterns.*** (a) Grilles on corridor wall, (b) sill outlets, (c) floor outlets, (d) ceiling slots.

It will be seen that all of these patterns require some consideration and planning, since the best equipment cannot cope with an impossible air circuit.

23.9 GRILLES

The air inlet grille is a device for converting static pressure to outlet velocity, having the required speed and direction to take the conditioned air across the room and entrain the surrounding air so as to reach the occupants at a suitable temperature.

Wall grilles have directional vanes in one or both planes, which can be set on commissioning. These need to be set by a competent person who is aware of the required room flow pattern. It is advisable that such a setting adjustment should be operated by special tools to prevent subsequent tampering.

Perimeter unit grilles should direct the air upwards and slightly away from the window to start the circulatory pattern. Fixed angles are preferable on these units, rather than the adjustable segments supplied with many packaged products. A common fault with such an installation is obstruction of inlet and outlet grilles by office equipment. It will be a definite advantage if the upper surface is sloped, to discourage its use as a shelf.

The geometry of ceiling grilles and slots can be fixed or adjustable. In the former case the flow pattern is set by the spacing and volume (=velocity), so site adjustment cannot always compensate for a faulty layout. Slots have a limited throw, of the order of 5 m at full volume, and the layout of supply and return slots must take into account any operating variation in this volume. The setting on commissioning of all these, especially those of adjustable geometry, should be left to competent hands and then locked.

For very large areas, such as assembly halls or sports arenas, jets of air will be required to obtain the large throw distances. Localised draughts may be unavoidable in such installations.

Many air-moving devices now have discharge grilles which move during operation, to distribute the air in a varying pattern, to avoid stagnation.

23.10 RETURN AIR

Air entering a return duct will be moved by the difference in pressure, the duct being at a lower static head than the room. Such movement will be radially towards the inlet and non-directional. At a distance of only 1 m from the grille this pressure gradient will be quite low, so return grilles can be located close to supply grilles, providing the overall circulation pattern ensures coverage of the space. In Fig. 23.17a, the return air grilles can also be in the corridor wall, if far enough from the inlets (see Fig. 23.18).

With ceiling inlet and extract systems, the opportunity is presented to remove heat from light troughs. This can reject a proportion of the cooling load, possibly as high as 20%, in the exhaust air. The recirculated air is also warmer, improving heat transfer at the cooling coil.

23.11 TEXTILE DUCTING

Textile-based air distribution, or 'air socks' as they are affectionately known, are still new to many in the industry although they have been around a long time. By means of strategically positioned nozzles or holes in specially selected materials virtually the whole spectrum of air distribution

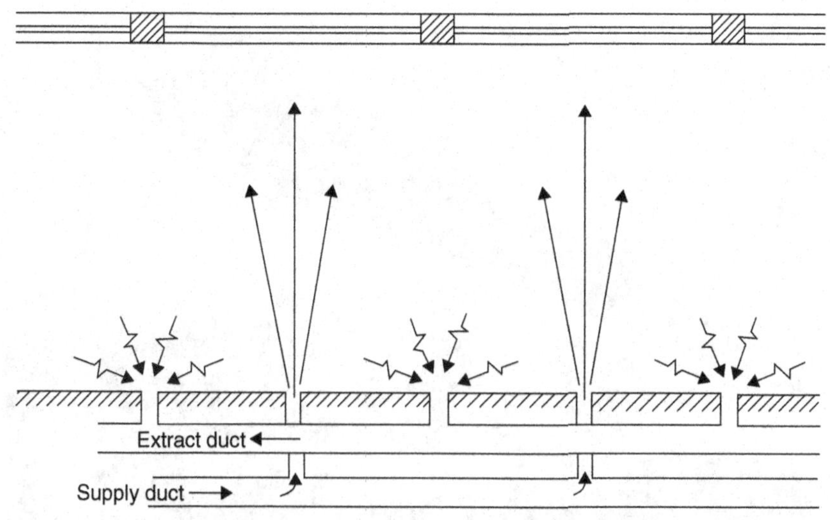

Figure 23.18 ***Discharge and return grilles on same wall (plan view).***

requirements can be satisfied for heating, cooling and ventilation with remarkable results. Air socks can provide uniform air distribution throughout a room and can be made pleasing and interesting to the eye as seen in the gym example in Fig. 23.19 where the colours and lighting are carefully coordinated. A cold room application for good air distribution for meat carcasses can be seen in Fig. 16.2.

Advantages of textile ducting are numerous and include freedom from draughts, quiet operation, low installation cost, removable for laundering, low weight, various shapes and colours that can be changed during refurbishments.

23.12 AIR FILTRATION

Ambient air contains many solid impurities, ranging from visible grit down to fine dusts, smokes and fumes. An air conditioning system will remove a proportion of these, depending on the application. There are three reasons for air filtration:

1. To remove impurities which may be harmful to a process, for example fine dust in a computer room, bacteria in a pharmaceutical packing room.
2. For the comfort of occupants and the cleanliness of papers and furnishings.

Figure 23.19 ***Textile ducting providing air distribution in a fitness centre (Business Edge).***

3. To keep the inside of the air-conditioning apparatus and ductwork clean. Impurities may be classified by size:

Pollens	9–80 μm
Mould spores	3–50 μm
Fine ash	0.7–60 μm
Bacteria	1–10 μm
Tobacco smoke	0.1–7 μm
Viruses	Up to 0.1 μm

Filtration apparatus is available to remove any size, but the very fine particles require a deep, bulky and expensive filter, which itself sets up a high resistance to airflow and therefore requires high fan power. A practical balance must be reached to satisfy the requirements:

1. To remove a high proportion of impurities in the air.

2. To hold a large weight of dust before having to be cleaned or replaced, so as to reduce the frequency of maintenance to an acceptable level (ie if maintenance is required too frequently, it may be neglected).

3. The filter must be cleanable or reasonably cheap to replace.

A high proportion of the weight of dust and fluff in the air is in large particles and so is fairly easy to trap. Filters for general air conditioning duty comprise a felt of glass or other fibres, used in a dry state and termed 'impingement filters'. Air passage through the fibres is turbulent, and dust particles strike the fibres and adhere to them. The filter material may be flat, but is more usually corrugated, so as to present a large surface area within a given face area. A typical filter in a comfort air conditioning system is 50 mm deep and may collect 95% or more of the impurities in the air, down to a size of 1 μm.

Increased dust-holding capacity can be obtained by making the filter material in a series of bags, which are normally about 400 mm deep, but also made up to 900 mm where maximum retaining capacity is required. Some bag filters are shown in Fig. 23.20.

Finer filtration is possible, down to 0.01 μm. Such filter elements are only used when the process demands this high standard. These fine filters would clog quickly with normal-size impurities, so they usually have a coarser filter upstream, to take out the larger dusts. They are about 300 mm deep, and require special mounting frames so that dirty air cannot escape around the edges.

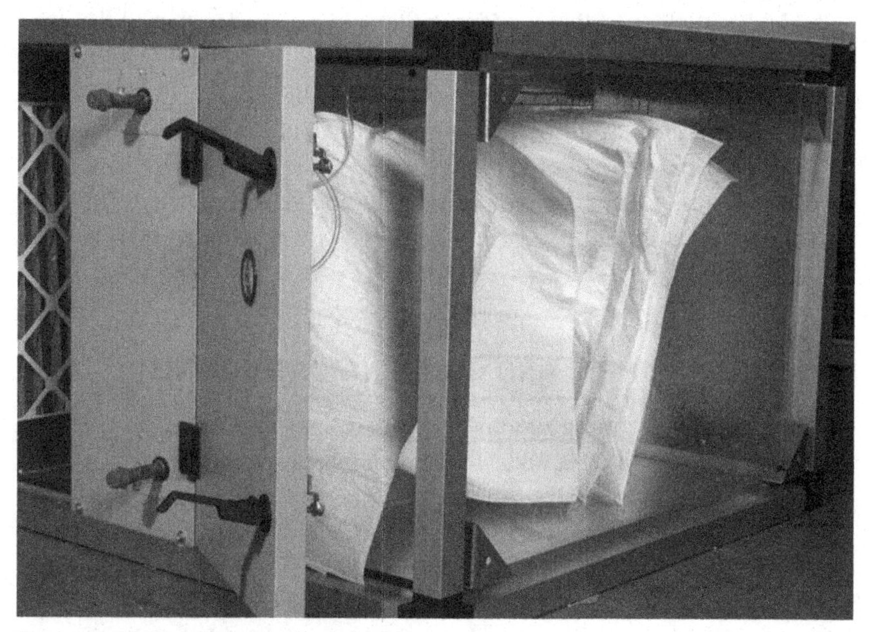

Figure 23.20 ***Bag filters (Business Edge).***

Very fine particles such as smokes can be caught by electrostatic precipitation. A high voltage is applied to plates or wires within the filter bank, to impart a static charge to dirt particles. These will then be attracted to earthed plates, and adhere to them. Impurities are generally cleaned off the plates by removing the stack and washing.

Electrostatic filters will not arrest large particles, and need to be backed up by coarser impingement filters for this purpose.

As a filter element collects dust, the air resistance through it will rise, to a point where the system airflow is impaired. Users need to have an objective indication of this limit, and all filters except those on small package units should be fitted with manometers (see Fig. 23.2). On installation, marks should be set to indicate 'clean' and 'dirty' resistance pressure levels.

Dry impingement filters cannot be effectively cleaned and will usually be replaced when dirty. Thin filters of this type are used on some package air-conditioners and much of the dirt can be dislodged by shaking, or with a vacuum cleaner. The problem of air filtration on small packaged units is the low fan power available and the possible neglect of maintenance. Since users will be reluctant to buy new filters when needed, some form of cleanable filter is employed. One such type is a plastic foam. Where replaceable filters are used, it is good practice to always have a complete spare set ready to insert, and to order another set when these are used. This avoids the inevitable delay which will occur if new filters are not ordered until the need is urgent.

Air filters are not used on cold store coolers, since the air should be a lot cleaner and small amounts of dust will be washed off the fins by condensate or by melted frost. Air-cooled condensers are not fitted with filters, since experience shows that they would never be maintained properly. In dusty areas, condensers should be selected with wide fin spacing, so that they can be cleaned easily.

23.13 CLEANLINESS AND CLEANING OF DUCTING

Filters in air conditioning systems do not remove all the dirt from the air, and this will settle on duct walls. There is an increasing awareness that ducting systems can harbour a great deal of dirt, and that this dirt will hold bacteria, condensed oils such as cooking fats and nicotine, fungi and other contaminants.

Where ducting cannot be stripped down for cleaning, it is strongly advisable to leave frequent access holes for inspection and cleaning.

CHAPTER 24

Air Conditioning Methods and Applications

24.1 INTRODUCTION

There are many ways of achieving the desired air conditions in the working space or defined zone. In all cases the air must pass through a process of some kind where its temperature, humidity, fresh air content and cleanliness is modified depending on the zone requirements. The heat gain is to be removed by a cooling medium which may be air, water, brine or refrigerant, or a combination of two of these (see Fig. 24.1). Many system types are available, starting with the basic window-fitted unit through to the very latest multi-split type equipment.

The choice of system type will depend on several criteria. For example, in a modern, design-conscious office aesthetic requirements may prove of greater importance than the number of control options provided. In a computer room close control of the air temperature and high reliability will be the critical factors. A client looking for the best life-cycle cost will need to balance capital cost with long-term operating costs, efficiency and predicted maintenance costs.

Air conditioning will always include the ability to reduce the temperature and humidity level of the air being processed. Heat pumps designed primarily for heating together with integral systems that provide heating and cooling are dealt with in chapter: Heat Pumps and Integrated Systems.

Heat passes from the occupants or equipment to the air within the space, and from there to the refrigerant or chilled water. Variations in the load due to number of occupants or operation of equipment must be dealt with by the control system. Variations in air temperature and humidity will also occur with the primary cooling medium flow changes. In the simplest case, on–off control of the compressor within an air-conditioning unit will allow the temperature to slowly rise whilst the compressor is 'off' until the compressor restarts.

The design engineer must consider the effect of load and cooling medium variations on the conditions within the space. This governs the selection of the cooling apparatus and method of control. The majority of

Refrigeration, Air Conditioning and Heat Pumps
http://dx.doi.org/10.1016/B978-0-08-100647-4.00024-3

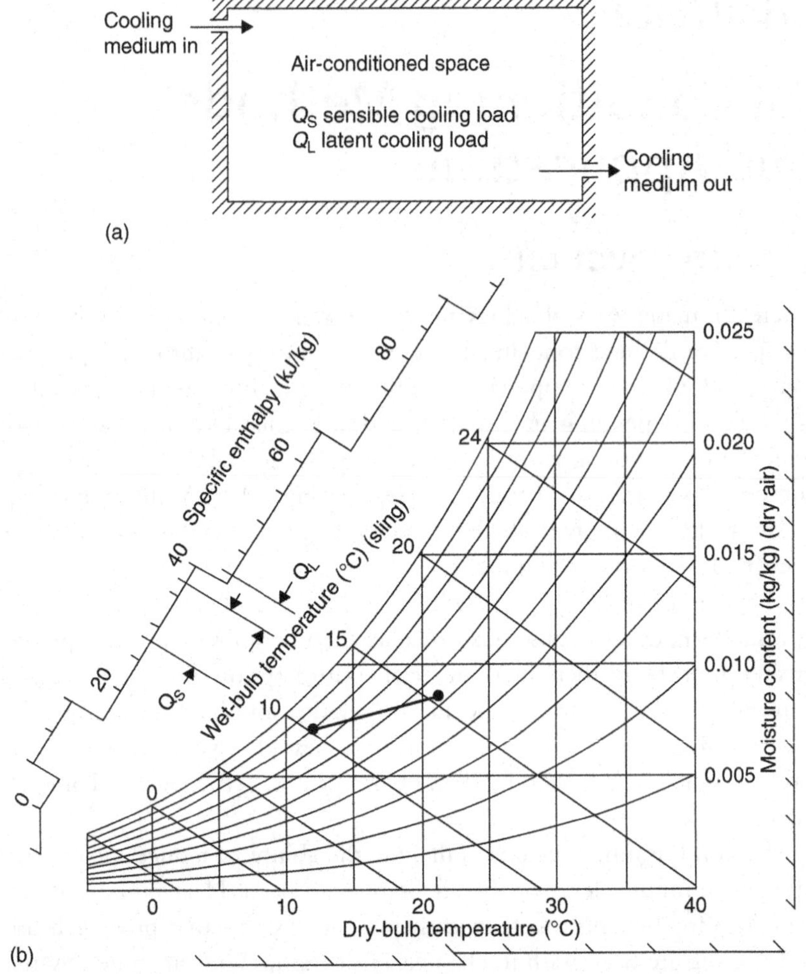

Figure 24.1 ***Removal of sensible and latent heat from conditioned space. (a) Flow of cooling medium, (b) process line.***

air conditioning systems are for ensuring the comfort of individuals in the conditioned zone and this is achieved by control of temperature and percentage saturation levels so that they remain within certain limits, typically as shown in Fig. 19.8.

Close control conditions may require more sophisticated air-distribution systems (see Fig. 24.2), and it may be necessary to restrict movements of human operatives in the sensitive area. Fully modulating coolant flow is advantageous.

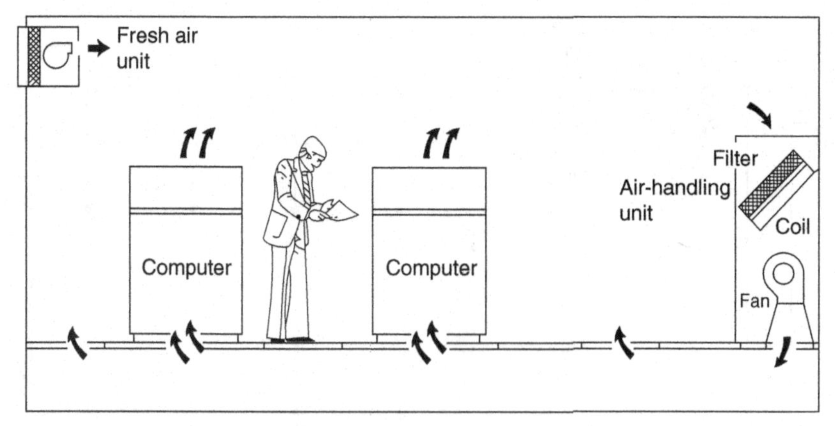

Figure 24.2 ***Raised floor computer room system.***

Regarding specific application groups, two are outlined in this chapter, data centres and dehumidification. Considerations for a full list of specific activities and building types are given in CIBSE Guide B0 Applications.

24.2 AIR HANDLING UNITS

Central air handling units, (AHUs), provide all the conditioned air that is circulated round the building. This type of system is often adopted for theatres, cinemas, factories and large open-plan offices. The units can be used for simple comfort cooling or can be very sophisticated for applications that require close temperature and humidity control, for example standards rooms and environmental test chambers. The basic schematic is shown in Fig. 24.3, and an example is pictured in Fig. 24.4.

The unit is normally built up from a number standard modules with elements such as heat exchangers, filters, etc., designed for specific requirements. It is possible to arrange for inclusion of any proportion of outside air up to 100%. This may be required for some applications, and in any case reduces the refrigeration load in cold weather. When outside air temperatures are suitable, mechanical refrigeration may not be needed. Size of ducting dictates that the plant is located away from the cooled space, and roof mounting is quite common. The cooling plant can be a direct expansion system or a chiller supplying water or glycol, and heat can be provided from direct electric heaters, hot water or steam coils.

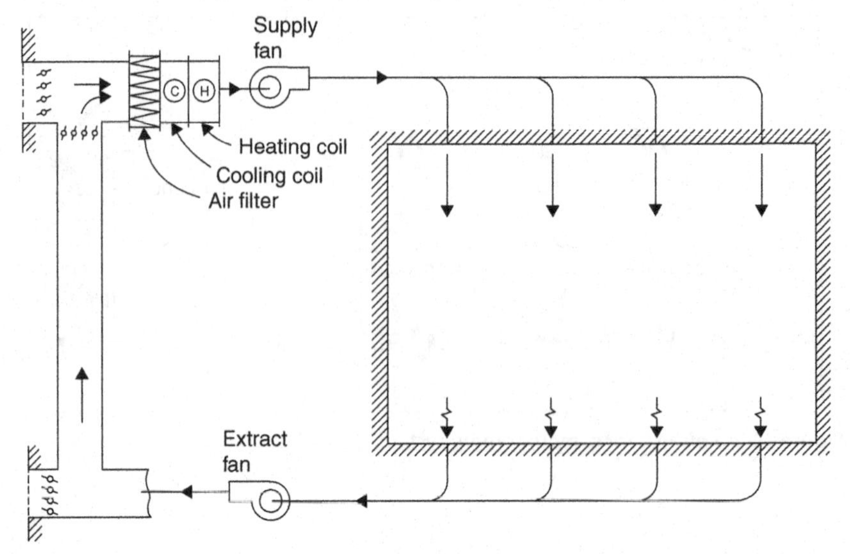

Figure 24.3 ***Central air-handling unit – basic circuit.***

The advantages of this type of unit may be summarised as follows:

- Major controls located together and close to the unit
- Ease of maintenance and fault diagnosis with all mechanical components within the AHU, and less need to work in occupied spaces
- Remote location reduces noise transmission and there is more space for noise control elements and anti-vibration fittings
- Flexibility of construction allows incorporation of many combinations of components and controls within a standardised framework
- Heat recovery in the form of fresh air to exhaust air heat exchangers can be incorporated in the unit.

On the other hand there are some disadvantages including the following:

- Large physical size and space required for ducting
- Zones with heat gain especially where solar gain moves round to different locations can require special controls, and zone heaters may be need to be incorporated into individual duct branches introducing inefficient re-heating of cooled air
- Difficult to alter room usage or change subdivisions
- Can be externally visually intrusive.

The distribution of air over a zone pre-supposes that the sensible and the latent heat loads are reasonably constant over the zone. As soon as large variations exist, it is necessary to provide air cold enough to satisfy the greatest load, and re-heat the air for other areas. Where a central plant serves

Figure 24.4 ***Air handling units providing air conditioning for production plant, total installed capacity 145 kW (Johnson Controls).***

a number of separate rooms and floors, this resolves into a system with re-heat coils in each zone branch duct (see Fig. 24.5). This is waste of energy and to be avoided.

To make the central air system more economical for multi-zone installations, the quantity of cooled air to the individual zones can be made variable and reduced when the cooling load is less. This will also reduce the amount of re-heat needed. An alternative is to supply the re-heat by blending with a variable quantity of warmed air, supplied through a second duct system (see Fig. 24.6).

In the first of these methods, the reduction in air mass flow is limited by considerations of distribution velocities within the rooms, so at light load more air may need to be used, together with more re-heat, to keep air speeds up. Within this constraint, any proportion of sensible and latent heat can be satisfied, to attain correct room conditions. However, full humidity control would be very wasteful in energy and a simple thermostatic control is preferred.

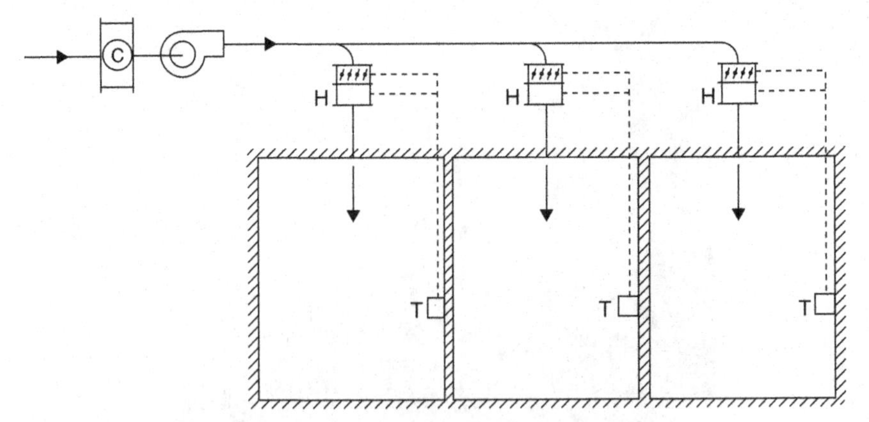

Figure 24.5 ***Variable airflow with re-heat to individual zones.***

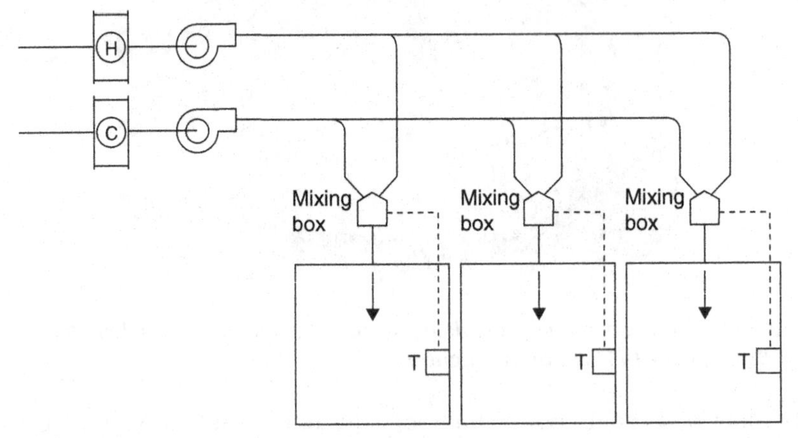

Figure 24.6 ***Dual duct supplying separate zones.***

Example 24.1

A room is to be maintained at 21°C, with a preferred 50% saturation, using air at 13°C dry bulb, 78% saturation and reheat. The load is 0.7 sensible/total ratio (see Fig. 24.7).

Air at the supply condition can be re-heated to about 18°C and will rise from 18°C to 21°C in the room, picking up the quantity of heat 'B' as shown. The final condition will be 50% saturation, as required (line abc).

Alternatively, supply air is used directly, without re-heat. It now picks up the quantity of heat 'A' (about three times as much) and only one-third the amount of air is needed. The final condition will be about 55% saturation. This is still well within comfort conditions and should be acceptable (line shown in Fig. 24.7).

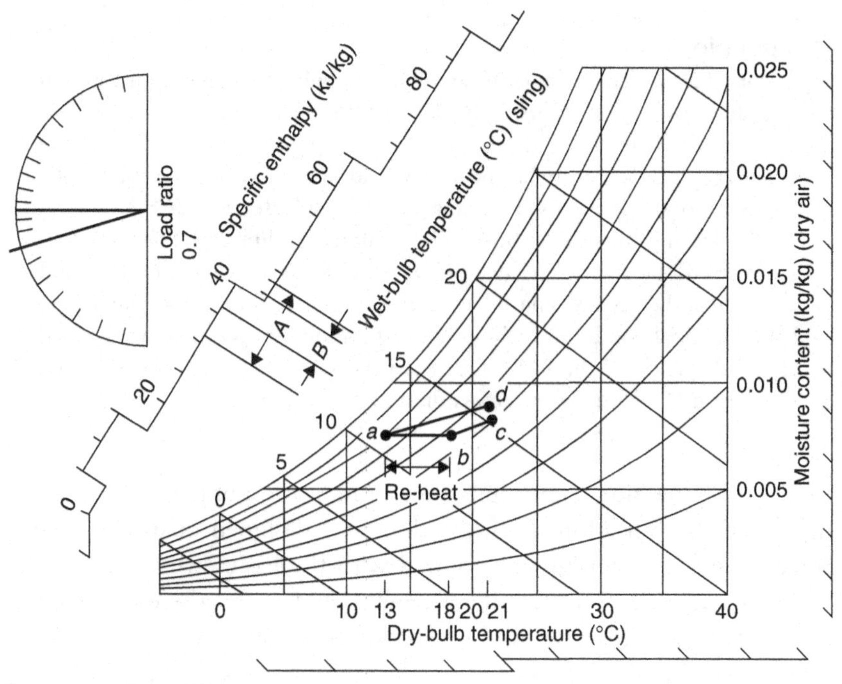

Figure 24.7 ***Zone differences with re-heat.***

With this variable volume method, the cold-air supply system will be required to deliver less air into the building during colder weather and must be capable of this degree of 'turn-down'. Below 30% of design flow it may be necessary to spill air back to the return duct, with loss of energy and, in some types, cold air in the ceiling void when trying to heat the room. If the final throttling is at the inlet grille, the reduction in grille area will give a higher outlet velocity, which will help to keep up the room circulation, even at lower mass flow.

The *dual-duct* system, having the second method of heating by blending cold and warm air, has reached a considerable degree of sophistication, normally being accommodated within the false ceiling and having cold and warm air ducts supplying a mixing chamber and thence through ceiling grilles or slots into the zone (see Fig. 24.6).

The blending of cold and warm air is thermostatically controlled, so that the humidity in each zone must be allowed to float, being lowest in the zones with the highest sensible heat ratio.

Example 24.2

A dual-duct system supplies air at 14°C dry bulb, 75% saturation through one duct, and at 25°C dry bulb, 45% saturation through the other. Two zones are to be maintained at 21°C and in both cases air leaves the mixing boxes at 17°C. Room A has no latent load. Room B has a sensible/total heat ratio of 0.7. What room conditions will result (see Fig. 24.8)?

Air leaving the mixing boxes will lie along the line *HA*. For these two zones it will be at *B* (17°C dry bulb). For room A, air will enter at *B* and leave at *C*, the process line being horizontal, since there is no latent heat load. The final condition is about 50% saturation. For room B, air enters at *B* and the slope of the line *BD* is from the sensible/total angle indicator. Condition D falls at about 56% saturation.

The aforementioned example gives an indication of the small and usually acceptable variations found with a well-designed dual-duct system. Since a constant total flow is required with the basic dual-duct circuit, a single fan may be used, supplying cooling and heating branches. Where variation of volume is employed, one or two fans may be used, as convenient

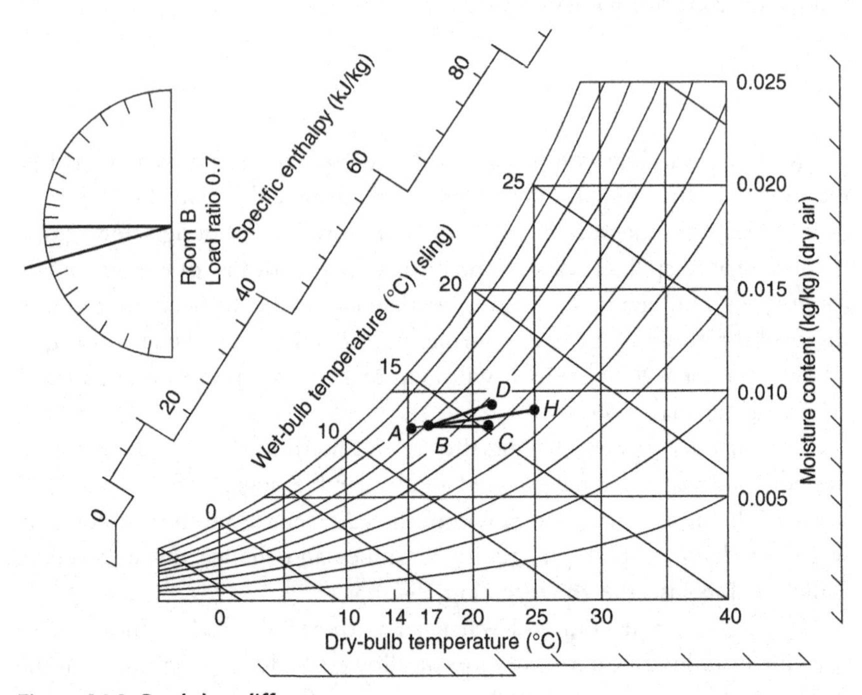

Figure 24.8 ***Dual-duct differences.***

for the circuit. In all cases an independent extract fan and duct system is required, so that the proportion of outside/re-circulated air can be controlled.

Since about 0.1 m^3/s of airflow is required for each kilowatt of cooling, the mass airflow for a large central station system is large and the ductwork to take this is very bulky. This represents a loss of available building space, in terms of both vertical feed ducts and the extra ceiling space to accommodate branches on each floor.

Reduction of duct size can be achieved by increasing the velocity from 3–6.5 m/s to 12–30 m/s. Such velocities cause much higher pressure losses, requiring pressures in excess of 1 kPa, for which ductwork must be carefully designed and installed, to conserve energy and avoid leakage. The use of high velocity is restricted to the supply ducts and is not practical for return air ducting.

With a supply system pressure of 1 kPa and another 250 Pa for the return air duct, the total fan energy of a central all-air system may amount to 12.5% of the maximum-installed cooling load, and a much greater proportion of the average operating load. This power load can only be reduced by careful attention to design factors.

If the conditioned space can be broken down into a number of zones or areas in which the load is fairly constant, then a single-zone AHU with localised ductwork may be able to satisfy conditions without re-heat in its branches. The success of such a system will depend on the selection of the zones. Large open offices can be considered as one zone, unless windows on adjacent or opposite walls cause a diurnal change in solar load. In such cases, it will be better to split the floor into arbitrary areas, depending on the aspect of the windows. Some local variations will occur and there may be 'hot spots' close to the windows, but conditions should generally be acceptable by comfort standards.

24.3 FAN COIL UNITS

As its name implies, a fan coil unit consists of a heat exchanger in which water is circulated and a fan assembly, incorporating a filter and simple controls, designed for wall perimeter units (Fig. 24.9) or ceiling units mounted within ceiling voids. Ceiling units can be configured as a cassette, drawing air into the centre and discharging at the periphery. Heating elements, electric, hot water or steam can be included.

The chilled water is fed to a number of units, each sized for a suitable zone, where the conditions throughout the zone can be satisfied by the

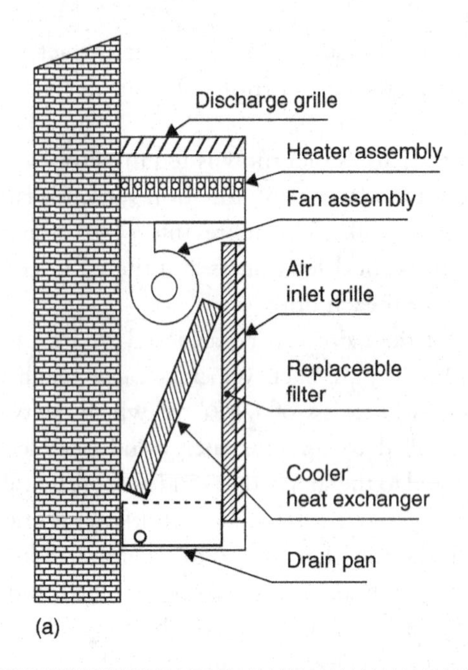

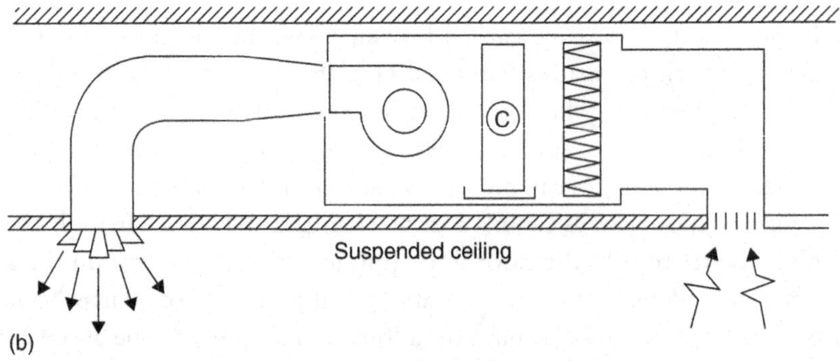

Figure 24.9 ***Fan coil units.*** (a) Wall mounted, (b) ceiling mounted.

outlet air from the unit. This offers a wide range of comfort conditions within the space, with units serving a single room, or part of a room. The coil is normally operated with a fin temperature below room dew point, so that some latent heat is removed by the coil, which requires a condensate drain. Multi-speed fans are usual, so that the noise level can be reduced at times of light load.

The advantages of this type of system are the following:

- Individual control for zone or office, including heating in some zones and cooling in others
- Relatively low cost of standard units

- Simple control system
- Built in standby capacity where several units are located in one zone.
  The disadvantages are as follows:
- Limited flexibility with standard units – all operating parameters fixed by the manufacturer
- Normally only simple dry bulb control is provided, although some specialist units incorporate electronic control systems and humidifiers
- Limited control of fresh air input, if any, so that advantage of free cooling cannot be taken
- Noise levels may be higher than central air handling units
- Limited ability to control air distribution
- Visibility of the units and connecting chilled water services may be an issue
- Maintenance and servicing can disrupt occupants of the room and there is a risk of damage to the room fabric by virtue of all services and connections within the room.

Primary (fresh) air supply can be used to induce room (secondary) air circulation. This air, at a pressure of 150–500 Pa, is released through nozzles within the coil assembly, and the resulting outlet velocity of 16–30 m/s entrains or induces room air to give a total circulation four or five times as much as the primary supply. This extra air passes over the chilled water coil. Most induction units are wall mounted for perimeter cooling (Fig. 24.10) but they have been adapted for ceiling mounting. With induction units, latent heat extraction can usually be handled by the primary air and they run with dry coils. Some systems have been installed having high latent loads which remove condensate at the coil.

24.4 INTEGRAL UNITS

A unit is defined as integral when all the system components are combined into one package. Most people are familiar with the 'through the window' units seen in individual offices, small shops and domestic locations. These units are usually in the capacity range 0.75 to 6 kW. Larger units of this type are frequently installed in warehouses and factories and may have capacities of up to 100 kW. The advantages of this system are low cost, space saving and ability to provide heating, whilst the disadvantages are large holes through the building structure, high noise level and limited humidity control.

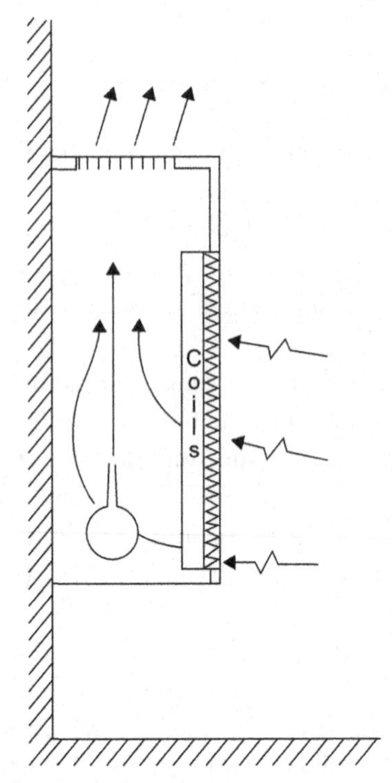

Figure 24.10 ***Fan coil induction unit.***

24.5 UNDERFLOOR SYSTEMS

A room containing heat-generating equipment such as computers may have a high cooling load, and require a high airflow to carry this heat away. If this amount of air was circulated in the usual way, it would be unpleasantly draughty for the occupants.

With computer cabinets on the floor, the general solution is to use fans to pressurise the space beneath a false floor with cool air and propel the air directly into the cabinets from below. Another supply may enter the room to deal with other heat loads. The air conditioning unit is floor mounted, taking warmed air from the upper part of the room (see Fig. 24.2).

Such units may use chilled water or direct-expansion refrigerant and will have the air filter at the top. If it is not possible to introduce outside air through it, the room will have a pressurised, filtered fresh air supply. Computer room units work with a very high sensible heat ratio of 0.95 or more, so they have large coils to keep the humidity near the dew point of the room air. Larger data centres are a specialised application (Section 24.9).

24.6 STATIC COOLING DEVICES

Chilled beams and chilled ceilings are static cooling devices in which chilled water is circulated to ceiling panels or beams. The beams can consist of a chilled water pipe with flat fins. An aluminium plate often encases each of two sides of the assembly and the bottom may have an aluminium plate with slots. Both the chilled ceiling and chilled beam provide sensible cooling only to the space, the heat exchange being by convection and radiation. In order to avoid the risk of condensation, the chilled water temperature is higher than for other methods, and whilst this improves the efficiency, it increases the surface area required for extraction of heat from the cooled space.

The advantages of static cooling are absence of noise, high-energy efficiency and low maintenance, whilst condensation risk is a potential issue.

24.7 PACKAGED UNITS AND SPLIT SYSTEMS

These air conditioners are characterised by being relatively small self-contained units, or split units that together make a complete functional unit, containing all the components necessary to provide space cooling (see chapter: Packaged Units). Larger packaged or split units become multi-splits when more than one evaporator is employed. Packaged and split air conditioners include their own integral controls and are therefore relatively simply installed wherever their cooling function is required.

Multi-split is a general name applied to packaged systems where two or more remote coolers are run from one condensing unit. Liquid from the condenser passes directly through an expansion device, and the resulting mixture of cold liquid and flash gas is distributed to each of the fan-coil units on the circuit. On–off control of the cold liquid to each room is effected by a solenoid valve within each indoor unit, controlled by a room thermostat. Return refrigerant gas passes through a suction trap and/or a suction/liquid heat exchanger in order to ensure that un-evaporated refrigerant is vaporised before entering the compressor. With this type of system both the outgoing and return refrigerant pipes to each fan-coil unit must be insulated.

Some method of shedding the excess cooling capacity is necessary for when some of the fan coil units do not call for cooling. Excess compressor capacity is sometimes shed by injecting hot gas directly from the compressor discharge into the return.

Inverter speed control of the compressor allows reduction to 40% of full speed or less. Some can also be run at higher speeds than normal, to deal with peak loads. As many as eight fan-coil units can be connected to one condensing unit, and the installed number can be nominally greater than the compressor capacity, on the basis that not all will be cooling at the same time.

For further information on variable refrigerant flow (VRF) systems please see chapter: Heat Pumps and Integrated Systems.

24.8 TRANSPORT AIR CONDITIONING

Automotive air conditioning is now almost universal. Mobile air conditioning for cars (MAC) systems generally use engine-driven compressors with refrigerant R134a, and these have separate category under the F-Gas regulations. This will result in the phase-out of R134a and new fluids having a GWP of less than 150 are being developed. Links to the regulatory requirements can be found on the ACRIB web site.

For rail vehicle air conditioning, specially designed split systems are used. There is at least one separate system for each vehicle. Semi-hermetic piston, hermetic scroll and small hermetic screw compressors are used. Horizontal scrolls have been developed which allow the compressors to fit inside roof pods.

24.9 DATA CENTRE COOLING APPLICATIONS

Data centres are remote servers communicating with local or distant computers or other media devices. They may be housed within a building or department within a building such as a University, or in a purpose built facility. Air is normally circulated to cool their electronics. Large dedicated data centres that handle the vast amounts of data now being generated are housed in secure, frequently anonymous buildings. In these buildings servers handle all the electronic communications materials, messages, attachments, videos, films including all the junk that is now part of our everyday lives. The data explosion has led to the term Big Data to describe data collections too large or complex to handle by traditional methods, yet increasing in size at an ever growing rate. It has been estimated in a 2011 report by Datacenter Dynamics that the world's data centres currently use 31 gigawatts of power, the equivalent of about half of the total peak electricity demand in United Kingdom.

With this scale of energy demand it is not surprising that the cooling methods are being actively developed to reduce the cooling power consumption element. Energy efficiency is important but still secondary to the key requirement for resilience of equipment and of services. The energy efficiency of a data centre can be expressed as *power usage effectiveness* (PUE)

$$\text{PUE} = \frac{\text{Total power input}}{\text{IT power input}}$$

Whilst this can be a useful benchmark, care must be taken with any PUE comparisons because of the following reasons:

1. Total power input includes all ancillary items such as lighting as well as the cooling power
2. No account is taken of outdoor conditions that may vary during the day and year, and are location dependent
3. Size dependency is not accounted
4. It does give information about the efficiency of processing of a specific volume of data

Earlier installations comprised Computer Room Air Conditioning units, known as *CRAC units* with cool air circulated via a pressurised void beneath the floor as described in Section 24.5. In recent years more powerful electronic chips have been developed, consuming more power and increasing the intensity of the heat energy released. This can give rise to a heat density of 50 kW/m^2 compared with a conventional CRAC unit capability of 1.5 to 4 kW/m^2. Data centre cooling is a rapidly developing application area, particularly with the following technologies.

Good airflow management prevents the recirculation of hot air exhausted from IT equipment and reducing bypass airflow. There are several methods of separating hot and cold airstreams, such as hot/cold aisle containment and in-row cooling units. This enhances efficiency and avoids reduced server performance or equipment damage due to hot exhaust air finding its way into an air inlet. Server spaces not in use should be blanked off. Atmospheric stratification can require setting cooling equipment temperatures lower than recommended. CFD techniques are being applied to optimise airflow patterns. A detailed study of air circulation metrics is given by Flucker and Tozer (2011).

Free cooling and evaporative cooling can be effective in PUE improvements. An outline of free cooling is given in chapter: Efficiency, Running Cost and Carbon Footprint, and evaporative cooling is in chapter: Practical

Air Treatment. Direct use can be made of outside air. Air below 24°C is drawn in and delivered, after suitable filtration to cold aisles from where it passes through the IT processors to the warm aisle from which it is exhausted to atmosphere. A conventional cooling system takes over when the outside air temperature is above 24°C. Another approach is to use an economiser coil on computer room air handlers or DX air conditioners, or a centralised cooling tower and heat exchanger that works alongside the chiller. Water can be used instead of air to provide a free cooling element and improve efficiency of conventional systems.

The aforementioned methods can be effectively used for heat densities up to about 5 kW/m^2. Above that, the following approaches are being developed.

In row cooling or direct-cabinet cooling is targeted cooling that precisely cools and conditions air in close proximity to the server cabinets. The coolers can be installed on the floor or suspended from overhead making them closer to the actual rack. This offers both capacity and efficiency gains. Neither cool air nor warm exhaust air has far to travel, allowing the units to dissipate high heat loads.

Carbon dioxide evaporators within the cabinets is being used as a method for localised cooling of the air stream. CO_2, pumped to the evaporators as a secondary fluid, is a very compact way of delivering energy absorbing capacity. Cooling of workstations on dealer floors where space precludes air ducts, can also benefit from this approach. The heat load generated per deck can reach over 2 kW and there can be up to 500 traders in a large dealer room (Fig. 24.11)

Liquid cooling uses the exceptional thermal conductivity of a special liquid to provide dense, concentrated cooling to targeted small surface areas. Dependence on fans and expensive air conditioning and air handling systems is drastically reduced. This enables over 45 kW densities per rack using water cooling and reducing power consumption. Liquid cooling solutions are installed either directly into enclosures or mounted into data centre spaces.

Systems are being researched that use a non-flammable non-conductive liquid coolant that can be in direct contact with electronics. A simple low energy pump, located at the bottom of the cabinet, pumps a secondary coolant (water) to the top where it cascades down throughout all the modules due to gravity. The secondary coolant terminates at heat exchangers within the cabinet for transfer of heat to a third and final coolant, on an external loop, taking the heat away for discard or re-use.

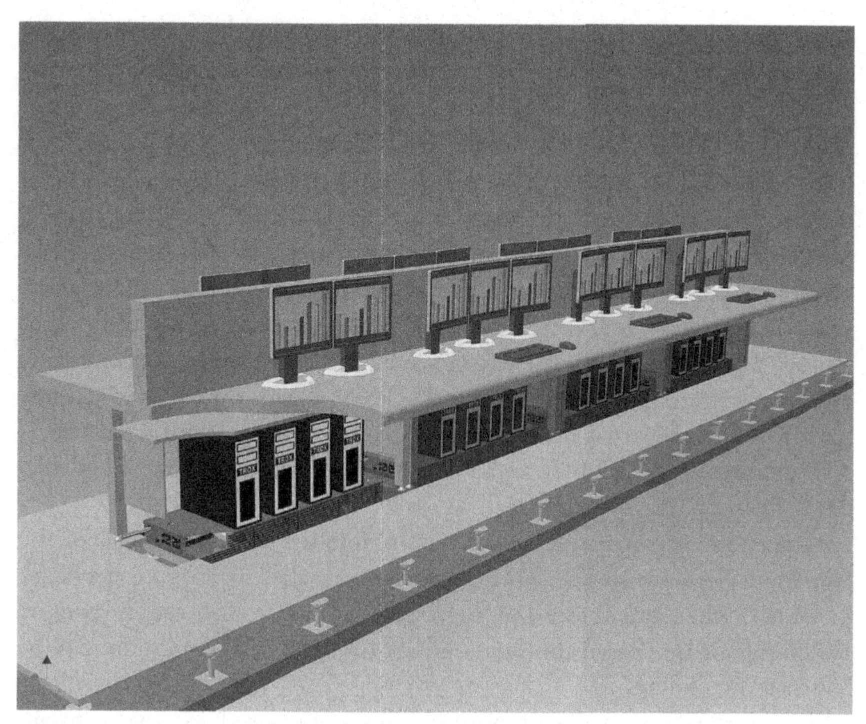

Figure 24.11 *Dealer desks with built in CO_2 coolers (Star Refrigeration).*

24.10 DEHUMIFICATION APPLICATIONS

Packaged one-piece dehumidifiers are used for the following:

1. Maintaining a dry atmosphere for the storage of metals, cardboard, books, timber, etc., that is, any material which is better preserved in low humidity.
2. Removal of moisture from newly constructed or plastered buildings, to expedite final decoration and occupation.
3. Drying out buildings which have been left unoccupied for some time, or have a condensation problem.
4. Removal of excess moisture from indoor swimming pools
5. Moisture control in cold stores (see chapter: Cold Storage and Refrigeration Load Estimation).
6. Some crop drying.

The drying load in a swimming pool varies throughout the year, and dehumidifiers built for this application may have an air cooled condenser for re-heating the air, and also a water-cooled condenser, so that some proportion of the heat may be used to warm the pool water. Automatic controls enable the condenser heat to be used the best advantage.

24.10.1 Drying of Compressed Air

If the air pressure is increased, the partial pressure of the moisture goes up in the same proportion, and more moisture can be removed without frosting the cooling surface. Air-drying evaporators for pressures above atmospheric are designed as pressure vessels, and take the form of shell-and-tube, shell-and-coil or plates. Such driers are found on compressed air installations, to remove moisture from the air which would otherwise settle in distribution piping, valves and pneumatic machines, causing corrosion which is accelerated by the high partial pressure of oxygen. By means of refrigerated driers, compressed air at 7 bar can be dried to a moisture content of less than 0.001 kg/kg.

Depending on the end use of the compressed air, some or all of the condenser heat can be used to re-heat the cold air. This may be necessary in winter, when distribution piping could be colder than the evaporator. When the air is released through a power tool, the final condition may be less than 5% saturation.

Unit driers for small compressed air systems need to have capacity control, in order to maintain a steady working dew point when there is a variation in air demand.

24.10.2 High-Temperature Dehumidifiers

The kilning, or accelerated drying, of newly cut timber requires higher temperatures than will usually be found in refrigeration and air-conditioning systems. Typically, the air is above 50°C and may be up to 80°C. Condensing temperatures of 85–90°C require use of R134a, or the use of a more specialised refrigerant.

Similar high-temperature dehumidification has been used in the drying of other fibrous materials and ceramics.

CHAPTER 25

Heat Pumps and Integrated Systems

25.1 INTRODUCTION

Refrigeration technology concerns the movement of heat energy from a lower to higher temperature level. Increasingly, familiar and sometimes less familiar techniques are being applied that utilise the high temperature heat for various purposes, mainly space and water heating. The term heat pump is used when the primary purpose of the system is to deliver heat. Sometimes the term *dedicated* Heat Pump is used to distinguish it from a reversible air conditioning unit because these units very often can provide heating, although cooling is the primary functionality. The industry has long been dealing with the heating and cooling of buildings as separate entities with separate groups of designers, specifiers and installers for each discipline. Heating would be provided by primary fuels such as oil, gas or electric resistance, and cooling with dedicated air conditioning systems discarding heat to atmosphere. This has even led to the absurd situation where both systems are operational, working against each other. Realisation that heat pumps can deliver heat with less carbon emissions and at lower cost than direct fossil fuel heating has given impetus to the industry. Installation costs can be a barrier, but may be offset by (frequently changing) government initiatives. Moreover, heat from refrigeration applications, previously discarded, is being usefully recovered. In this chapter the techniques and constraints are considered, together with some details of specific systems and potential development possibilities.

25.2 HEAT RECOVERY AND REVERSIBLE AIR CONDITIONERS

Mechanical heat recovery systems utilise heat rejected from a cooling process and thus provide useful cooling and heating simultaneously. The term *heat recovery* is also used to describe passive systems that recover heating or cooling which would otherwise be lost due to the need to maintain ventilation.

Refrigeration, Air Conditioning and Heat Pumps
http://dx.doi.org/10.1016/B978-0-08-100647-4.00025-5

Opportunities for heat recovery may arise when the following happen:

1. Cooled or warmed air is exhausted from a building or process and might be used in heat exchangers to pre-cool or pre-warm fresh air brought in to replace it.
2. Cooled or warmed liquids leave a process, and heat exchangers can provide the means of pre-cooling or pre-warming fresh liquids entering.
3. Hot discharge gas from a refrigeration circuit can be used to heat water.
4. Condenser heat can be diverted into a building, for heating in winter.

Passive heat exchanger equipment for air heat exchange is described in Section 22.4 and shown in Fig. 22.5. This can only be used where the ducts are adjacent. Other methods are the following:

1. The rotating wheel heat exchanger. The wheel has a rotating matrix that picks up heat from one duct air flow and transfers it to the other. If the matrix is coated in a hygroscopic material, there can also be some transfer of moisture, see chapter: The Refrigeration Cycle, desiccant cooling. The thermal wheel is illustrated in Fig. 25.1
2. Coils in the two ducts, with a fluid circulated between them. It is important to get the fluid in counterflow in the two airstreams. If fresh air is to be heated in winter, the fluid must be a non-freeze solution.
3. Heat pipes between the two ducts. These comprise a coil made with closed pipes, filled with a volatile liquid. This liquid will condense in one coil and evaporate in the other, at the same pressure and therefore at the same temperature.

All these methods will transfer heat in either direction, providing heat recovery in summer and winter. All devices using air should be protected by filters, or they will choke with dirt and become ineffective. Heat exchangers for liquids may be double-pipe, shell-and-tube or plate type. Waste fluids

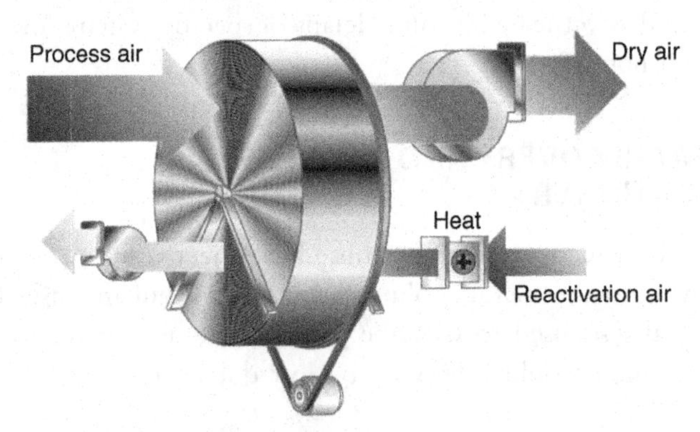

Figure 25.1 ***Thermal wheel (Business Edge).***

may be contaminated by the process, and heat exchangers for such fluids must be cleanable, and kept clean. All heat exchange equipment should be fitted with temperature sensors at the inlet and outlet of both fluids, so that operation can be monitored.

If the flow of refrigerant in a cooling vapour compression cycle is reversed, the heat exchanger which was the evaporator becomes the condenser and vice versa. A *reversing valve* is used to enable a system to operate in either mode. In Fig. 25.2a, a simple reversible air/air system is shown in cooling mode. Hot, high pressure gas from the compressor is directed to the outdoor unit acting as a condenser, rejecting heat to the atmosphere. The condensed liquid flows through the check valve and into the indoor unit via the expansion valve, where it evaporates at low pressure cooling the indoor air. When the piston in the reversing valve moves up (Fig. 25.2b) the ports are changed so that the flow in the system is reversed and the unit heats the building.

Equipment which can provide either cooling or heating includes air/air, air/water and water/water types. Some air/air packaged systems have the ability to provide simultaneous heating and cooling and an example of this type is described in Section 24.2. Air/water chillers may use water circuits and fan coil units that can operate either on chilled or heated water. Water/water systems cover a very large size range, and installations may have several packaged chillers with the water circuits piped to provide heating and cooling simultaneously to various sections of a building on demand. An outstanding example of an installation of this type is described by Macklin (2006).

25.3 GROUND SOURCE HEAT PUMPS

The HP type and performance is characterised by the nature of the source, which is the medium from which heat is extracted by the evaporator, and the sink, which is the distribution of heat in the building.

Ground source heat pumps (GSHP) use ground (sometimes termed geothermal) or water as a source of heat. Water is probably the best source, if available, and comes under the general heading of ground source. Some notably successful HPs use nearby lakes or sea as a source. Water (or preferably containing some glycol to prevent freezing) is circulated through a coil submerged in the lake and it picks up heat so that it arrives at the evaporator in the HP at a temperature close to the lake temperature. The advantage of water as a source is that the temperature is quite stable and usually above 10°C. Lake water or ground water may be fed directly to the evaporator but the effect of contaminants must then be considered. Geological conditions

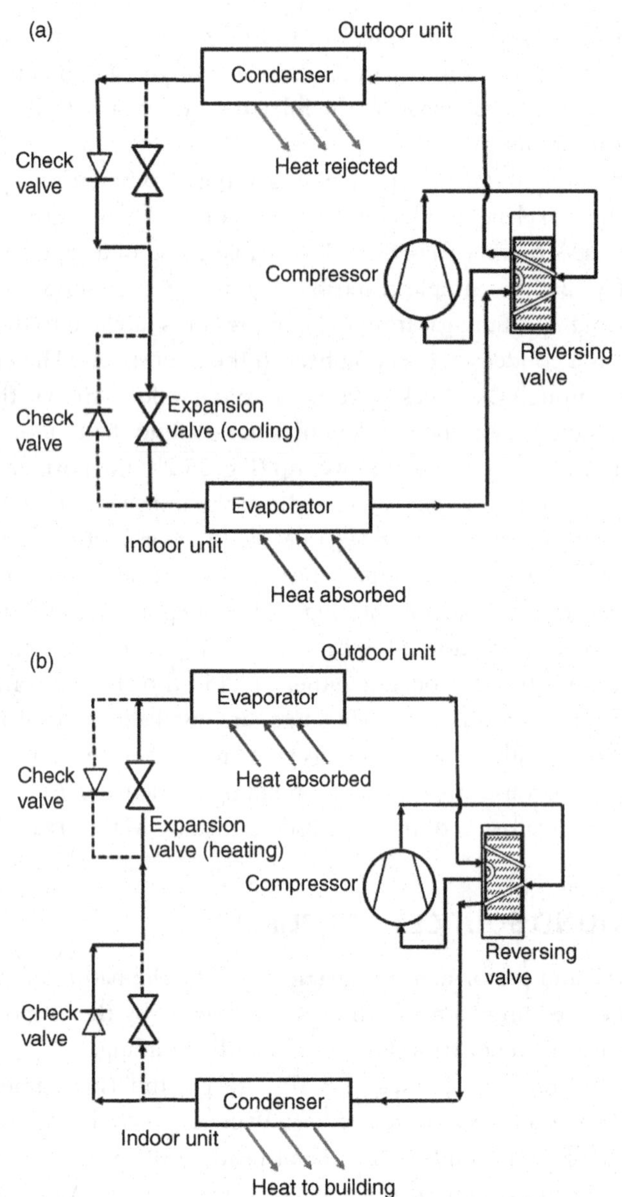

Figure 25.2 ***Reversible circuit – basic principle*** (a) cooling mode, (b) heating mode.

may allow drilling for a water source and for return of the ground water, but again, contaminants in the water may present problems. With any water source, care must be taken to observe local regulations.

Below a certain depth, ground temperatures remain relatively stable and there are a number ways in which to extract heat from the ground. The

GSHP has a typical source temperature of 5–10°C and is the most commonly quoted technology when heating-only systems are being considered. One way of extracting the heat is to circulate a glycol solution through plastic pipes, laid in trenches (Fig. 25.3), and the ground source then behaves in a very similar way to the water source.

An alternative method, which may be used where space is restricted, is to drill a bore hole and to circulate brine (again in a sealed loop) to sufficient depth to gain the necessary heat. Drillings of 50 m depth and upwards are typical, but this is still within the region where ground temperatures are primarily determined by average air temperatures. GSHPs capture solar heat stored in the ground. It is also possible to circulate the refrigerant directly by inserting the copper tube evaporator in the drilling. This eliminates one of the heat barriers, but increases refrigerant charge and there is always the danger of underground refrigerant leakage, which can be costly to rectify. The factory-assembled HP unit itself is quite compact and contains the complete refrigeration circuit and pumps for circulating the cold and warm secondary fluids (Fig. 25.4). HFC refrigerants are generally used, moving now to low GWP fluids. Note that the compressor and hot discharge piping are insulated to minimise heat loss to atmosphere.

Figure 25.3 ***Trench for GSHP showing pipe through which glycol solution will be circulated to collect heat (Kensa).***

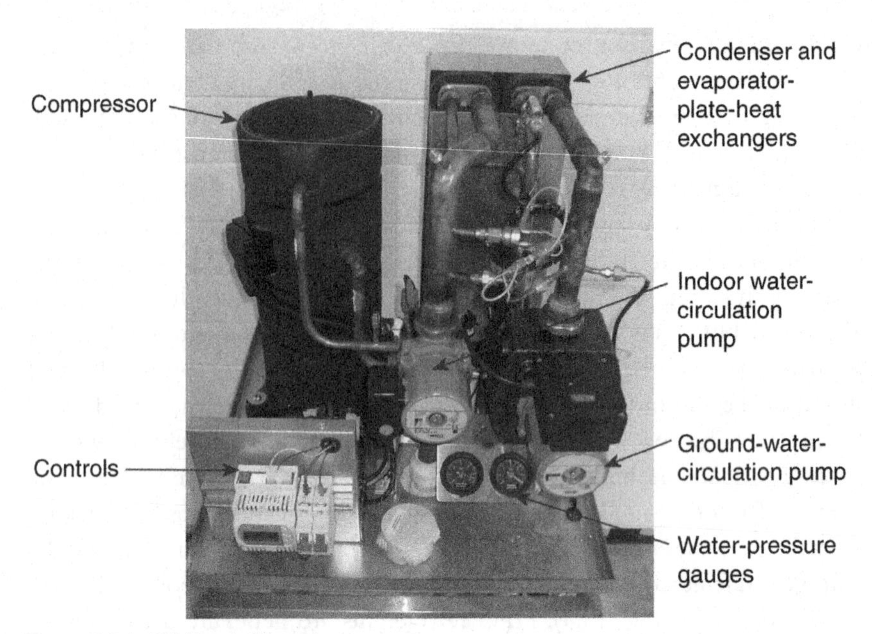

Figure 25.4 *GSHP unit (Kensa).*

25.4 AIR SOURCE HEAT PUMPS

In the United Kingdom air is an obvious and very suitable heat source. Air temperatures are within the range 5–15°C for a large part of the heating season, higher than with ground source for much of the year, resulting in excellent performance. With no drilling or trenching the installed cost of such a unit is usually much less, and there is also much less disruption. The ASHP in Fig. 25.5 contains the complete refrigeration circuit and is similar in this respect to the GSHP, but with an air coil evaporator and fan instead of a plate heat exchanger and water pump.

However there are some drawbacks with air source. Outdoor temperature variation significantly affects the capacity and efficiency. In particular, low outdoor temperatures result in poor performance and bring the additional problem of ice formation on the outdoor coil. Defrosting can be done by using a reversing valve to temporarily deliver hot gas to the outdoor coil, although this may have the undesirable effect of suddenly cooling the hot water supply! Another method is to divert some discharge gas to the evaporator using a hot gas bypass valve. The duration of very low outside air temperature conditions may be relatively short, but their occurrence coincides with the most acute need for heating, and so it is critical that the ASHP can cope. Sufficient capacity requires a larger unit than

Figure 25.5 ***Wall mounted ASHP delivering 8 kW heating to a farmhouse (HeatKing).***

would be needed for ground source, or an alternative is to use an additional heat source during cold spells. This is sometimes termed *back-up heating*.

The temperature to which the heat has to be raised or 'lifted' is dependent on both the required temperature at the point of use and the distribution system. Space heating can be delivered at less than 25°C if the air is directly heated, but the necessity to keep the room units small demands a relatively high condensing temperature. Refrigerant to air heat exchangers require larger temperature difference than refrigerant to liquid heat exchangers.

25.5 HEAT DELIVERY IN RESIDENTIAL AND COMMERCIAL HEAT PUMPS

One of the most efficient ways of delivering space heating is an under-floor water distribution system. This consists of plastic coils embedded in a solid floor, above a layer of insulation. With this system, the water may be circulated at a relatively low temperature of say, 35–40°C provided that there is sufficient floor surface available. Under-floor is less suited to buildings with suspended floors and is almost exclusively applied to new build. Wall-hung radiators result in lower COPs than under-floor coils because higher water temperatures are required.

The provision of domestic hot water (DHW) in addition to space heating necessitates a control system so that the HP can deliver lower temperature heat for space heating once the DHW demand is satisfied. Alternatively a top-up electrical immersion heater or inline heater can bring the water temperature up to 60°C.

It can be seen that there are a number of possible HP configurations and each application requires careful consideration of source, sink and use pattern. The performance is primarily governed by the effectiveness of the heat source and heat sink rather than the rated performance of the unit itself. There is a danger of cutting costs by reducing collector area, or by installing insufficient heating surface in the building. Both lead to excessive temperature lift and lowering of capacity and COP. The building insulation should be the first consideration when designing a heating system. It will usually be cheaper to bring the insulation up to standard than install additional HP capacity – and this is quite apart from running cost considerations.

Rating points for HPs are given in European Standard 14511, 2013, but the actual operating conditions may be considerably different for much of the year. Water flow temperature reduction, sometimes termed *compensated control* (see Fig. 25.6), brings considerable performance benefit during periods of lower heating load. In order to determine a more meaningful efficiency parameter the concept of seasonal COP or SCOP, sometimes called *seasonal performance factor* (SPF) is being introduced. It specifies the ratio of annual heat output to annual power input, with local weather pattern data to establish the heating load, and in the case of air source, the HP

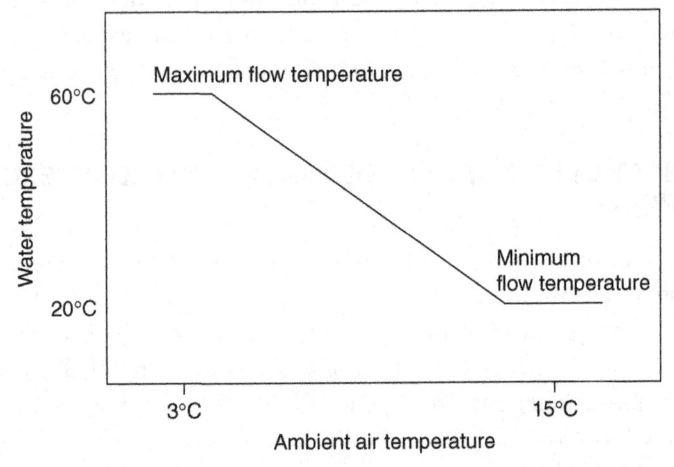

Figure 25.6 ***Compensated control – water flow temperature is varied with outside air temperature.***

performance. This is the parameter that should be used in the evaluation of the running cost benefits of a HP when compared to traditional heating methods. An example of HP benefits in terms of fossil fuel use, carbon emissions, and running costs is given in Section 28.4.

25.6 VRV/VRF SYSTEMS

Variable refrigerant volume (VRV) or variable refrigerant flow (VRF) systems extract or deliver heat to individual zones as required using refrigerant to air heat exchangers. They have multiple indoor units with a refrigerant flow distributor control device. This device controls the refrigerant with a PID control loop with the inputs being sent and received on a communication system. These systems generally offer high efficiency through the effective use of speed-controlled compressor, heat exchangers and controls. Heat recovery regimes allow rejected heat to be transferred within a building or stored as hot water. Typically there are 20–40 units on one system, although some manufacturers can allow as many as 64 indoor units. Individual indoor units can provide heating or cooling as required using either a three or two pipe system. Specialised heat recovery systems employ refrigerant distribution units with vapour and fluid separation and flow control. Some systems can provide simultaneous heating and cooling. Heat recovered from units providing cooling is transferred to units providing heating. With an extensive choice of indoor units these systems are suitable for larger buildings and offer installation flexibility.

Maximum interconnecting pipe work lengths are specified and it is vital to fully comply with the manufacturers' instructions to ensure reliable operation. In Europe installations must comply with BSEN 378 and must be carried out by a fully qualified F-gas registered technician. Safety limits apply to the volume of refrigerant that could displace oxygen in the event of a significant leak, and leak detection systems are available.

Many control options are available, from simple wireless controllers to powerful PC software packages. All the units are linked, providing a fully networked air-conditioning system. A simplified circuit diagram for a three-pipe system of this type is shown in Fig. 25.7. The outdoor unit contains the modulating compressor and heat exchangers that may work as either evaporators or condensers depending on the position of the reversing valves. The mode of each indoor unit is controlled by a branching controller. The detail of the branching controller is shown in Fig. 25.8.

In Fig. 25.7 the system is primarily providing cooling, three indoor units B, C and D are cooling and unit A is heating. The compressor is delivering

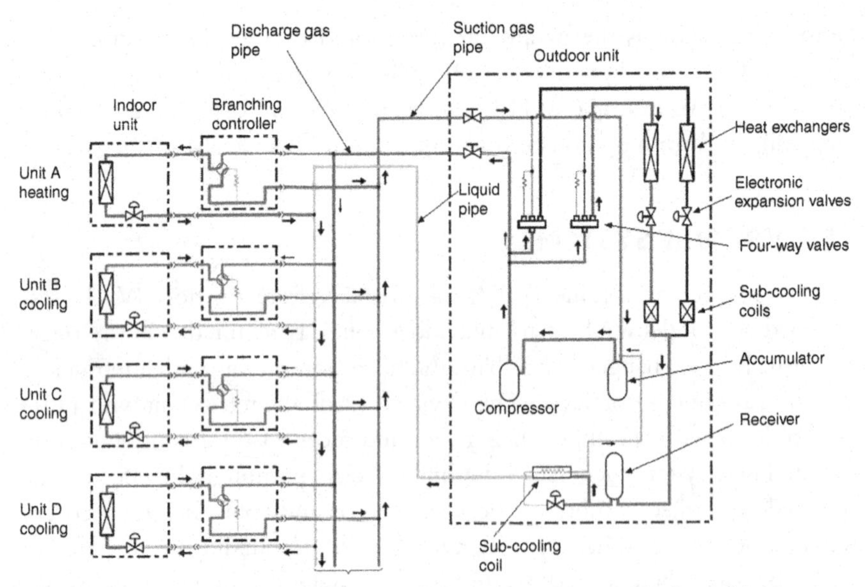

Figure 25.7 ***Three-pipe VRF system – simplified circuit diagram (Business Edge).***

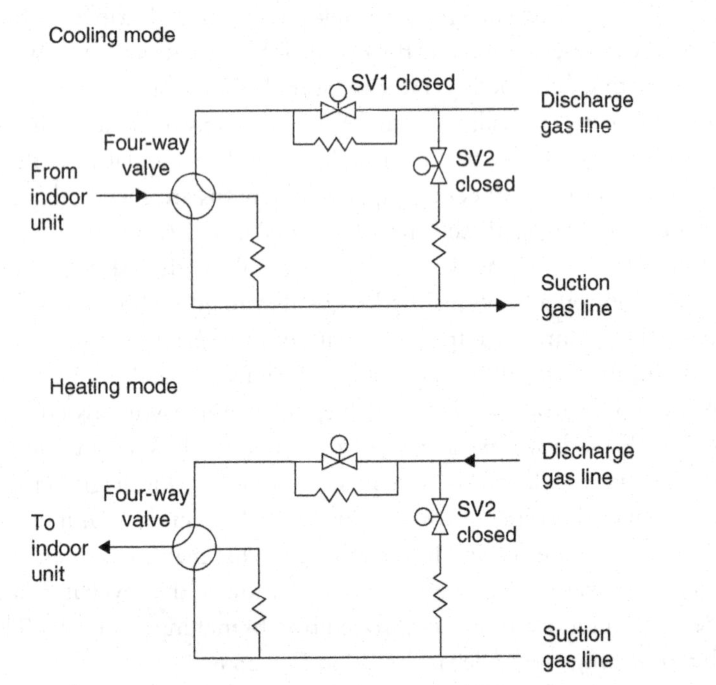

Figure 25.8 ***Branching controller function for three-pipe VRF system.***

hot high-pressure gas to the heat exchangers, acting as condensers. The liquid from the condensers flows via the receiver to the liquid line and hence to the indoor units B, C and D where the individual electronic expansion valves regulate the evaporation and cooling. The resulting suction gas flows through the suction pipe back to the outdoor unit. Unit A however is receiving some of the compressor discharge gas and condensing it (reducing the outdoor unit condenser load). The rate of condensation is determined by the rate at which liquid is allowed to feed back into the liquid line – controlled by the individual electronic valve in the unit (which acts as an expansion valve in cooling mode).

When the system is primarily providing heating, the solenoid valves SV1 in most of the branching controllers are open, and the indoor units are supplied with hot discharge gas instead of liquid. The flow in the liquid pipe is reversed and the heat exchangers in the outdoor unit work as evaporators, regulated by their electronic expansion valves. The four-way valves are switched so that the resulting liquid flows to the accumulator, where it is joined by liquid coming from any indoor units that happen to be cooling. There can be a 'balanced' condition where cooling matches heating and the outdoor heat exchangers are not in use.

Additional functionalities include defrost and oil return. Assistance with oil return is provided by periodically opening valve SV2 (Fig. 25.8) in the branching controller in order to raise the velocity in the suction line. Individual flow restrictors are also placed in the circuit where shown. These systems are fully engineered and specified by the manufacturers, and proven within the specified maximum line lengths. The line length allowances have been increased with the move to R410A and other high pressure refrigerants – the pressure drop effect is less because the pressures are higher.

VRF systems are suitable for a range of building types. There is flexibility of mixing a range of indoor units (eg floor, wall, ceiling-mounted and ducted units) with the correctly matched outdoor unit. The modular format lends itself to phased installations, and compared to traditional air-conditioning systems VRF is easier to design and install.

25.7 TWO-PIPE SYSTEMS

It is possible, with the correct selection of components to arrange a two-pipe circuit which will heat in one indoor unit and cool in another. The outdoor unit is connected to a distribution device local to a group of fan-coils, and the direction of gas and liquid flow will be determined by the overall balance of load, whether cooling or heating. If most indoor units call for cooling, the flow will be as follows. Discharge gas from the compressor passes through a four-port valve into the outdoor coil and is partly condensed to a high-pressure mixture of liquid and gas. The liquid is separated in the distributor unit and passed through two stages of pressure reduction, evaporates in the indoor unit and is returned as a low-pressure gas, through a controlling solenoid valve to the compressor. If any room calls for heating, the solenoid valves change over – the 'cooling' solenoid closing and the 'heating' solenoid opening. This admits hot gas from the top of the separator to the coil, where it gives up its heat to the room air and condenses. This liquid flows into the liquid header in the distributor and can then pass directly to a 'cooling' unit.

If the greater demand is for heating, the flows are reversed. The hot gas from the compressor goes directly to the solenoid valves, and so to each unit, to give up its heat, liquefy and pass back to the compressor unit, to be evaporated in the outdoor coil. Now if any unit wishes to cool, the solenoid valves on this item changes over and liquid will be drawn from the heater in the distributor, and pass as a gas, through the separator and back to the compressor.

It will be seen that the operation of this two-pipe cooling and heat pump circuit is delicately balanced and requires electronic control of the compressor and outdoor fan speeds in order to pump the right amount of gas, and still maintain working pressures.

25.8 HYBRID HEAT PUMPS

Hybrid heat pumps are designed to combine gas boiler heating and ASHP heating in a single unit so that the heat pump always operates under favourable conditions. This avoids the main disadvantages of air source mentioned above, and the ASHP size can be smaller. This goes a step further than a back-up electric heater, taking advantage of lower cost per kWh for gas, and at the same time providing two levels of heat, one for underfloor or radiator heating and a higher level for DHW.

An electronic control system determines the heat requirements and optimises the operation of the two sources to minimise the running cost. The compressor is usually speed controlled, and the water flows and be regulated. These constitute outputs from the control logic along with heated water temperature. The principle behind the logic is illustrated in Fig. 25.9. The curve represents the annual load profile, the area under the curve being the annual kWh heat load. It can be seen that the 'boiler only' element of heating only supplies a small fraction of the annual heat load.

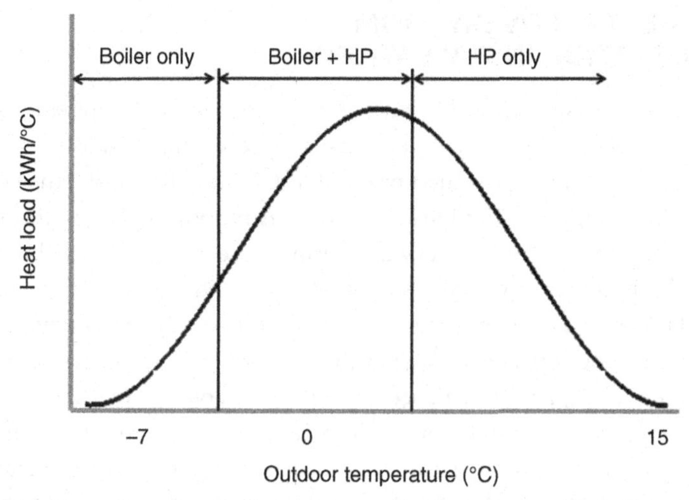

Figure 25.9 ***Operation zones for heat pump and boiler in a typical hybrid system.***

25.9 ENGINE-DRIVEN AND HEAT POWERED HEAT PUMPS

Gas engine-driven HPs offer the bonus of additional heat from the engine coolant that is at a higher temperature than the refrigerant condenser. The engines are normally conventional spark ignition or diesel IC type adapted from vehicle versions. An open compressor is required and since the installation is static, and required to run for comparatively long periods without attention, drive assemblies must be robust to withstand the extra vibration and should be separate from the rest of the circuit. Gas engine-driven HPs are generally applicable to large buildings and VRF systems are starting to appear in commercial applications.

Ammonia–water absorption heat pumps available that can extract heat from an air, water and ground source. They make use of the absorption cycle (chapter: The Refrigeration Cycle) and use burning gas to drive the refrigerant from the sorbent. High temperature hot water for heating and DHW can be supplied. The three main refrigerants used for these HPs are water, methanol and ammonia. Each has advantages and disadvantages. Water is low cost with high specific heat but operates at very low pressures. Ammonia is more practical for air source with positive pressures at the lowest likely temperatures, but lower latent heat. Methanol lies somewhere between the two but decomposes at high temperatures.

Adsorption technology is also being developed for heat pumps. A development using active carbon adsorbent is published by Critoph and Metcalf (2012).

25.10 HEAT RECOVERY FROM REFRIGERATION INSTALLATIONS

Industrial heat pumps can utilise a number of heat sources to upgrade heat for various large scale heating requirements. A good example is district heating. Although district heating is uncommon in the United Kingdom, the concept well developed in some European countries. Ammonia, R717, with its good thermal properties and high critical temperature is proving to be a very effective heat pump refrigerant. It is best suited to medium to large applications where the installation is separated from building occupancy. District heating schemes can require water temperatures of up to 90°C in order to transmit the required heat quantity with acceptable capital cost networks, and this also enables heat pumps to replace oil or gas boilers on existing networks. In order to achieve these high temperatures it is necessary to condense the ammonia at pressures up to 50 bar abs. Conventional compressors are limited to around 26 bar which limits the condensing temperature to 60°C.

By taking advantage of the single screw compressor's balanced radial and thrust rotor loadings, see chapter: Compressors, these machines can be designed for high pressures. Ductile iron or cast steel is used for the casing instead of cast iron. Heat can be recovered from a variety of sources including existing refrigeration systems, industrial process waste heat or river/seawater. Capacity control is facilitated by sliding valves as described in chapter: Compressors and the use of separate valves to adjust the size of the two discharge ports enables the volume ratio to be closely controlled, ensuring best energy efficiency at part load conditions when lower heating temperatures suffice. Fig. 25.10 is a schematic illustrating the process of heat recovery from an existing cooling plant.

In this example the heat recovered is maximised by heating the water in stages, taking it first through a subcooler, followed by oil cooler, desuperheater, and finally condenser. For temperatures above 71°C, two stage recovery systems can provide water temperatures of up to 90°C. This is achieved by utilising heat rejection at different stages in the compression process as indicated in Fig. 25.11. This sequential water heating process uses a desuperheater for the final temperature lift enabling a lower condensing temperature of approximately 80°C to be maintained, and enhancing the COP. A study of a 15MW district heating installation that extracts heat from seawater at 8°C and delivers heat in stages to raise the incoming water from 60 to 90°C is given by Hoffmann and Pearson (2011).

Innovations are also taking place with heat recovery from commercial refrigeration systems. Supermarkets have a fairly steady cooling load throughout the year but space heating is only required in winter. In the concept

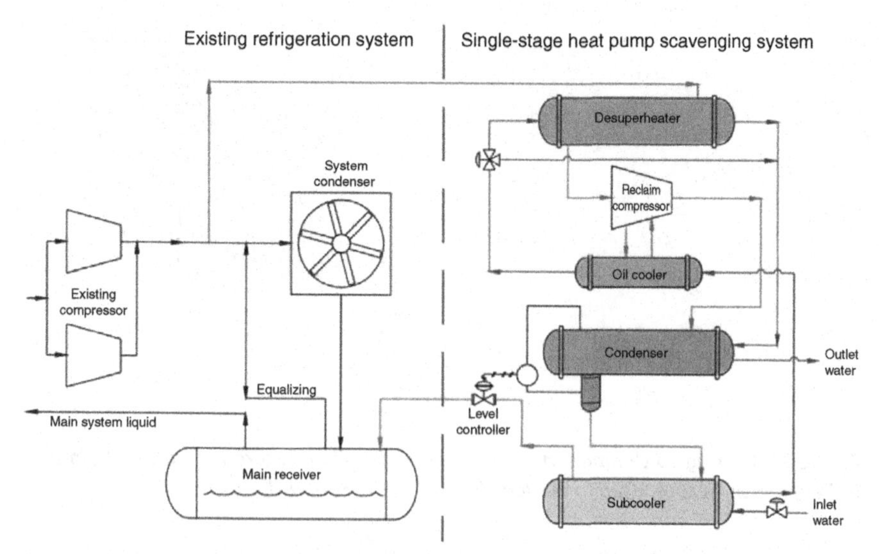

Figure 25.10 ***Heat reclaim process. Ammonia from the compressor discharge of an existing plant taken through a high pressure single screw compressor heat recovery system (Emerson Climate Technologies).***

illustrated in Fig. 25.12, angled drilling (left) gives access to a large volume of underground heat storage that is used to store heat rejected by the booster refrigeration system serving the low and medium temperature cooling loads (Fig. 25.13, and see also Section 2.5). A water based secondary refrigerant is circulated to the closed loop borehole heat exchangers. The introduction of

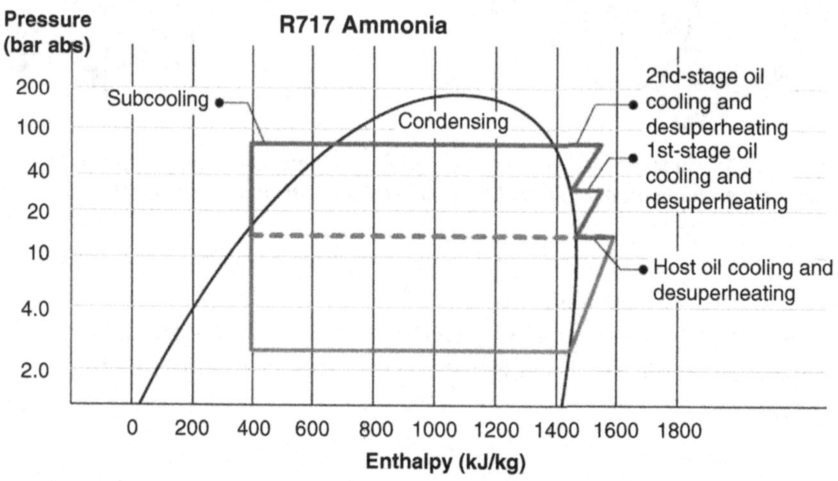

Figure 25.11 ***Two-stage heat recovery process P–h chart, not to scale (Emerson Climate Technologies).***

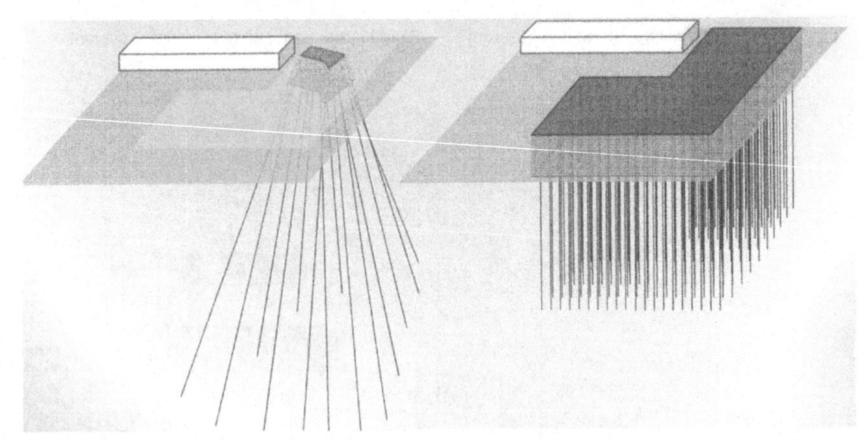

Figure 25.12 ***Angled drilling left, maximises underground storage space with limited drilling footprint compared to vertical drilling, right (Hubbard).***

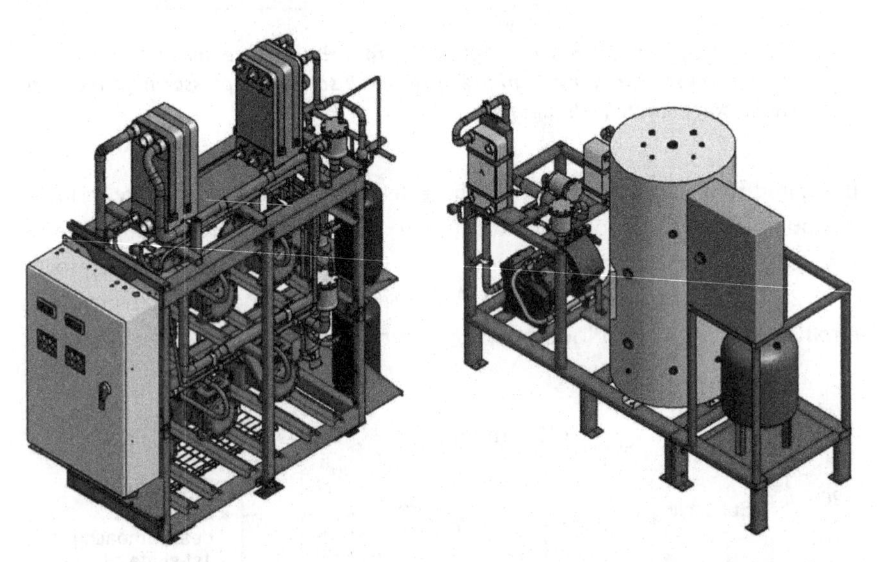

Figure 25.13 ***Supermarket heat recovery packages, left: heat pump showing brine plate heat exchangers, and right: LT booster package with liquid receiver (Hubbard).***

angled drilling gives enables the storage capacity to be accessed from a limited footprint area such as an on-site car park.

In addition to storing the rejected heat the ground serves as a heat source for GSHPs. The result is that both systems see significantly improved efficiencies.

CHAPTER 26

Control Systems

26.1 INTRODUCTION

Control systems can be classified into four categories:

1. Controls that make the system safe.
2. Controls necessary for correct and efficient operation of the system.
3. Controls that ensure the correct conditions within the conditioned space or of the product are maintained.
4. Controls that optimise the operation – for example provide the necessary cooling whilst using minimum energy.

Safety standards and codes such as EN378 specify certain safety requirements according to the type and size of the installation, and some of these, such as high-pressure cut-out switches will be self-actuating and independent of any control systems. Where the appropriate sensors are in place, a control system is quite likely to incorporate additional safety elements which are activated prior to any local cut-outs.

Correct operation includes, for example, ensuring that the compressor is operating within its allowable working conditions and with adequate oil pressure. Protection devices such as these are not necessarily safety critical, but will help to ensure plant reliability. Efficient operation may require that some compressors are switched off instead of running multiple compressors part loaded. This type of sequencing will normally be incorporated in the main controller.

There are two types of devices that report on conditions. These are sensors and detectors. Sensors provide a variable output according to the value that they are measuring. By contrast detectors give only a present or absent signal.

26.2 THE CONTROL LOOP

The control system for a sensor-based system consists of a loop, with sensor, controller and controlled device. The communication between these parts of the loop can be electric, pneumatic or mechanical (see Fig. 26.1).

The controller action can have three elements. These are proportional, integral and derivative. Such a controller is known as a three-term controller,

Refrigeration, Air Conditioning and Heat Pumps
http://dx.doi.org/10.1016/B978-0-08-100647-4.00026-7

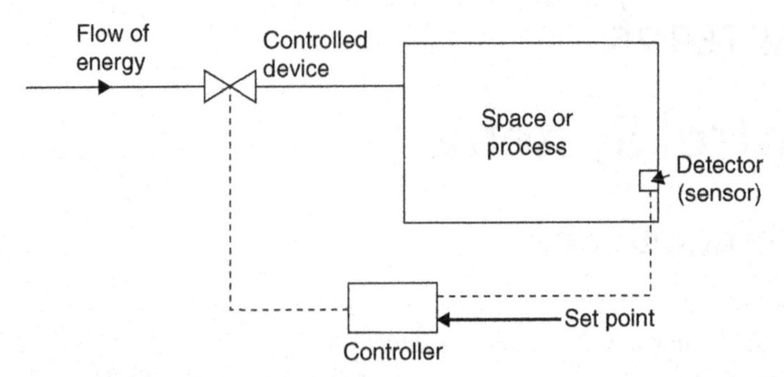

Figure 26.1 ***Basic control loop.***

or a PID controller. The proportional element creates part of the output for which the reaction is proportional to the deviation from set point or 'error'. The integral element takes account of the history of the errors and creates part of the output based on the sum of the errors. The differential element creates part of the output based on the rate of change of the error. Most refrigeration control uses proportional and integral elements, setting the derivative part to zero. Experience has shown that using a significant amount of derivative in the control algorithm leads to instability.

26.3 DETECTORS AND SENSORS

In principle detectors are two-position (on–off); however, in many cases they have some hysteresis. This means that there are three levels at which they operate. They switch in one way when their input is above a certain level and switch the other way when their input is below a lower level. Between these two levels they maintain their previously arrived at state. Using a simple detector as the control mechanism will cause the controlled condition to overshoot to some extent, since all devices have some time lag in operation. The amount of the overshoot depends on the time lag of the detector and the extent to which the rate of supply of energy to the process exceeds the load (see Fig. 26.2). The range of the control will therefore be the differential of the detector along with the upper and lower overshoots under load.

Where the two-position sensor is also the controller, it provides only two outputs: maximum or zero. A narrower dead band gives tighter control, but results in more rapid cycling, which may be undesirable, particularly if compressors are being switched on and off.

A two-position detector can be used to operate a floating control. At the upper limit it will operate the control in one direction, and if it reaches the

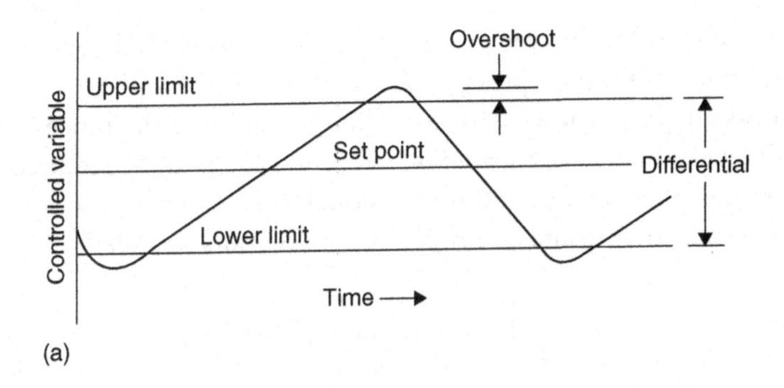

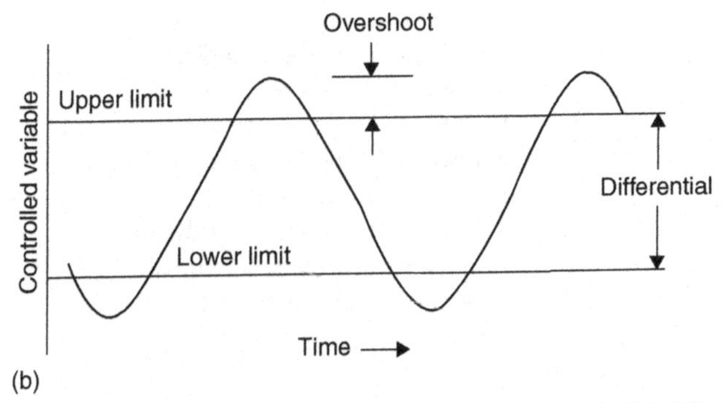

Figure 26.2 ***Limits of controlled variable with two-position control. (a) Capacity closely matched to load, (b) capacity much greater than load.***

lower limit it will operate the control in the other direction. Between the two limits the control is not actuated (see also Section 26.4).

Two-position detectors can be classified according to the purpose.

Thermostatic	• Bimetal • Liquid expansion • Solid expansion • Vapour pressure
Pressure	• Diaphragm • Bellows • Bourdon tube
Fluid flow	• Moving vane • Rotating turbine
Time	• Clock • Bimetal and heater • Dashpot
Humidity	• Dimensional change of hygroscopic element
Level	• Float • Light refraction

Many of these devices are direct acting on the controlled device and do not require a controller to process the signal.

Sensors are proportional detectors and they measure the process condition, which can then be compared by the controller with the required value. They are not direct acting, and need a controller to convert the signal to a working instruction to the controlled device. Sensors include:

Temperature	• Those mentioned previously, plus • Electrical resistance of a metal or a semiconductor • Thermocouple • Infrared radiation
Pressure	• Those mentioned previously
Fluid flow	• Those mentioned previously, plus • Electronic, Doppler effect
Time	• Those mentioned previously, plus • Electronic timing devices
Humidity	• Those mentioned previously, plus • Resistance of a hygroscopic salt
Level	• Float with impedance coil • Ultrasonic

26.4 CONTROLLERS

Various implementations are possible using a controller with an on–off detector. In the first, it functions only as an amplifier to transmit the detector signal to the controlled device. This is fixed control.

In the second and third implementations floating control is achieved. For this two detectors are required with different on–off settings. In the second implementation, the output of the controller is fed to a motor which drives a camshaft. If the signal to the detectors is above them both, the motor is driven in one direction. If it is below them both it is driven in the other direction. If it is between them, there is no drive to the motor. The cams on the shaft cause on–off signals to the controlled devices, for example a row of pumps in parallel. Thus a signal in one direction causes the number of pumps running to increase and a signal in the other direction causes the number running to decrease. The rate at which the number running changes is dependent on the gearing of the motor. It is possible to implement this functionality with an electronic device taking either two input signals or one analogue signal from the parameter to be controlled.

In a further implementation, the operation is similar in that it is all done electronically in the controller. The rate at which the output value changes are dependent on the amount by which the input signal differs from the upper or lower setpoint.

In all cases the controller has as many outputs as there are controlled devices or it has one analogue output which feeds a speed control device.

Some detectors are combined in the same instrument with a suitable transducer which can perform some of the functions of a controller. For example, for pneumatic systems the primary sensing element actuates a variable air jet, thus modulating an air pressure which is transmitted to a further controller or direct to the controlled device. Electric and electronic detectors such as the infrared detector include the sensing and amplifying circuits of the instrument.

Controllers generally for use with proportional detectors measure the displacement of the signal from a set point and transmit a signal to the controlling device(s) on signal on the basis of PID algorithm. Any one, or two, of the three PID terms can be zero. In refrigeration, it is normal for the D-term to be zero so that the control is PI – proportional and integral. A controller may be arranged to accept input signals from more than one detector, for example the flow temperature of a hot water heating system may be raised at the request of an outdoor detector if the ambient falls, or may start the heating earlier in the morning to pre-heat the building before it is occupied; a servo back pressure regulation valve (Fig. 9.8) can respond both to evaporator pressure and to load temperature. With the advent of microcomputer devices almost any combination of signals can be processed by an electronic controller, providing the output signals can be made coherent and not conflicting.

Pneumatic controllers, which may include part of the sensing instrument, are supplied with compressed air at 1 bar gauge which is allowed to escape from an orifice controlled by a detector. The resulting pressure modulates about 0.4 bar and is used in a servo piston, diaphragm or bellows to actuate the controlled device (see Fig. 26.3).

26.5 CONTROLLED DEVICES

Controlled devices commonly consist of an actuator, which accepts the signal from the controller and works the final element. Typical examples are as follows:

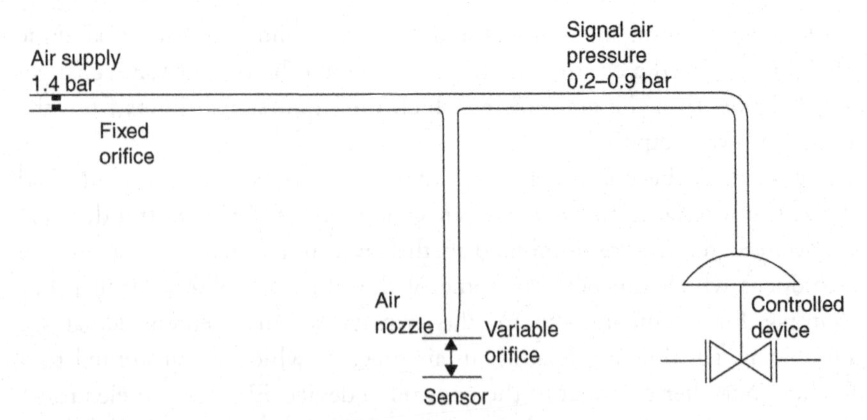

Figure 26.3 ***Pneumatic operation of controlled device.***

1.	Electric relay	Operating	Contactor Motor Motorised valve Dampers
2.	Electric solenoid	Operating	Solenoid valve
3.	Modulated electronic signal	Operating	Magnetically positioned valve Thyristor power control Pulsed expansion valve
4.	Pneumatic pressure (and hydraulic)	Operating	Pneumatic relay Valve positioner Damper positioner

The effect of a controlled device may not be proportional to its movement. In particular, the shape of valve plugs and the angle of opening of dampers will not give a linear result, and the signal from the controller may take this into account. Alternatively, the feedback system will ensure that the device moves to the correct position to give the desired output.

26.6 CONTROLS COMMUNICATIONS

Communications today are mainly via electronic networks and suppliers of equipment have built a number of proprietary protocols which enable individual devices to communicate. The industry is moving away from proprietary control schemes and centralised systems. Manufacturers are using open operating systems and parts to build products that feature improved reliability, flexibility, system cost and performance. The local operating network (LON) and MODBUS protocols are examples of an open access protocols designed to be used in general purpose control network solutions. The protocols refer to the pattern of signals that are used in communication. These patterns can be transmitted via many different media. Examples are the following:

- Mains voltage electricity, usually 230 V but sometimes 115 V
- Low voltage, mainly 24 V a.c.
- Low voltage, electronic circuits
- Telephone cable
- Pneumatic
- Optical fibre
- Optical – infrared
- Radio
- Internet

Input signals to controllers can be transmitted in a number of ways. These include the following:

- Mechanical means – rods, levers and cables
- Fluid pressure
- Low voltage, thermocouple (microvolts) or thermistors (millivolts to volts)

Mechanical devices need careful installation to ensure that there is no distortion of the parts. This is especially the case with damper mechanisms, which need maintenance and periodic inspection to ensure they are working correctly.

Where fluid pressure is carried by a capillary tube, such as with the thermostatic expansion valve or pressure switches, the tube should be installed with due attention to the risk of it chafing against metal edges and wearing through. Tubes to manometers are usually in plastic, but may be copper. These must be carefully tested for leaks, as they are transmitting very low pressures.

Mains voltage communications must be run according to IEE Wiring or the appropriate safety regulations. In particular, these may cause interference with telephonic, computer and other electronic signals carried in or near the same conduit. In the same way, electronic control signals may suffer interference. Thermocouple signals are very low DC voltage and should be run as far as possible with unbroken conductors. Terminal boxes should be compact and insulated from sudden temperature changes. Terminals must be tight.

Pneumatic controls are used widely in hazardous situations such as chemical plants and oil refineries. The same risk of chafing applies as with capillary tubes. Pneumatic tubing is usually in copper and is correctly secured.

Optical fibres are not yet very much in use, but there is no interference between them and electrical signals of any sort. For this reason their use will probably become more widespread. Line-of-sight optical signals require that no obstruction is inserted at any time. Such points are easily noticed when installing and commissioning, but are not so obvious if a malfunction occurs at a later date.

Remote plant is sometimes controlled or monitored by radio link. This is now largely superseded by internet systems. On-line systems allow users to display control panels for remote systems in their browser for monitoring operation or making control adjustments.

26.7 CONTROL SYSTEM PLANNING

Control systems can quickly escalate into unmanageable complexity, and the initial approach to the design of a suitable control system should examine the purpose of each item, and the effect on others, to eliminate those which are not essential.

The action of a control may combine two or more of the purposes, as set out in Section 26.1, which may then be interdependent. It is more informative to consider the action of a control and examine what purpose it may serve in the circuit. Controls should ensure that functions are shut down when not needed (the boiler in summer and the chiller in winter). Optimum start controls now complement the starting clock, to advance or retard the starting time according to the ambient conditions.

In planning a control system, a flow diagram is needed to indicate what may influence each item of plant. In many diagrams it will be seen that complexity arises and two items work in conflict. A typical instance is the cooling and dehumidifying of air, to a room condition lower than design, with concurrent operation of a humidifier.

Since most controls will be electrical and largely of the two-position type, it is a convenient notation to set out the initial control scheme as an electrical circuit and in 'book page' form, that is from left to right and line by line, to indicate the sequence of operation, with the controlled device always in the right-hand column. This analysis should indicate the different items which might act to produce a final effect and bring errors to light.

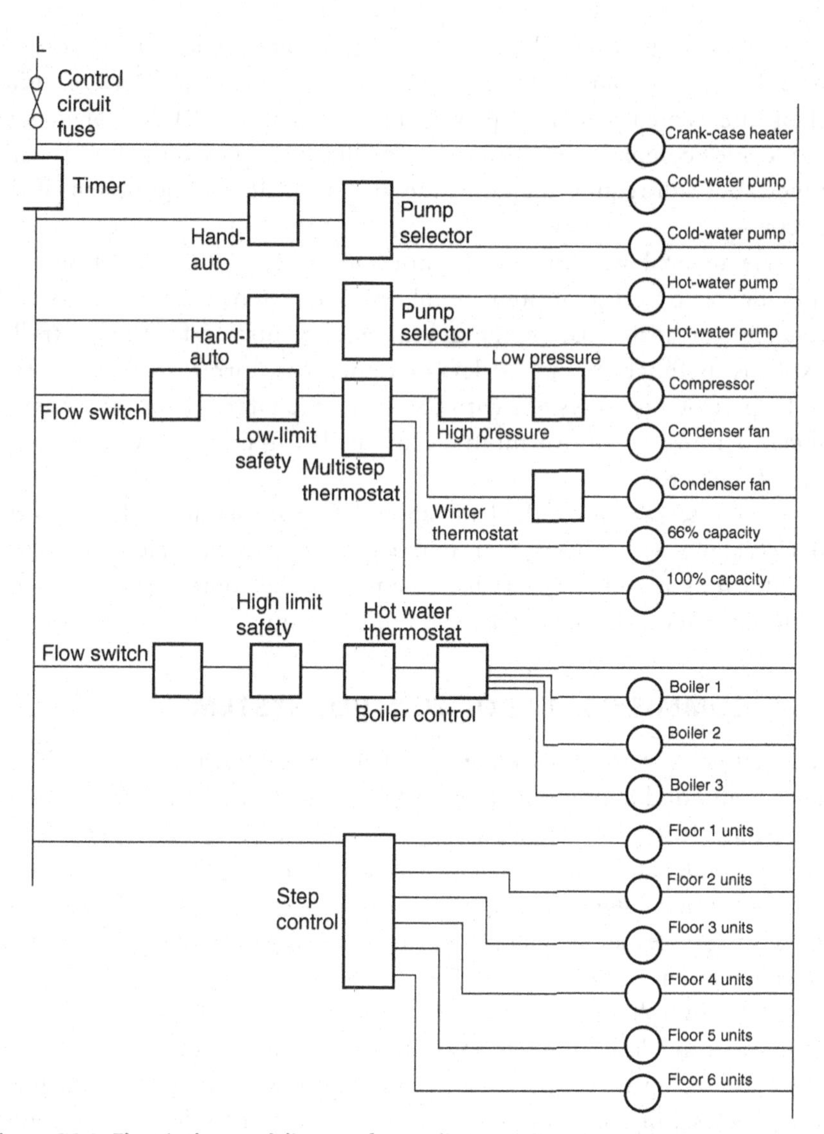

Figure 26.4 ***Electrical control diagram for small air-conditioning system.***

Fig. 26.4 is a simplified control circuit for a small air-conditioning system. Non-electrical items can be shown on the same initial scheme, possibly with dotted lines to indicate a non-electrical part of the system. The possibilities of abnormal operation should be examined, and grouped as system not working, system unsafe and system dangerous, and protected accordingly. The last category requires two independent safety controls or one control and an alarm.

Complex timing and logic controlling, monitoring and indication can now be carried out with programmable computer-type devices, using algorithms stored in some type of read only memory (ROM). These save the former complex arrangements of sequencing and interlock relays and timers, but still require the same attention to planning and design of the circuit.

The normal way in which information about these controllers is provided is by a list of 'item numbers'. Each item number refers to a settable or readable parameter and the documentation for the controller explains its function and its default value. It is normal for such system to be part of a total system with parts, like the safety chain that must be intact before a pump (or motor) can start being made-up of a string of on–off detectors.

In all cases, a copy of the basic control diagram should be left with the device, along with a list of item numbers for any sophisticated controllers, to inform users and service staff of the plan of the control system, and any subsequent modifications updated on the diagram.

26.8 COMMISSIONING OF CONTROL SYSTEMS

The setting up, testing and recording of all control functions of a refrigeration or air conditioning system must be seen as part of the commissioning procedure. It requires that all items of equipment within the system are in working order and that the function of each item of control is checked, initially set at the design value (if this is known), readjusted as necessary during the testing stages, and finally placed on record as part of the commissioning documentation.

Most controllers have adjustments, not only to the set points but to differentials, time delays and response rates. It is of paramount importance that these are set up by an engineer who completely understands their function. Such settings should be marked on the instrument itself and recorded separately.

26.9 THE INTERNET OF THINGS

The internet of things (IOT) is about connecting devices over the internet. Controlling energy use in buildings can be assisted by smart meters that enable, for example, systems to be turned on remotely, adjust temperatures according to conditions, and monitor motion sensing cameras. Some heating and cooling systems have the capability to integrate forecast weather conditions for the specific location, and use the information to program advance preparatory operations such as backing up heat to increase capacity during a cold spell.

These straightforward examples are just an introduction to what could become a wide ranging scenario as more and more components are developed with the ability to connect and communicate. For this to happen the infrastructure has to be open; any device, from any platform, must be able to connect and communicate with others.

CHAPTER 27

Commissioning and Maintenance

27.1 INTRODUCTION

The commissioning of a refrigeration or air conditioning plant starts from the stage of static completion and progresses through the setting-to-work procedure and regulation to a state of full working order to specified requirements. Commissioning is the completion stage of a contract, when the contractor considers that the plant is in a correct state to hand over to the purchaser for acceptance and payment. Since the final object of commissioning is to ensure that the equipment meets with a specified set of conditions, all parties should agree this specification when a contract is placed.

Maintenance is the effort required to ensure that a commissioned plant continues to deliver correct performance, and includes inspections designed to detect signs of deterioration prior to any noticeable effects.

The key points are summarised in this chapter. The details given are by no means exhaustive; they are used to illustrate the procedures which should be followed. For small packaged equipment such as a simple split system, the requirements may be relatively simple with many items pre-set, in contrast to a large custom built industrial installation that requires much more detailed attention. The key points apply equally to heat pumps.

27.2 SPECIFICATION

A final contract specification should state the following (but not be limited to these data):

1. The medium or product to be cooled, or the area to be cooled.
2. The total required cooling capacity, both at the extreme (design) conditions and at nominal (average) conditions.
3. Maximum design and nominal ambient or coolant temperatures.
4. The power input of the compressor and auxiliaries at design and nominal conditions.
5. Part load requirements, if any.
6. The required limits of control.
7. The conditions of the refrigeration system during nominal operation including condensing, evaporation, superheat and sub-cooling.

Refrigeration, Air Conditioning and Heat Pumps
http://dx.doi.org/10.1016/B978-0-08-100647-4.00027-9

If no such specification exists at the time of commissioning, some basis of acceptance must be agreed between the parties concerned. Basic flow diagrams should be available and if not, the commissioning engineer must draw them up, against which actual plant performance can be checked.

27.3 COMMISSIONING

The work of commissioning must be under the control of a single competent authority, whether it be the main contractor, a consultant or the user. It can happen that when building specifications are written, they split the responsibilities to sub-contractors with the presumption that everything will be exactly as specified after commissioning, which it rarely is. With no one responsibility for the total operation of the cooling system it can be found that it operates, but in an unsatisfactory manner – for example, frequent cycling instead of steadily operating at reduced capacity.

Complex systems need to have follow-up on the initial commissioning at low, medium and high load (warm, medium and cold ambient) to ensure proper and energy efficient operation. To include this as a responsibility in the contract is very cost-effective.

The commissioning engineer will require details and ratings of all major items of the plant and copies of any manufacturer's instructions on setting to work and operating their products. If this information is not to hand, the work will be delayed. The following stages may be identified in a commissioning process.

27.3.1 Documentation

The commissioning documentation should include the following:

- Specification, including cooling capacity, operating conditions and limits
- Refrigeration and electrical circuit diagrams
- Refrigerant charge and operation conditions
- Set points for all the controls and safety devices
- Commissioning and operating information for all major components
- Site tests performed such as pressure, tightness, vacuum and electrical insulation tests
- For larger systems, sub-system controls done by different sub-contractors for running currents, flows and pressure drops.

27.3.2 Initial Checks

The installation should be initially checked to ensure that it is in accordance with the specified design. For example, components are those specified, electrical equipment is suitable for the pressures and temperatures, pipework correctly installed and adequately supported, cleanliness of heat exchangers, water circuits and filters, compressor mountings correctly installed, safety and pressure controls correctly connected, non-return and pressure regulating valves correctly positioned, correct wiring and control sequence.

27.3.3 Preset Controls

The next stage is to preset as many controls and protection devices as possible. Only after all possible static checks and adjustments have been made to the system should it be started for the first time. These precautions will prevent most of the common types of failure occurring during the initial running period. Typically settings include service shut-off valves open, water controls set, control switches preset as accurately as possible, temperature control and cut-outs set. Also compressor rotational direction checked and defrost timers set.

27.3.4 Operational Checks

Start pumps and fans without refrigeration machinery and evaluate flows from available indications such as flow meters, pressure differences over pumps, fans, filters depending on what is available and check for air in water/brine systems. For secondary systems with freeze protection, it must be ensured that they are free of air before the temperature is lowered and the systems are left to operate for longer periods. Air in these systems can be virtually impossible to remove if the gas is distributed as micro bubbles in the system due to operation without proper degassing. This will affect the performance of the system during tests and operation.

The refrigerant charge and system operation should be checked and set for the complete range of load and ambient conditions. Follow-up tests may be required to cover a range of ambient conditions. The refrigerant charge should have been added according to the weight specified, and additional charge should not be required. When the system has been operating for a sufficiently long period, conditions in the cooled space can be checked against specification, and other typical points to watch for include evaporator superheat and refrigerant distribution, excessive pressure drops, compressor oil levels, correct condensing pressure. The operation of pressure controls such as evaporator pressure regulator and condensing pressure controls together with defrost systems can be checked.

It may not be possible to operate at the extreme conditions and whilst it may be possible to achieve high-condensing pressure by turning off fans, this will not be realistic, as this creates the same condensing pressure but not the same sub-cooling and the load will be different. A more realistic test of the refrigerant cycle is to temporarily enclose the condenser and force the air to re-circulate without limiting the flow. It is best to have follow-up controls at different seasons in the first year.

The whole system is now left to run for a shake-down period, which may be from a few hours to several days, depending on the size and complexity. During this time, all components are checked for vibration, leaks or other malfunction, and remedial action taken.

For large low-temperature systems and cold stores, temperatures should be reduced slowly, to allow for shrinkage in the structure. A fall of 5 K per day is reasonable, moving more slowly through the band 2°C to −2°C. At the end of the shake-down period all strainers and filters are cleaned ready for the final test.

27.3.5 Performance Check

In the final commissioning stage, readings are taken and recorded and compared with the specification and design figures. Some final adjustments to airflows, secondary fluid flows, etc. may be necessary. The following measurements, as applicable, should be considered as the absolute minimum to be taken and recorded:

- Ambient conditions, dry and wet bulb
- Refrigerant pressures and temperatures at expansion valve inlet, evaporator outlet, and compressor suction and discharge
- Secondary fluid temperatures at heat exchanger inlet and outlets
- Pump, fan and filter pressures
- Settings of all adjustable controls
- Electric motor currents

Further details of performance measuring techniques are given in the next section. It is probable that a full load cannot be obtained during the final test, for reasons of low ambient or lack of completion of other equipment for the process. In the absence of analysis tools, the commissioning engineer must make an estimate of the system performance, on the basis of time run, or otherwise interpret the figures obtained. In such cases it may be advisable to agree to a tentative acceptance of the plant and carry out a full, minimum and part load tests at a later date.

27.3.6 Hand Over

A complete set of plant documentation and commissioning records should be left on site for future reference. These should include the following:

- A copy of the commissioning log
- Flow, control, electrical and layout diagram and drawings
- The system refrigerant type, and charge; and oil type and oil charge
- Operating instructions
- Maintenance instructions
- Copies of instructions and manuals for all proprietary items of equipment
- A list of recommended spare parts
- Declaration of conformity and any other information as required to complete the health and safety file

27.4 PERFORMANCE MEASUREMENTS

Field performance measurement is known to be notoriously difficult. In the first place, it is necessary to ensure that the system is stable. The traditional method is then to meter the secondary flow, usually on the chilled side, for example the chilled water flow rate, and this together with the inlet and outlet temperatures gives the cooling capacity. Orifice plates or flow meters are required and they need to be carefully installed to obtain acceptable accuracy. Additionally the result is very sensitive to small temperature differences. Because the measurements are on the secondary side, these techniques are classed as external methods.

The internal method uses the measured refrigerant pressure and temperature values to plot the actual vapour compression cycle on the P–h chart. With an adjustment for the heat losses from the compressor shell this enables the COP to be determined. Measurement of the compressor power input provides the information necessary to determine the refrigerant mass flow rate, and once this information is known, the secondary side flow rates can be found from temperature measurements. The internal method is described by Berglöf (2004) and manual readings of pressures and temperatures listed in Section 27.3, under 'performance check' may be used. Many systems have most of the required temperature measurements included in their control systems, and it may only be necessary to add a few additional channels and sensors to collect the required data. The advantage of this method is that there is no intrusion into the system for flow meters and there is less sensitivity to errors of measurement than with the external method.

Figure 27.1 ***Field performance analyser connected to a chiller, showing visual performance display (Climacheck).***

The internal method is greatly enhanced by the use of a computer-based analyser which performs all the necessary calculations using refrigerant properties to generate an output which gives immediate information on key parameters such as superheat, sub-cooling, cooling capacity and COP. This makes the information easy to access and interpret. The computer software can be run on the user's building management system (BMS) computers. Claimed accuracy with sensors that are commercially available at reasonable cost is ~5% COP and ~7% capacity. Where the BMS does not convey the necessary parameters, or where an independent analysis is to be carried out, a field inspection kit containing a logger and all the necessary sensors can be applied (see Fig. 27.1). This is also very useful for commissioning.

27.5 PLANNED MAINTENANCE

Planned maintenance should be of preventative nature, and may be carried out by the user or by a contracting company. The objective is to ensure that the system remains leak tight, to maintain efficient operation and to reduce the incidence of breakdowns. There are minimum inspection requirements for all HFC systems depending on refrigerant charge, see Section 27.9, and leak testing procedures are covered in Section 11.3. For small self-contained sealed water/water systems with hermetic compressors and pre-set controls, maintenance will be little more than a visual check and control of available information such as energy consumption. A check of system pressures and the water side and temperature differences should be included whenever possible. Air heat exchangers and cooling towers introduce the need to check for cleanliness, and large complex systems require a maintenance schedule for sustained reliable operation. The following points have been found by experience to be prime sources of performance deterioration.

Accumulated dirt on air filters increases the resistance and leads to reduced airflow. This is by far the most frequent cause of malfunction of air-conditioning equipment. An example of the effect of reduced air flow is given in Example 27.1.

Example 27.1

An R410A direct expansion coil evaporates at 3°C when cooling air from 20°C to 11°C. Condensing is at 35°C. If the airflow is reduced by 15% because of a dirty filter, what is the approximate increase in running cost?

Approximate calculation:

$$\text{Air entering coil} = 20°\text{C}$$

$$\text{Air off coil at full air flow} = 11°\text{C}$$

$$\text{Evaporating temperature at full air flow} = 3°\text{C}$$

$$\text{LMTD at full air flow, } \frac{17-8}{\ln(17/8)} = 11.94\,\text{K}$$

$$\text{Air off coil at 85\% air flow, } 20 - \frac{20-11}{0.85} = 9.41°\text{C}$$

$$\text{Coil performance at 85\% air flow, } (0.85)^{0.8} = 0.88$$

$$\text{LMTD at 85\% air flow, } \frac{11.94}{0.88} = 13.6\,\text{K}$$

See the cooling curves in Fig. 27.2. The evaporating temperature falls to about 0.2°C. Compressor manufacturers' software shows 10.3% loss in duty and 1.5% less power at the new condition – a power increase of 9% to sustain the same load, assuming the compressor cycles off for shorter intervals. A more accurate estimate can be obtained by calculating a new basic rating for the reduced airflow. This results in about 8% extra power. The provision of a manometer across the filter to indicate the pressure drop will give a positive indication of the need to clean or replace.

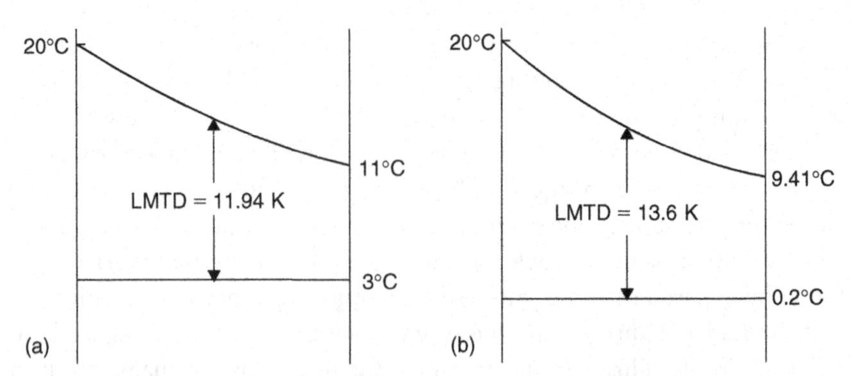

Figure 27.2 ***Effect of airflow reduction.*** (a) Clean filters, (b) dirty filters.

Fouling of a condenser, either water or air-cooled, has the effect of raising condensing temperature, and this directly increases compressor power consumption. This may easily go unnoticed until the compressor trips on high pressure cut-out when the outdoor temperature reaches an unusually high level. A typical condenser before and after cleaning is shown in Fig. 27.3.

Incorrect refrigerant charge is another fault that can result in prolonged operation at sub-optimal conditions. Loss of refrigerant causes a reduction in the wetted surface in the evaporator and reduction in evaporating temperature, leading to a similar situation to the blocked air filter. If charge is being lost the effect on performance may be delayed if extra refrigerant is held in a liquid receiver, or if the operating conditions are such that the full charge is not required. Clear liquid, without bubbles, in the sight glass upstream of the expansion valve can be a useful indication of adequate charge but it is affected by operating conditions, and it is insufficient indication for an economised system where sub-cooling effect is lost if vapour is entering the sub-cooler expansion device. Overcharging can also occur as a result of adding charge unnecessarily. In severe cases this has been known to cause rapid short cycling as a result of high condenser pressure caused by excess liquid in the condenser.

Erratic expansion valve operation can occur as a result of poor adjustment and this can give rise to excessive fluctuations in pressures and temperatures, particularly at low-load conditions. Unstable operation can give rise to reliability problems, particularly if it results in low superheat and potential liquid damage to compressors. It is likely to be difficult at the design stage to predict system dynamic behaviour, and changes to expansion valve specifications and settings may be necessary on commissioning

Figure 27.3 ***A typical condenser before and after cleaning (Advanced Engineering).***

or during maintenance. Manufacturer's documentation for packaged components such as chillers should carry clear instructions regarding superheat settings at appropriate running conditions to enable commissioning and maintenance technicians to subsequently check operation. High superheat results in poor performance due to low evaporation pressure and there is also a risk that the compressor discharge temperature becomes too high.

A planned maintenance schedule should be designed to ensure that the above faults are detected and remedied, together with cleanliness of secondary circuits, water and airflow checks, water treatment, etc., the details of which will be specific to individual plants.

27.6 REPAIRS AND MODIFICATIONS

Depending on the extent to which a system is disturbed by major changes of components, it will be necessary to adopt the procedures for evacuation, leak tightness and pressure testing applicable to new installations (see chapter: Installation and Construction). Similarly some repeating of the commissioning procedures will be necessary.

There may be an additional risk of moisture ingress during repair activities. Moisture in halocarbon circuits is indicated by the colour trace on the sight glass, where this is fitted. The drier should be changed and the sight glass monitored for reversal of the colour to 'dry'. Severe cases of contamination may need a second change of drier. Oil analysis tests can be taken to validate the moisture content in the oil and if hygroscopic oils are badly contaminated it may be preferably to replace the oil. If the liquid line leaving the drier or strainer (if separate) is colder than the inlet, this indicates pressure drop due to blockage and a new drier, or cleaning of the strainer is required.

In the maintenance of R744 systems care must be taken to avoid liquid being trapped in sections isolated by closure of valves. Subsequent temperature rise can cause pressure to rise above safe levels. Venting of liquid of atmosphere can give rise to very low temperatures or solidification of the refrigerant causing a valve to become blocked by solidified refrigerant. Technicians should receive R744 training prior on working on these systems.

27.7 FAULTFINDING

System faults fall into two general classes: the sudden catastrophe of a mechanical breakdown and the slow fall-off of performance which can be detected as a malfunction in its early stages, but will also lead to a breakdown if not rectified. Identification of the first will be obvious. To track down the cause of a malfunction is more complicated.

Fault tracing is seen as a multi-step process of deduction, ending in normal operation again and a record of the incident to inform other operatives. The steps are as follows:

1. Detection, that is detection of abnormal operation.
2. Knowledge of the system to track down the cause.
3. Observation of exact operating conditions.
4. Identification of the fault.
5. Decision: What to do? How? When? Can it be left?
6. Action to rectify the fault.
7. Test: is it now normal?
8. Record note in log, for future information.

An experienced technician will often know where to look, and more sophisticated tools, which give immediate indication of superheat, and sub-cooling, for example, are now becoming available. Often refrigeration systems show dynamic characteristics that are difficult to follow if not visualised in graphs. In Fig. 27.4 the poor operation of the expansion valve at part load (superheat too low) is made clear, and the improvement in isentropic efficiency (and hence COP) when compressor goes to full load is can also be seen.

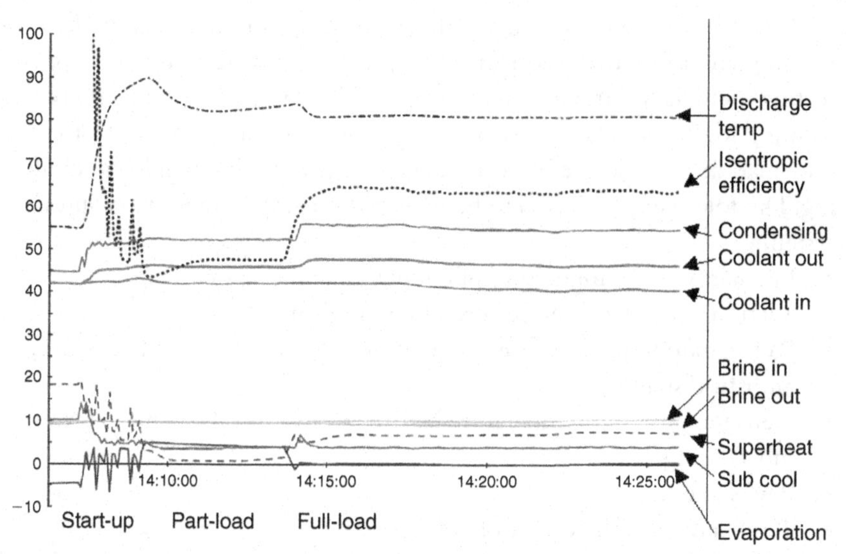

Figure 27.4 ***Industrial chiller start-up, part-load and full-load operation (Climacheck).***

Frequently the symptom of a fault is failure of a compressor to run. When the obvious faults such as electrical supply have been verified, this frequently comes down to contaminants in the system or overheating causing loss of lubrication. Motor winding insulation failure or "burn-out" can be the result. Oil analysis will point the way, but before any rectification starts it is necessary to thoroughly clean the system. Traditionally this could be a complex operation involving dismantling of pipework and flushing components with a solvent. More recently sophisticated flushing kits with built in recovery capability enable refrigerant to be re-used and a flushing cycle to be operated that can reach all parts of the system. The principle is based on the miscibility and solubility of refrigerant and oils, coupled with that of acids and water, together with refrigerant/oil/acid/water decanting and separation through evaporation.

27.8 RUNNING LOG

The detection of abnormal operation can only occur if normal operation is monitored. The skilled operator or the visiting service technician will have a working knowledge of the pressures and temperatures to be expected, but will not be able to make an accurate assessment of the actual conditions without previous measurements for comparison. The commissioning log will show readings taken at that time, but only at one set of running conditions.

It is therefore essential on a plant of any size to maintain some kind of running record, so that performance can be monitored with a view to detecting inefficiency and incipient troubles. The degree of complexity of this running log must be a matter of judgement, and a small amount of useful information is to be preferred to a mass of data which would be confusing. The following are basics to be included for simple vapour compression system:

- Inlet air/water temperature of condenser and evaporator
- Suction and discharge pressure of compressor(s)
- Refrigerant temperature at the inlet of compressor and outlet of evaporator (if distant)
- Refrigerant liquid temperature before expansion valve. This determines sub-cooling
- Power input
- Temperature at the outlet of compressor

With these measurements all relevant parameters for a suction gas cooled compressor can be calculated, that is COP, capacity, sub-cooling, superheat, compressor efficiency together with evaporator and condenser temperature differences.

Experience shows that few of running logs taken today have a quality that allows evaluation of performance and that there is frequently a lack of instruction or advice on how such logs should be taken. The log should be taken with consideration to the stability of the system. One proposal is a series of three sets of 'records' with 2 minutes in between after the system has been made to operate at stable conditions as far as is practically possible. This would take 6 minutes and cannot be seen as a costly requirement but it would allow any person evaluating the data to judge how stable the system is, and judge if the data is reliable and worth an evaluation.

Much of this recording is now automated as part of the BMS system. There is a danger of being overwhelmed with data and it still requires competence to interpret the information and identify potential problems.

27.9 LEAK TESTING

Loss of refrigerant can have a direct impact on global warming will eventually cause inefficient and unreliable operation. The EU F-gas Regulation No. 842/2006 directs that operators of HFC equipment must prevent leakage, ensure leak checks are carried out and repair any leaks as soon as possible as well as arranging proper refrigerant recovery. The detailed requirements concerning frequency of mandatory checks, qualifications for personnel and other matters can be obtained from the Institute of Refrigeration (IOR). Certain equipment will need instruction manuals containing information about the HFC in use.

Methods for identifying leaks are given in chapter: 'Installation and Construction' and a study of types of leaks and methods of minimising them has been made by Bostock (2007). For ongoing detection fluorescent additives can be added to the system, and when they are carried round by the oil, a leak can be pinpointed with an ultraviolet lamp. The plant must be cleaned after use and any additions of dye to the system must be noted because unlimited use could impair oil properties.

27.10 MANDATORY INSPECTIONS

The F-Gas regulation and the Energy performance of Buildings Directive (EPBD) cover inspection requirements.

The F-Gas regulation deals with minimisation of refrigerant emissions and requires operators of systems containing more than 3 kg HFC refrigerants to implement a leakage inspection schedule, the nature of which is dependant on the size of the system. Full details and guidance can be found from the ACRIB website.

The EPBD requires a performance inspection of all air-conditioning systems over 12 kW rated output installed in buildings, and this may include the combined output of several air-conditioning systems. This has been implemented in the United Kingdom by a revision to the Building Regulations 2000. The inspections are intended to give building owners information about the performance of their equipment and identify energy saving opportunities. European Standard EN15240, 2008 provides guidelines for inspection of air conditioning systems and CIBSE TM44, 2012 is used as the guidance for air conditioning inspections in the United Kingdom. The guidance is primarily intended to support inspections which are carried out for compliance with the EPBD.

27.11 TRAINING AND COMPETENCE

Refrigeration and air conditioning equipment requires specialised training for the people operating and maintaining equipment. The basic skills of those entering the industry at this level should be the ability to read an engineering drawing and to read and understand flow, circuit and electrical diagrams. A route map for refrigeration and air conditioning qualifications can be found on the IOR website, together with guidance notes and links to further sources of information.

REAL Zero was developed by the UK Institute of Refrigeration to provide a series of guidance notes based on research in a variety of sites on the impact of refrigerant leakage on systems and end users.

Many of the larger companies run their own training schemes, and courses both with and without practical content, part- and full-time, and of various standards, are run by technical colleges and colleges of higher education. Training organisations offer short courses, which cover essential requirements and skills, including certification where necessary, as well as specialist courses covering specific types of system and technologies. Tailor made and in-house courses are also offered.

The F-Gas regulation requires that technicians who carry out service and maintenance work on stationary equipment containing fluorinated gases (HFCs) must be certified. The exact requirements for certification can be found from the ACRIB website.

Air conditioning plant comes under the Building Regulation Act for England and Wales under 'controlled services'. Competent Person Schemes were introduced to allow organisations and individuals to self certify their work as being in compliance with the Building Regulations. The term 'Competent Person' in this case refers to a company and not an individual.

CHAPTER 28

Efficiency, Running Cost and Carbon Footprint

28.1 INTRODUCTION

The environmental impact of energy consumption is coming under close scrutiny, and refrigeration is no exception. Nearly half (46%) of the final energy consumed in the United Kingdom is used to provide heat and contributes to approximately 31% of United Kingdom total CO_2 emissions. Domestic and commercial building heating accounts for 67% of the total heat demand and refrigeration, air conditioning and heat pump (RACHP) accounts for 19% of UK electricity demand or 7% of total CO_2 emissions (derived from a UKERC report by Chaudry et al, 2014). Efficiency improvements can potentially make a significant impact on reduction of carbon emissions.

At present, there is no definitive method for the calculation of the carbon footprint of a refrigeration system, but this does not prevent calculation of the lifetime carbon emissions using the total equivalent warming impact (TEWI) concept as explained in chapter: Refrigerants. The energy consumed in manufacture and disposal is likely to be small in comparison with energy consumption over a typical 15-year life, and so energy efficiency along with leakage of global warming substances must take top priority when considering carbon footprint.

Several additional benefits accrue from improvements in energy efficiency, including reduced running cost, and less primary fuel consumption, that is gas, oil and coal, but these additional benefits are dependent on a number of different factors and can change according to circumstances.

Integrated systems (chapter: Heat Pumps and Integrated Systems) can be an effective way of providing heating and cooling, but this in itself does not guarantee energy efficiency. It is the performance of the system when dealing with the necessary load that determines its efficiency.

28.2 ENERGY EFFICIENCY, COP AND SEI

COP is the parameter most used to quantify efficiency of refrigeration and heat pump systems (chapter: The Refrigeration Cycle), but this is not a true efficiency indicator and it is strongly dependent on operating conditions.

Refrigeration, Air Conditioning and Heat Pumps
http://dx.doi.org/10.1016/B978-0-08-100647-4.00028-0

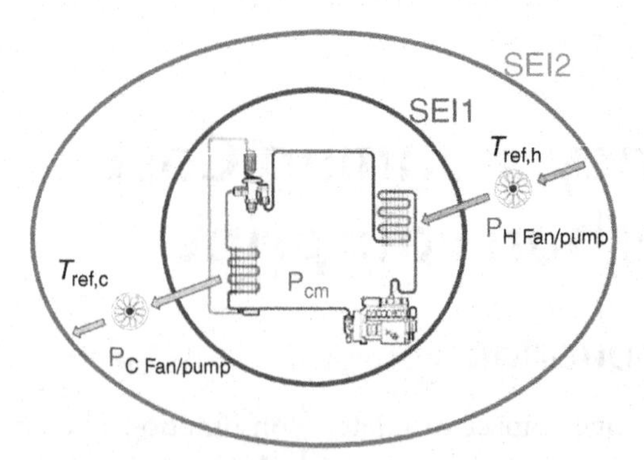

Figure 28.1 ***System boundaries for efficiency comparisons, compressor power input Pcm, pump/fan power inputs P_C and P_H.***

The system efficiency index (SEI) has been independently proposed by Maidment et al. (2007) and by the German engineering association, VDMA. The ideal or Carnot COP provides the ultimate reference, consistent with the laws of thermodynamics, for a process of transferring heat energy to a higher temperature level. The COP achieved through design or through field measurements is divided by the ideal COP:

$$\text{SEI (cooling)} = \frac{\text{Actual COP}}{\text{Carnot COP}} = \frac{\text{Cooling capacity}}{\text{Power Input}} / \frac{T_{\text{ref,c}}}{T_{\text{ref,h}} - T_{\text{ref,c}}}$$

$$\text{SEI (heating)} = \frac{\text{Actual COP}}{\text{Carnot COP}} = \frac{\text{Heating capacity}}{\text{Power Input}} / \frac{T_{\text{ref,h}}}{T_{\text{ref,h}} - T_{\text{ref,c}}}$$

where $T_{\text{ref,h}}$ and $T_{\text{ref,c}}$ are the secondary fluid temperatures at the reference conditions. Fig. 28.1 illustrates the general concept of system boundaries where SEI1 refers to the refrigeration cycle itself, SEI2 accounts for power input to fans and pumps required to operate the cycle. This is the power required to circulate the secondary medium (air or other fluid) through the refrigerant heat exchangers. Where the pump also circulates fluid through distribution circuits the relevant portion of the pump power is included in SEI2. Further boundaries can be added if it is desired to include pumping or fan power of distribution systems.

In addition to the system boundary, the reference temperatures must be defined. It is preferable to use mid or mean values where possible. For secondary flows in condensers and evaporators this would suggest mean of fluid entering and fluid leaving temperatures. However, for practical reasons and due to the difficulties of measuring condenser exit air temperatures, the entering air temperature is generally used. In order to analyse the performance of a system further and to find reasons for poor performance, breaking down the SEI into sub-efficiencies can be very useful. The sub-efficiencies indicate efficiency of the components of the refrigeration process. The SEI is the product of these sub-efficiencies. By analysing the sub-efficiencies it is possible localise where in the process the performance problem exits.

Refrigeration cycle efficiency: This takes account of losses inherent in the refrigeration cycle itself, and can, for example be used in the design process to compare effects of different refrigerants, or the effects of the use of an economiser.

Compressor efficiency: For cooling cycle this is the compressor isentropic efficiency (IE). For heating the effect of IE is slightly less since the compressor losses result in heat addition to the refrigerant and form part of the useful heat output.

Pressure drop in refrigerant lines: This effect is probably most significant for large refrigeration systems typically with several long pipes.

Heat exchanger efficiency: This includes the heat exchangers effectiveness and is important in evaluating or comparing condenser and evaporator temperature differences.

Fluid transfer efficiency and non-useful heat loss/gain: This includes 'non useful' heat pick up.

All of the aforementioned parameters can be identified during the design process or from field measurements. A full description and analysis can be found in the SP report, 2014. In summary, COP answers the question about how much energy is used for production of cooling and heating at a specific measurement point. The indicator is specific for the measurement point and it is difficult to relate to other circumstances. SEI answers the question how efficient the process is in the same point. The measured value can be compared with values for other conditions. It shows the potential for optimisation and the quality of COP.

28.3 ENERGY EFFICIENCY, PRACTICAL CONSIDERATIONS

The following points have been prioritised by the Institute of Refrigeration when designing for energy efficiency.

28.3.1 Avoid Refrigeration

There is no point designing an efficient system if the load is unnecessary and so the starting point is to ensure that the load is an absolute minimum.

This requires a critical energy survey of the process, the cold store operation or the building to be air conditioned. The purpose of this survey is to determine the elements of the cooling/heating load, and ways of reducing each item to a minimum, consistent with the cost of doing so. Such an evaluation may not be exact, but the need at this stage is to put these cooling loads into some order of magnitude to find what proportion of the total load may be reduced by any form of treatment. The load elements are following:

- Reduction in temperature of the product
- Heat conduction through building structure; consider better insulation
- Direct solar radiation into the cooled space; consider shading
- Convection heat gains/losses from infiltration
- Heat input from auxiliaries – fan motors, pumps, defrosting
- Lighting – efficient equipment and switching schedule. This energy may be paid for twice – to put it in and then to take it out again

Further, find out if free cooling can offset any portion of the load. The term *free cooling* is used to describe any method that does not require mechanical refrigeration although it may not be totally free. It can potentially be applied whenever the wet-bulb temperature is below the required temperature level of the cooled space. Recognising that evaporative cooling may not always be practical, there are still opportunities for free cooling throughout a percentage of the year when outdoor temperatures are low and heat needs to be removed from buildings. Increasing the fresh air proportion is one way to reduce the load with air distribution systems.

Chilled water for circulation in the building may be pre-cooled by the outside air prior to being cooled by the refrigeration process. An example of a free cooling chiller air conditioning circuit is shown in Fig. 28.2.

The water to be chilled passes from the inlet at the top to the evaporator (shown at the bottom of the diagram) via a three-way valve. During normal operation, the three-way valve directs all the water to the evaporator. In this example, there are two identical condensers, and each condenser has three refrigerant coils and an additional water coil. When a falling ambient temperature reaches a certain point, water is directed into these coils by the three-way valve and becomes pre-chilled by the ambient air prior to passing through the evaporator. Eventually the air temperature may become sufficiently low to enable all the cooling to be performed directly from the cold outside air, and the refrigerant circuits become idle. The savings are visualised in Fig. 28.3.

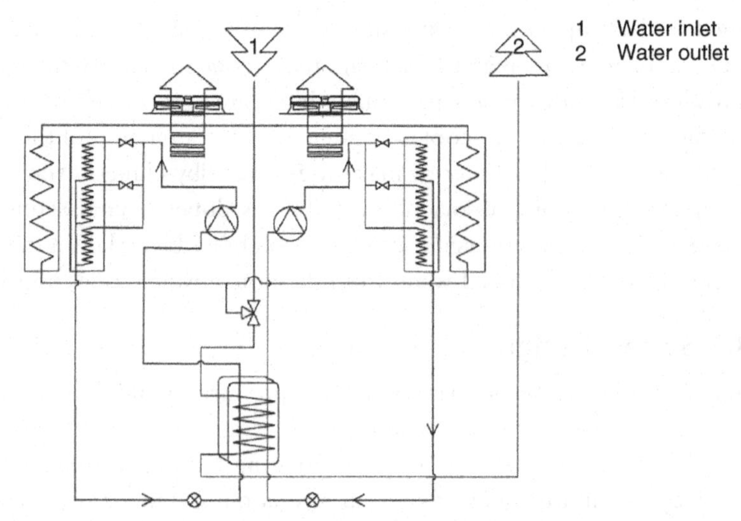

Figure 28.2 *Dual-circuit free cooling air-cooled chiller flow diagram (Airedale).*

Chilled water may be produced without running the compressors or fans with *thermosyphon system* as described by Pearson (1990) and Blackhurst (1999). When the ambient is at a lower temperature than the chilled water, condensation can occur at a pressure below the evaporating pressure in the chiller. By positioning the condenser above the chiller, the liquid can be made to return to the chiller by gravity. The chiller needs to be carefully selected but does not need to be special. Plate and shell chillers or flooded shell and tube chillers can be used. Direct expansion tubes in shell coolers

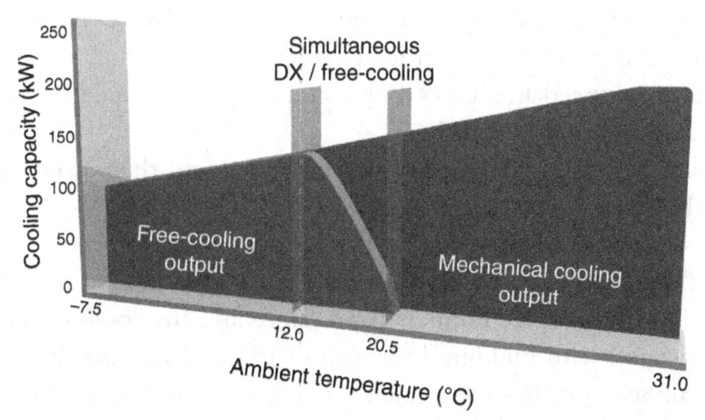

Figure 28.3 *Illustration of free cooling output for a chiller with additional air/water coils (Airedale).*

designed for TEV operation are not suitable because the refrigerant will not boil and circulate naturally in them. Evaporative condensers cool the refrigerant towards the ambient dew point allowing more thermosyphon cooling during the summer than air-cooled condensers. One attraction of the thermosyphon system is the ability to make use of standby chillers even when other chillers have compressors running. This is done by connecting the chillers in series on the waterside with the lead chiller being last in the line, and receiving partially chilled water from the non-powered ones upstream.

28.3.2 System Design

It is mostly well understood that condensing pressure should be minimised, and allowed to float downwards as ambient air temperature reduces. Savings can also be made by maximising the evaporating temperature, and in this respect loads should be split where separate cooling temperatures are required to avoid all compressors working at the suction pressure corresponding to the lowest load temperature. For large temperature lifts consideration should be given to enhanced cycles, for example two-stage or economised cycles that give an inherently better efficiency.

The design should focus on the operating conditions at which most running hours occur. Optimisation at these conditions will be more beneficial than a single extreme design point calculation. For high efficiency, single-circuit chillers with two compressors can utilise the full surface area of heat exchangers in part load conditions.

28.3.3 Control Philosophy

Some points to consider here include the following:

- Avoid fixed set head pressure control to ensure low condensing temperatures are available at part load operation.
- Avoid fixed speed auxiliaries. Pump and fan power should be reduced at low loads wherever possible.
- Ensure defrost is only applied when needed and then as efficiently as possible.

For comfort air conditioning systems a single comfort condition is often specified. This may be, for example 21°C and 50% saturation, but variations within the accepted comfort band can reduce the cooling and heating costs associated with building heat gain or loss and also for the treatment of the proportion of fresh air which is needed. For example, summer fresh air reduced to 21°C, 60% saturation, imposes 4 kJ/kg less cooling load, and winter air raised to 40% saturation requires 4 kJ/kg less heating than at 50%

saturation. These changes in humidity cannot be detected without instruments and have a small effect on personal comfort.

A further consideration is to allow the indoor condition to vary slightly depending on ambient conditions. People dressed for summer conditions with an outdoor temperature above 27°C will generally be comfortable with a room temperature above 21°C. For example the design maximum for the London area may be taken as 27°C dry bulb, but exceeds this for an average of 25 h/year. The increased cooling power to hold 21°C under such conditions is considerable and experience has shown that short-term rises can be permitted. Automatic controls can have a bias set point imposed by a high ambient.

28.3.4 Component Optimisation

Here we consider the main components of the system with a view to ensuring choices that benefit system efficiency. There may be a cost trade-off, but in many cases it can be more simply a case of using components best suited to the application.

- Compressors and condensing units – select for normal running conditions, whilst at the same time having adequate capacity for the design point.
- Evaporating units such as food display cabinets, freezing tunnels and similar components should be checked for evaporating temperature at which they are required to work, and the amount of fan and lighting power consumed.
- Consideration of temperature differences for choice of heat exchangers.
- A refrigerant appropriate to the application should be chosen. The impact of the refrigerant on efficiency is likely be less than 5%, assuming all other design parameters are optimised.

28.3.5 Operation and Maintenance

Maintenance and its effects on energy consumption are considered in chapter: Commissioning and Maintenance. A study made in Sweden by Prakesh (2006) provides an insight into the potential for energy saving in existing installations. Over 150 systems of various types were analysed and the following is a list of main faults, by frequency occurrence, all of which contribute to excessive power consumption:

- Incorrect system charge – mostly undercharge, which causes low evaporating temperature.
- Incorrect adjustment of expansion valve.
- Secondary flow problems.

This is after attending to the obvious points such as dirt on filters, coils and fans, etc. Only 13% of systems were functioning in an optimised manner, and it was estimated that on average 10% energy savings could be achieved if all installed systems were operating correctly. This study supports the widely held belief that there are a large number of installations operating in the field at energy efficiencies well below the desired and practically achievable levels.

Historically, low price has often been the only interest from equipment purchasers/owners and is still often the most important criterion, making it difficult to sell on claims of long-term savings. An additional issue is that after installation there is rarely any follow up to see if the plant installed gave the performance estimated in the calculations. There is little risk of exaggerating performance in quotations, for example by excluding accessory power or not taking actual operation conditions into account. On-site measurement of electricity consumed, kWh, over a period is relatively straightforward, but climate and variations in usage pattern can easily mask equipment performance shortfalls.

28.4 RUNNING COST

The aforementioned energy efficiency measures directly affect reducing running costs, but there are additional points to note which concern actual cost of operation, but not necessarily energy consumption.

Prime mover – the vast majority of equipment is electric motor driven, but savings may be made by applying absorption systems where waste heat is available, or by using IC engine drive, particularly where there are opportunities for heat recovery. Gas engine driven variable refrigerant flow (VRF) air conditioning systems that utilise engine waste heat for heating are available on the market. Combined heat and power (CHP) systems can generate sufficient electricity to run the cooling equipment, or may use waste heat for absorption cycle. This is termed 'Tri-generation' and is the subject of research.

Tariff structure – running equipment at certain periods during the day can save costs by taking advantage of off-peak electricity tariffs. Ice banks or other phase change materials are applicable here (see chapter: Distributed Cooling and Heating) and can be used to reduce running costs. If the tariff includes a maximum demand penalty, automatic load-shedding demand limiters may be applicable.

28.5 LIFE CYCLE COST ANALYSIS

Lifecycle cost analysis is used to support the selection of energy efficient products by demonstrating he real cost of ownership. In its simplest form it accounts the true cost over, say, a 5 year period by estimating the operating hours, power consumption and energy tariff to derive a running cost. Lifecycle calculation frequently shows that over a given period, refrigeration equipment running costs can be several times greater than the initial equipment purchase price. LCCAs demonstrate the effect of overlooking running costs in favour of short-term investment savings. Fig. 28.4 illustrates a typical result.

For example, two refrigeration condensing units are available to achieve a desired room temperature. Unit A is £200 cheaper to purchase and appears to be the best option. However, LCCA shows that its energy consumption is 500 W greater than the more expensive unit B. In real terms, this translates to an additional £300 on the annual electricity bill when using unit A. As such, the initial £200 saving offered by unit A is more than offset in the first year of use. Unit B could potentially save the end user as much as £1,300 in energy costs over 5 years.

More sophisticated LCCA models take account of depreciation, disposal costs, and projected energy and maintenance costs to determine optimum replacement period.

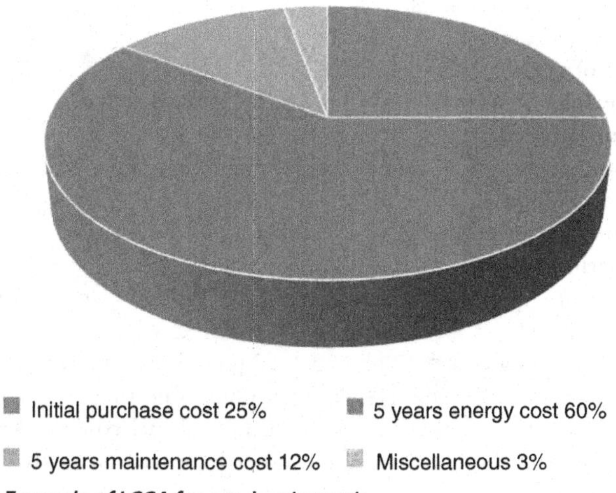

Figure 28.4 ***Example of LCCA for condensing unit.***

28.6 HEAT PUMP/BOILER COMPARISON

Example 28.1

A building requires 15 kW of heating, currently provided by a gas boiler having an efficiency of 85%. It is planned to replace this with a heat pump having an average or seasonal COP (SCOP) of 3. Will this result in more or less primary fuel consumption and will the carbon emissions attributable to the operation of the system be increased or reduced? What about the running costs?

Gas consists mainly of methane, CH_4. When it burns in the boiler, it reacts with oxygen in the air to form carbon dioxide and water. For every kWh of heat 0.19 kg of CO_2 is liberated. The boiler is 85% efficient, and hence 17.65 kW equivalent of gas is required.

For the boiler

Rate of primary fuel use = 17.65 kW

Rate of CO_2 emission = 17.65 × 0.19 = 3.35 kg/h

Running cost = 17.65 × Gp per h of heating (where Gp is gas price, pence per kWh)

The heat pump will use 5 kW of electrical energy, and the primary fuel consumption required to produce this 5 kW can be as low as 10 kW, taking into account generation and distribution losses. This value is based on the latest gas turbine combined cycle generation technology, in other words burning the gas in a modern power plant instead of locally at the point of use. For electricity drawn from the grid, there is an average carbon emission factor of 0.49 kg CO_2 for every kWh of electricity supplied at the user's meter (Section 3.4). This takes account of the generation mix of coal, oil, gas, nuclear and renewables.

For the heat pump

Rate of primary fuel use = 10 kW

Rate of CO_2 emission = 5 × 0.49 × 2.5 kg/h (based on average carbon emission factor)

Running cost = 5 × Ep per h of heating (where Ep is electricity price, pence per kWh)

Running cost comparisons are complicated by an array of suppliers' energy tariffs and regional variations. For example

Gas (Gp) = 3 p/kWh, Electricity (Ep) = 9 p/kWh (mix of peak and off peak)

Boiler running cost = 17.65 × 3 = 53 p per hour of heating

HP running cost = 5 × 9 = 45 p per hour of heating

Note that it is necessary to establish the real fuel cost, per kWh, taking into account service charges and other local factors, to obtain a true comparison.

28.7 EFFICIENCY STANDARDS, CERTIFICATION AND LABELLING

The introduction of reliable testing procedures where results can be independently verified has enabled refrigeration, air conditioning and heat pump products to be 'rated' for energy efficiency. Domestic refrigerators are energy labelled and performance labelling has been extended to household air-conditioners below 12 kW rated cooling output. Larger packaged or split units are not required to be labelled, although a similar performance rating system is used where the performance of such products is published by Eurovent Certification. Eurovent is a voluntary initiative of the industry and certifies the performance ratings of air conditioning and refrigeration products according to European and international standards. A similar scheme for refrigeration compressors is operated by ASERCOM.

The revision of Part L of the Building Regulations published in 2006 introduced, for new buildings, a method of compliance based on a calculation of building carbon dioxide emissions. A Non-Domestic Heating, Cooling and Ventilation Compliance Guide (2006) specifies the minimum acceptable efficiencies for air-conditioning equipment to be used in new construction and when replacing existing systems.

The European Seasonal Energy Efficiency Ratio (ESEER) is a weighted formula, which takes into account the variation of system COP, as applied to a chiller, with the load rate and the variation of air or water inlet condenser temperature. It is becoming widely adopted as a criterion for chiller performance and details can be found on the Eurovent website.

In the United Kingdom heat pumps may be eligible for grants under the UK Microgeneration Certification (MCS) scheme. The requirements cover performance and installation. Enhanced capital allowances (ECA) are available for the purchase of products that meet defined efficiency criteria and these are published by the Carbon Trust on the Energy Technology List (ETL). Refrigeration products include compressors, condensing units, heat pumps, chillers and cellar cooling equipment.

The European Heat Pump Association (EHPA) has introduced a quality label which shows the end user a quality heat pump unit or model range on the market. Testing by an accredited centre is required. The EU Ecodesign Directive (Directive 2009/125/EC) is superseding some of these earlier initiatives. It is a framework directive that obliges manufacturers of energy consuming products to reduce the energy consumption and sometimes also other negative environmental impacts occurring throughout the product life cycle for products that are within scope of the many separate 'implementing directives' under the framework. The Directive is complemented by the Energy Labelling Directive (Directive 2010/30/EU). Products are considered under 'Lots' or groups of similar products. The scheme covers all industrial products. Included are refrigeration products - display cabinets, condensing units, room air conditioners, unitary heat pumps, and process chillers. Implementation progress and changes on these schemes is on-going, see relevant websites and IOR Guidance Note "Energy Related Products Directive" for further information on this topic.

28.8 COMMITMENT TO ENERGY SAVINGS

A positive energy policy needs to be a company decision, taken at boardroom level and backed by boardroom authority, since it cuts across departmental boundaries and may conflict with departmental financial targets. Typical issues are

- The capital, operating, maintenance and fuel costs come from four separate budgets, possibly accounted for by four different managers, so these budgets need to be coordinated.
- Documented system analysis and/or energy metering may be needed to prove the savings.
- There may be some disruption to normal working whilst energy investment schemes are being investigated and implemented.
- Staff may need to be released for training schemes.
- The improvements may require changes in operating methods previously considered as adequate.

It is important to be able to quantify the results of an energy conservation programme. It is an ongoing process with all concerned alert to the possibilities of further improvements.

CHAPTER 29

Noise and Vibration

29.1 INTRODUCTION

Unwanted sound is noise, and the sound generated by cooling and heating equipment potentially causes annoyance indoors and outdoors. This makes it an essential consideration for system design in many situations. The detailed theory of sound is beyond the scope of this book, and can be found in specialist books on the subject. Useful references include the Audio Engineer's Reference Book, for air conditioning CIBSE Guide B4, and for standards, BS EN 12354, parts 1–6. Here a brief summary of relevant noise sources, transmission, perception and attenuation is given, together with some indication of how to interpret data. Some basic points concerning vibration are also included.

Building occupants may be troubled by noise generated within the building by its air conditioning system or other cooling or heating equipment. Occupants of offices, hotel bedrooms and similar 'quiet' locations will be disturbed by relatively low sound levels. The acceptance level for noise in plant rooms is higher with limits subject to regulation, see Controlling Noise at Work (2005) for details. Noise transmitted or originating outside the building can cause annoyance to neighbours and its control comes under the responsibility of the local authorities which may require the noise to be abated. Conditions may apply to planning consents in order to protect neighbours from nuisance caused by noise. Noise-level prediction requires information about the noise emission from the source, and about the transmission and attenuation. In addition, flow-generated noise emission from duct systems may need to be considered. The accuracy of the prediction will depend on the reliability of the input data. Manufacturers' data, based on British or International standard test procedures, should always be used.

29.2 NOISE SOURCES

Mechanical equipment produces noise over different frequency ranges as illustrated in Fig. 29.1, which also gives a typical subjective descriptions of the sound. Relevant noise sources include fans, duct components, grilles and diffusers, compressors, cooling towers, condensers and pumps.

Refrigeration, Air Conditioning and Heat Pumps
http://dx.doi.org/10.1016/B978-0-08-100647-4.00029-2

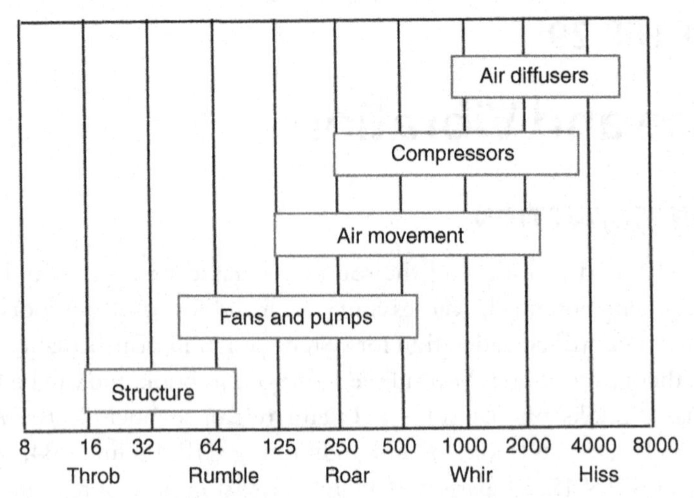

Figure 29.1 ***Typical frequency ranges, Hz, for various sound sources.***

Centrifugal fans produce most of their noise at low frequencies, whereas axial types generate higher-frequency noise. Fans with high tip speeds will generate noise levels that may require attenuation. Fan noise control depends on choosing an efficient operating point, design of good flow conditions, ensuring that the fan is vibration isolated from the structure, and that the fan is flexibly connected to the duct.

Noise generated by a room air terminal unit located in a ceiling void arises where conditioned air pressure is reduced. The pressure-dropping valve is noisy as a result of turbulent flow losses and a noise attenuation element can be included. Control of air velocity and flow conditions is the key to reducing noise arising from grilles and diffusers. Fan coil units are located close (1 m) to the occupants so that the prime sound level will be that from the direct sound path, and the primary source of noise will be the fan.

Compressors generate tonal noise linked to a fundamental tone based on the rotation speed. Different compressor types (reciprocating, centrifugal, scroll, etc.) have different sound characteristics:

1. Reciprocating compressors have a relatively low frequency fundamental tone. They produce more noise in the low frequency bands than the rotary types.
2. The noise produced by compressor motors is located in the mid-frequency bands.
3. Scroll and screw compressors have strong tones in the octave bands between 250 and 2000 Hz.

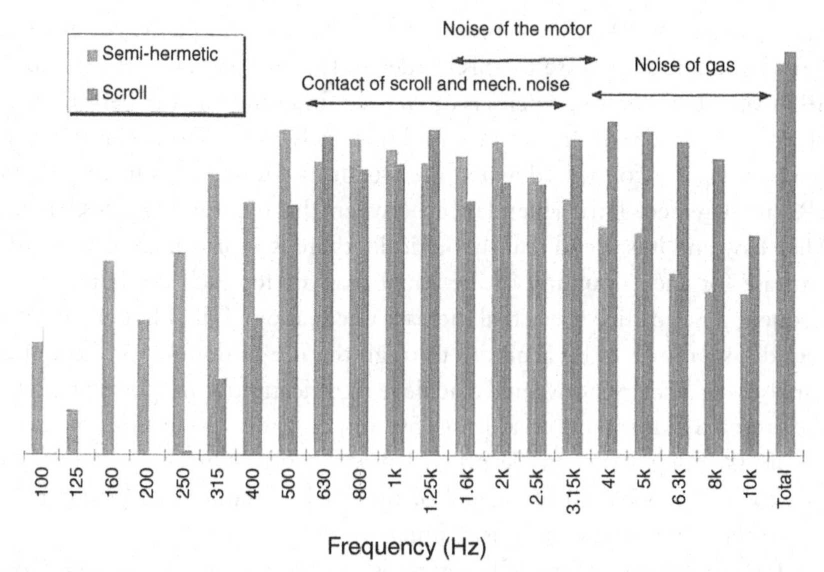

Figure 29.2 ***Comparative sound power levels, dBA, for a typical scroll and semi-hermetic reciprocating compressor type (Emerson Climate Technologies).***

4. The noise in the high frequency bands is due to the gas compression and flow.
5. The noise level will be dependent on operating conditions.

Sound jackets or acoustic enclosures may be used to provide attenuation at source.

In Fig. 29.2 the frequency banded noise levels for a typical scroll and semi-hermetic reciprocating compressor are shown. These are general trends only and manufacturer's data should always be consulted for specific information.

Pumps produce external noise from the motor, fluid borne noise from the impeller and vibration into both the structure and the pipes. Condenser, chiller and cooling tower noise arises from the component fans and compressors.

29.3 NOISE TRANSMISSION AND ATTENUATION

When airborne noise is generated in a room, by a compressor or another machine, it can be transferred to adjacent rooms via a number of transmission paths – walls, floors, gaps between the rooms, doors and the general building structure. The sound insulation performance of a material structure is termed sound reduction index (SRI) [sometimes called sound transmission loss (dB)]. The larger the SRI is for a partition, the better is the sound insulation. Because the SRI is dependent on frequency, the averaged SRI over a band of frequencies is often used for evaluating the performance of different materials. In simple one-leaf panels, where both exposed surfaces are rigidly connected, when the weight per unit area is increased, the SRI is increased. This dependence between the mass and the SRI is the Mass Law and it is noted that theoretically there is approximately 6 dB SRI increase for each doubling of the panel mass or for each doubling of the frequency. In practice the actual increase rate is about 4 dB. Higher values of sound insulation can be achieved through the use of double (or more) leaf panels with a cavity between. To achieve significant low-frequency benefits the cavity width should be at least 150 mm. In the cavity, porous materials should be hung, for example mineral wool or glass fibre mats. It is noted that for achieving high efficiency of the total structure, good sealing between the panel connections is required.

The normal treatment of fan noise is to fit a lined section of ductwork lined with an absorptive material on the outlet or on both sides of the fan. Such treatment needs to be selected for the particular application regarding frequency of the generated noise and the degree of attenuation required. Control of noise arising from air flow through air ducts depends on good flow conditions for the air stream, isolating vibrating components, and incorporation of in-duct silencers.

A passive attenuator contains narrowed air passages. Both rectangular and circular attenuators are used. The cross section of the attenuator is often significantly greater than that of the duct in which it is located. Longer attenuators have increased attenuation but cause additional pressure loss, although this is not necessarily a linear increase. Flow generated noise occurs at both the inlet and outlet of an attenuator as a result of the momentum change as the air speeds up on airway entry and slows on exit. This new noise sometimes (inaccurately) termed 'regenerated noise'.

Active attenuators detect the noise travelling in the duct, generate and add an opposing noise, of opposite phase which cancels out the original noise. They are most effective in the low frequencies, where passive attenuators have limited performance.

Transmission of noise to outdoors occurs mainly from equipment located outdoors such as chillers, air handling units, condensers and cooling towers. Screens or enclosures can be used to provide attenuation, and louvres with aerodynamic blades are employed where air circulation is required (Fig. 29.3). They offer a pressure loss dependent on face velocity.

Figure 29.3 ***An acoustic louvre enclosure (CIBSE reproduced courtesy of SRL).***

29.4 NOISE PERCEPTION

The human ear detects sound pressure levels expressed as decibels (dB) above a reference level of 20 μPa. The sound spectrum defines the sound in terms of frequency bands it may be octave bandwidth or narrower as in Fig. 29.2. The description of the spectrum as a single number is obtained by adding a weighting number to each octave band and logarithmically adding the octaves together. The weighting recognises that the ear is more sensitive to sound in the range 1–4 kHz than at higher or lower frequencies. The resulting single number is given as A, B or C weighted sound level. The A weighting is widely used. A dBA-filter makes a sound level meter less sensitive to very high and very low frequencies. Measurements made with this scale are expressed as dB(A).

The human comfort level depends on both frequency and sound pressure level. Noise rating (NR) curves are used to determine the acceptable indoor environment for hearing preservation, speech communication and annoyance. The curves for different sound pressure levels are plotted at acceptable sound pressure levels at different frequencies. Acceptable sound pressure level varies with the room and the use of it. Different curves are obtained for each type of use. Each curve is designated by a NR number. For example a quiet location such as office might require NR25 and a supermarket NR40. Spectra that fall under a particular curve are said to meet that curve. NR curves can be found in the Audio Engineer's Reference Book, and other reference works.

29.5 SOME SIMPLE RULES

1. For each doubling of distance from the source there is a 6 dB decrease of the sound pressure level.
2. Addition and subtraction of sound levels is logarithmic. The total combined the sound level (either power or pressure) of two separate machines having sound levels L1 and L2, when they are working together is given by:

$$10\,\mathrm{Log}(10^{0.1*L1}+10^{0.1*L2})\,\mathrm{dB}$$

   For the case of two identical machines, the addition of the second machine increases the sound level by 3 dB and addition of a third machine would increase the sound level by a total of 4.8 dB.
3. If the source is mounted on a smooth surface, the sound level will be equivalent to the free field level + 3 dB (see Fig. 29.4). In the case where the source is also close to a wall there will be a 9 dB increase in the sound level radiated in comparison with the free field case.

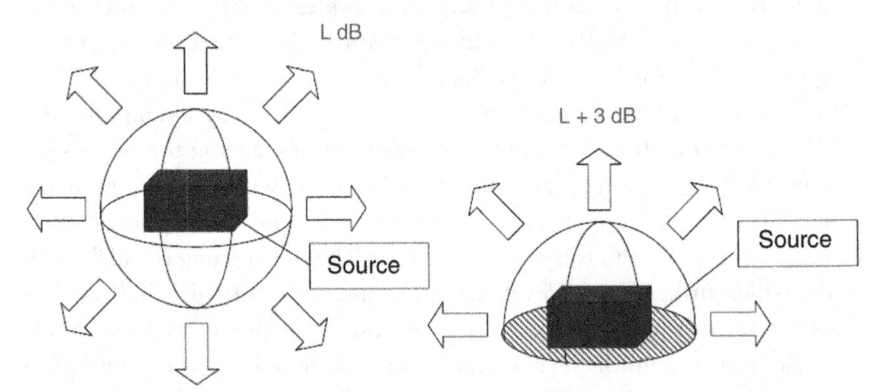

Figure 29.4 ***Effect of placing a sound source on a reflective surface (Emerson Climate Technologies).***

29.6 PUBLISHED INFORMATION

Manufacturers publish sound levels for their products and such figures should be scrutinised and compared as part of a purchasing decision. When comparing published noise levels it is essential to check that the figures are on a like-for-like basis. The main considerations are given as follows:

- Is the noise level a sound power level or sound pressure level?
- If sound pressure levels are quoted, what is the distance from the source?
- Are the quoted levels 'A' weighted?

Sound power is the rate of sound energy output of a source and its units are watts. Sound power is an inherent property of the source while sound pressure is also dependent on the surroundings (distance, reflection, absorption, transmission). Sound pressure is measured directly by a pressure sensitive microphone sensor; sound power is found from sound intensity measurements – sound intensity is the amount and direction of flow of acoustic energy per unit area at a particular location. 'A' weighted data is adjusted so that the individual frequency levels match the sensitivity of the human ear.

29.7 AIR SYSTEM NOISE

All air systems have a noise level made up of the following:

1. Noise of central station machinery transmitted by air, building conduction and duct-borne
2. Noise from air flow within ducts
3. Grille outlet noise

The first of these can be reduced by suitable siting of the plantroom, anti-vibration mounting and possible enclosure of the machinery. Air flow noise is a function of velocity and smooth flow. High-velocity ducts usually need some acoustic treatment. Grille noise will only be serious if long throws are used, or if poor duct design requires severe throttling on outlet dampers. Apart from machinery noise, these noises are mostly 'white', that is with no discrete frequencies, and they are comparatively easy to attenuate.

Where machinery of any type is mounted within or close to the conditioned area, discrete frequencies will be set up and some knowledge of their pattern will be required before acoustic treatment can be specified. Manufacturers should be able to supply this basic data and offer technical assistance towards a solution.

Where several units of the same type are mounted within a space, discrete frequencies will be amplified and 'beat' notes may be apparent. Special treatment is usually called for, in the way of indirect air paths and mass-loaded panels.

29.8 NOISE PREDICTION SOFTWARE

Prediction software is becoming increasingly used to calculate both internal room noise levels and external noise impacts. Care must be taken to check source data and noise pathways, and the input data must be in the right format. Commercially available computer prediction packages exist which provide a reasonably high level of accuracy. It is desirable to check against known situations to gain confidence.

Computer prediction models which have no specified level of uncertainty, should be used cautiously. Input of 'worst case' operating conditions is a good approach. Considerations for use of noise prediction software include the following:

- Validation of results
- Ensure that the software employs reputable and published algorithms
- An uncertainty estimate should be provided for the outcome and input uncertainties should be accommodated
- Comparison of results obtained from different software enhances confidence greater level of confidence.
- Variation of the input parameters helps in assessing uncertainty.

Familiarity and expertise in the subject matter enables the user to make critical judgments about the validity and robustness of the final results.

29.9 VIBRATION

Whereas airborne sound can be attenuated by the use of barriers and walls placed physically between the noise and the receiver, structure borne sound is generated by vibrations induced in the ground and/or structure. These vibrations excite walls and slabs in buildings and cause them to radiate noise. This type of noise cannot be attenuated by barriers, or walls but requires the interposition of a resilient break between the source and the receiver.

Quite apart from sound transmission, excessive vibration can reduce the service life of structures, and may interfere with proper functioning of equipment and human comfort. The best form of vibration control is avoidance, achieved by careful design, and selection and location of low vibration equipment. Vibration control may still be necessary by reducing vibration input through the mounting points of the vibration generating components to the building. The mechanical isolation of reciprocating and rotating components is via mount and hanger systems and flexible connectors. The helical spring is the most commonly used, most reliable and most predictable device employed in vibration isolation. They may be open, caged or enclosed and restrained and can also be used as pipe or duct hangers. Pads having elastic properties are easily applied and widely used but performance is somewhat unpredictable. The uncertainty of the loads also makes selection of vibration isolators difficult. Compressor and fan manufacturers' information should be followed where practical.

Measurement of vibration takes account of four parameters, displacement, velocity, acceleration and frequency. They are related in such way that given the frequency and any one of the other parameters, values of the others can be calculated. Frequency is an important parameter and is likely to produce a disturbing frequency coincident with that speed of rotation. For variable speed machines it may be necessary to investigate behaviour over a speed range and sometimes it is necessary to avoid certain speeds due to resonance effects. Sound and vibration attenuation is a specialist field and remedial situations may be require expert advice.

CHAPTER 30

Renewable Energy Innovations

30.1 INTRODUCTION

This short chapter concerns refrigeration technologies that have the potential to eliminate their carbon emissions arising from their operation. They are in various stages of development, and yet to be commercially proven.

30.2 RENEWABLE ENERGY STORAGE AND AVAILABILITY

A major challenge of renewable energy is efficient and cost-effective storage. Most forms of renewable energy, such as wind, tidal and solar radiation convert the source energy into electricity. Electricity can 'stored' in batteries or hydroelectric pumping schemes, but these are really conversion into another energy form, chemical reaction in the case of batteries, and gravitational potential in the case of hydro. Losses arising in the conversion – both ways – together with cost factors limit applications of these technologies. Any refrigeration or heat pump system consuming only energy directly from a renewable source could be described as '100% clean', in the sense that zero carbon emissions are generated as a result of its use (excluding any direct emissions from leakage of GWP refrigerants). Likewise, the same applies when drawing energy from a stored source derived from renewable such as battery. Another challenge of renewable energy is its low density. It is free, but requires large-scale collection and conversion facilities.

The vast majority of systems use energy from the grid. The mix of electricity generation types, fossil fuel, nuclear and renewable, in determining the carbon emissions associated with this energy source is represented by the CO_2 emission factor (chapter: Refrigerants, Section 3.4).

Geothermal energy is a carbon free and steady source of heat from rock in the earth's crust. It can arrive directly at the surface in the form of hot springs, but in most locations drilling is necessary. The concept is then for cold water to be injected down one borehole, pick up heat from the rock and return to the surface at temperatures up to about 180°C. High-temperature water could be used to power turbines for electricity generation. Lower temperatures are practical as a source for heat-powered cycles.

Refrigeration, Air Conditioning and Heat Pumps
http://dx.doi.org/10.1016/B978-0-08-100647-4.00030-9

This energy source should be distinguished from the energy stored near the surface and used for ground source heat pumps or integral systems with energy storage. These energy reservoirs are replenished by solar heating or cooling-system heat rejection.

30.3 EFFICIENCY

It is necessary to qualify statements of COP with the reference points or boundary conditions as explained in chapter: Efficiency, Running Cost and Carbon Footprint (Section 28.4), COP is energy delivered/energy consumed, and may take on a different significance depending on the energy source. For example, an absorption system will have a low COP compared to a vapour compression cycle, but the energy input is low-grade heat.

Example 30.1

Compare the COP of chilling water at a rate of 10 kW using an absorption cycle, to that of a vapour compression cycle both having a refrigerant evaporating temperature of 5°C and a condensing temperature of 40°C. Both are solar powered, with COP based on useful cooling/solar heat input.

For the absorption cycle, typical COP = 0.7, hence heat power input = 14.3 kW

For the vapour compression cycle, typical COP = 4.5, electrical power input = 2.22 kW

To achieve the 2.22 kW power input using a with solar PV panel with average efficiency 15% requires 14.8 kW solar energy, hence overall COP = 10/14.8 = 0.68

Storage of the solar power may be needed to cover variations in availability. To store the solar power production as hydrogen and re-convert back to electricity via electrolysis at the point of use would be about 50% efficient, so a vapour compression cycle using this source would have a COP of 0.34 based on the solar energy input. Likewise the absorption cycle may require heat storage if the source is intermittent, resulting in additional losses.

30.4 EXAMPLES

The heat-powered technologies described in chapter: The Refrigeration Cycle have the potential to operate directly from solar radiation. Solar-powered absorption units may take directly heated hot water as a source

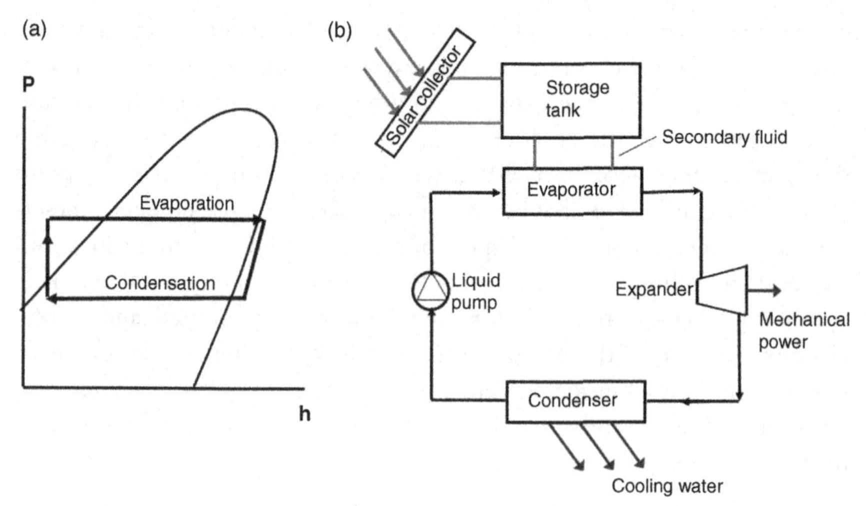

Figure 30.1 ***Organic Rankine Cycle, basic principle. (a) P–h diagram; (b) circuit schematic.***

for an ammonia/water cooling system to provide air conditioning during summer months. A study by Bajpai (2012) demonstrates the viability of such a system. In addition, two further innovative examples are:

Heat-powered mechanical refrigeration – solar or geothermal mechanical refrigeration uses a conventional mechanically powered vapour compression system where the mechanical power is produced with a heat-powered cycle. The Rankine cycle, similar to that used in steam-powered turbines is the cycle usually the one considered. It is termed *Organic Rankine Cycle* (ORC) because the fluid chosen is similar to a refrigerant, and suited to the working temperatures. It is vaporised at high pressure by heat exchange with another fluid heated by solar or geothermal collectors. A storage tank can be included to provide some high-temperature thermal storage. The vapour is expanded through an expander to produce mechanical power. An expander can be a piston, screw, turbine or scroll configuration, working in reverse (compared to a refrigeration compressor). The fluid leaving the expander is condensed and pumped back to the evaporator (boiler) pressure as shown in Fig. 30.1. It is necessary to choose a suitable refrigerant and the cycle can benefit from a liquid/vapour heat exchanger, similar to the refrigeration cycle suction/liquid exchanger. Detailed analysis and recent developments in expander technology are beyond the scope of this book, and some references are given.

Liquid air and liquid nitrogen expanders – the potential of cryogenic energy storage has started to be recognised, particularly with the advent of the cryogenic engine. This is an innovative expander which generates power in an efficient way in an expander and delivers useful cooling at the same time. Use of the expander to power a conventional refrigeration cycle together with the cooling produced by evaporation of the cryogenic fluid can provide a compact system with no carbon emissions at the point of use. Cryogenic power has become a candidate for replacement of diesel-driven cooling systems in transport applications and in remote locations where there is limited access to grid electricity. In assessing overall efficiency the liquefaction energy requirement should be accounted, and carbon emissions are zero if the liquefaction is powered by a renewable source, or maybe sometimes considered zero if the cryogenic fluid is generated as a by-product. Further information can be found on the Birmingham Centre for Cryogenic Energy Storage and IMechE websites.

APPENDIX

Notes on Units of Measurement

In this book, the SI system is used and numerous reference sources for definitions and conversions are available. Visual conversion illustrations can provide a quick way of relating quantities. The pressure and energy flow rate examples shown in Figs A1, A2 give relationships between commonly encountered units, the arrow points towards the larger quantity.

Imperial units, previously used in the United Kingdom, have now been superseded by international SI units, but are still used in the United States. These together with some other idiosyncrasies deserve special mention:

Btu and Btu/h – British Thermal Units are very deeply rooted in the world of heating and cooling. Data originating in the USA and some parts of Asia are likely to be expressed in terms of Btu. The rate of thermal energy is Btu/h, commonly abbreviated verbally as simply 'Btu'. The Fahrenheit temperature scale is likely to be used in these situations.

Tons refrigeration (TR) – Defined as 12,000 Btu/h. Originally one ton-day was the amount of heat removed to make on US ton (2000lb) of ice in one day.

Horse power (HP) – It is a defined rate of thermal energy (746 Watts). Refrigeration equipment of a specific size can deliver cooling and absorb power at very different rates depending on refrigerant and operating conditions. This makes it very difficult to differentiate equipment physical size by, for example wattage. Commonly a piece of equipment will be described in terms of HP, for example '5hp condensing unit'. This has no quantitative meaning except that at one time a nominal 5hp motor would be necessary to drive it, so it gives an idea of the size.

Coefficient of performance (COP) – It is the dimensionless ratio of thermal energy rate and power input rate (mechanical or electrical). The same units are used for both, normally W or kW. COP may apply to cooling or to heating, depending on whether a system is delivering useful heat or useful cooling. For a simple system:

$$\mathrm{COP}(\text{heating}) = \mathrm{COP}(\text{cooling}) + 1$$

Normally it is quite clear from the context which COP is intended, and occasionally suffixes are used, for example COP_R, COP_H.

Refrigeration, Air Conditioning and Heat Pumps
http://dx.doi.org/10.1016/B978-0-08-100647-4.00038-3

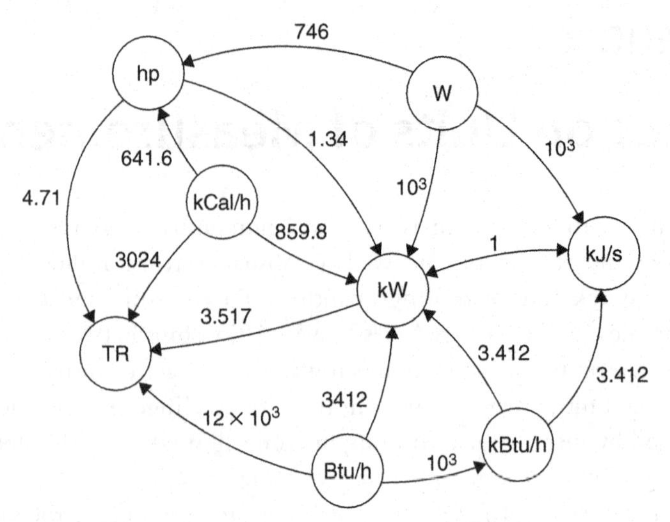

Figure A1 ***Energy flow rate conversion.***

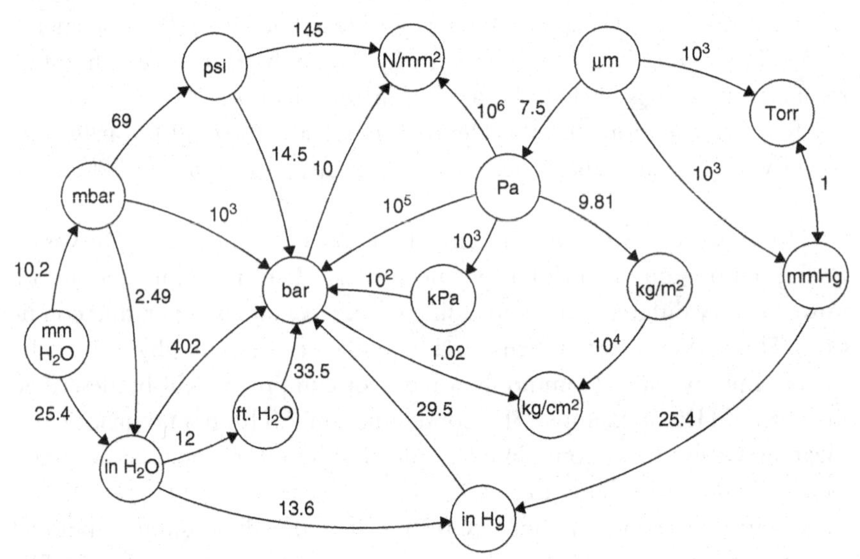

Figure A2 ***Pressure conversion.***

Energy efficiency ratio (EER) – It is the Imperial Units version of COP. It has the dimensions of Btu/h/W, and is found in US documentation and standards. Seasonal efficiency (SEER) is a benchmark rating for air conditioners in the USA. To convert EER to COP it is necessary to divide by 3.412. Somewhat confusingly, EER is used instead of COP in Europe for some air-conditioning units including chillers. Instead of using the normal Btu/h/W definition, EER for a chiller is defined as kW cooling capacity/kW power input (including fans). Thus the chiller 'EER' value is the dimensionless ratio normally expressed as COP.

Pressure, bar – Pressure is shown in absolute values unless otherwise stated. Traditional pressure gauges measure the difference between system pressure and atmospheric pressure, and it is normal to refer to this reading as bar gauge, or bar g. For a fixed absolute pressure in the system, the gauge pressure reading will vary slightly depending on atmospheric conditions and altitude, but this is normally ignored and the absolute pressure is obtained by adding 1.103.

Pressure drop, K – Compressor capacity data is given with reference to evaporating and condensing temperatures (see Section 10.5) and refrigeration pressure gauges are provided with temperature scales for various refrigerants (Figure 9.6). Therefore, pressure drops in suction and discharge lines are frequently referred to in terms of temperature differences, for example '2°C pressure drop', or more correctly, '2K pressure drop'.

LIST OF SOME SUGGESTED SOURCES OF FURTHER INFORMATION

Chapter	Reference
1–3, 12, 15, 19, 22, 23	ASHRAE Handbook – Fundamentals (latest edition).
10, 12, 14–17, 19	ASHRAE Handbook – Refrigeration (latest edition).
4, 7–9, 13, 23–25	ASHRAE Handbook – HVAC Systems and Equipment (latest edition).
24, 25, 28	ASHRAE Handbook – HVAC Applications (latest edition).
1–12, 14, 18, 19	Dossat, R.J., Horan, T.J., 2002. Principles of Refrigeration. Prentice Hall, Upper Saddle River, NJ.
1, 2	Gosney, W.B., 1982. Principles of Refrigeration. Cambridge University Press, New York, NY.
1, 2	Rogers, G.F.C., Mayhew, Y.R., 1992. Engineering Thermodynamics, Work & Heat Transfer. Pearson Higher Education, United Kingdom.
2, 3	Bullard, C., 2004–5. Transcritical CO_2 systems – recent progress and new challenges. IIR Bulletin 2004-5.
2, 3	Beating the Ban – is CO_2 a viable alternative?, 2000. 9th Annual IOR Conference.
3	Pearson, A.B., 2014. CO_2 as a refrigerant. IIR Guide.
3	Guideline Methods of Calculating TEWI, Issue 2, 2006, British Refrigeration Association.
3	Putting into Use Replacement Refrigerants, 2015, British Refrigeration Association, Reading UK
3	Refrigerant Report 18, 2014 (or later edition), Bitzer Kühlmaschinenbau Gmbh.
3	Pearson, A.B., 2007–8. Possibilities and pitfalls in carbon dioxide refrigeration, IOR 2007–8.
4	Winandy, E., 2012. New compressor technology trends for heat pumps, IOR.
4	Hundy, G.F., Kulkarni, S., 1996–7. The refrigeration scroll compressor and its application, IOR.
4	Creed, J., 2011. Compressor innovation - part load performance single screw compressors asymmetric unloading. IOR.
5	Sundaresan, S.G., 2015. Compressor Lubrication – the Key to Performance. International conferences on compressors and their systems, London.

Refrigeration, Air Conditioning and Heat Pumps
http://dx.doi.org/10.1016/B978-0-08-100647-4.00039-5

6, 7 Kays, W.M., London, A.L., 1964. Compact Heat Exchangers. McGraw Hill, New York, NY.

11 Safety code of practice for refrigerating systems using A1 refrigerants, 2015, IOR.

11 Safety code of practice for refrigerating systems using A2 and A3 refrigerants, 2015, IOR.

11 Safety code of practice for refrigerating systems using carbon dioxide, 2013 (or later) IOR.

11 Safety code of practice for refrigerating systems using ammonia, 2009 (or later) IOR.

11 Appointing and managing refrigeration contractors, Food and Drink Federation Initiative, Guide 1, 2007. Available from: IOR.

11 European standard EN 378, 2008 (or later), Refrigerating systems and heat pumps – safety and environmental requirements, Parts 1–4, BSI, London.

11, 27 Service Engineers' Technical Bulletins, IOR, Service Section.

11, 27 Code of practice for minimization of refrigerant emissions from refrigerating systems, 2009, IOR.

11, 27 Code of practice for refrigerant leak tightness in compliance with the F-gas regulation, 2007, British Refrigeration Association.

12 Pearson, S.F., 1993. Development of improved secondary refrigerants, IOR.

12 Melinder, Å., Granryd, E., 1992. Secondary Refrigerants for Heat Pumps and Low Temperature Refrigeration – A Comparison of Thermodynamic Properties of Aqueous Solutions and Non-Aqueous Liquids. Department of Applied Thermodynamics and Refrigeration, The Royal Institute of Technology, Stockholm, Sweden.

14 Duiven, J.E., Binard, P., 2002. Refrigerated Storage: New Developments, IIR Bulletin.

15 Cold Store Code of Practice, 2015, IOR.

16 Department of Health Cook-Chill/Freeze Guidelines.

16 Brown, T., James, S.J., 2006. The effect of air temperature, velocity and visual lean (VL) composition on the tempering times of frozen boneless beef blocks. Meat Science 78 (4), 545–552.

16, 17 Garnett, T., 2007–8. Fridge magnetism: an exploration of refrigeration dependence in the context of the UK food system and its contribution to climate changing emissions, IOR.

17 Guide to Refrigerated Transport, 1995, IIR.

17 Commère, B., 2002–3. Controlling the cold chain to ensure food hygiene and quality, IIR Bulletin.

17 European standard EN441, Refrigerated display cabinets, 1995, BSI, London.

17	BS EN ISO 23953:2005 Refrigerated display cabinets, Part 1: Vocabulary, and Part 2: Classification, Requirements and Test Conditions. BSI, London.
17	Stera, A.C., 1999. Long Distance Refrigerated Transport Into Third Millennium. 20th International Congress of Refrigeration. IIF/IIR. Sydney, Australia. Paper no. 736.
18	Stoecker, W.F., 1998. Industrial Refrigeration Handbook, McGraw Hill, New York, NY.
20–21	CIBSE Knowledge Series 20 KS20, 2012, Practical Psychrometry.
21	CIBSE Knowledge Series 19 KS19, 2012, Humidification.
20–26	Jones, W.P., 2000. Air Conditioning Engineering, Butterworth Heinemann.
23	BS EN 15780, 2011. Ventilation for Buildings: Ductwork – Cleanliness of Ventilation Systems.
23, 31	CIBSE TM42, 2006, Fan Application Guide.
23	Building and Engineering Services Association. Ductwork publications
23	Energy Savings in Fans and Fan Systems, 2004, Good Practice Guide GPG383, The Carbon Trust.
23	Cory, W.T.W., 2005. Fans and Ventilation, A Practical Guide. Elsevier, Saint Louis, Missouri.
24	ACHPI bulletin 33, 2012, Introduction to data centre cooling requirements, IOR.
25	CIBSE/IOR Guide to Variable Refrigerant Flow Air Conditioning (to be published 2016)
25	Doing cold smarter, Birmingham Energy Institute 2015
25	European standard EN 15450, 2007, Heating systems in buildings, design of heat pump systems. BSI, London.
25	Heap, R.D., 1983. Heat Pumps, second ed. E&FN Spon, New York.
25	Wang, F., et al., 2006–7. Ground source heating and cooling in the UK, IOR.
27, 28	Operational efficiency improvements for refrigeration systems, 2007, Food and Drink Federation Initiative, Guide 3. Available from: IOR.
27	Gartshore, J., 2007. Improving service and maintenance to reduce emissions, IOR conference.
28	Improving refrigeration system efficiency, 2007, Food and Drink Federation Initiative, Guide 5. Available from: IOR.
28	Purchase of efficient refrigeration plant, 2007, Food and Drink Federation Initiative, Guide 2. Available from: IOR.
28	Reducing refrigeration running costs, 1997, IOR conference.
28	Berglöf, K., 2007. Getting it right in practice – energy management case studies, IOR conference.
29	Audio Engineer's Reference Book, 2001, Edited Michael Talbot-Smith.

29	European standard EN 12354, 2000, parts 1 to 6, Building Acoustics. BSI, London.
29	Comparin, R., 1996. Coping with noise and vibration from compressors, Annual IOR Conference, Safe and reliable refrigeration.
29	Pearson, A., 1996. Coping with noise and vibration from fans, pumps and fluid flow, Annual IOR Conference, Safe and reliable refrigeration.
30	Smith, I.K., Stosic, N., Kovacevic, A., 2014. Power Recovery from Low Grade Heat by Means of Screw Expanders, Woodhead.
30	Klein, S.A., Reindl, D.T., 2005. Solar Refrigeration, Supplement to ASHRAE Journal, S26.
30	Nusiaputra, Y.Y., et al., 2014. Thermal-economic modularization of small, organic rankine cycle power plants for mid-enthalpy geothermal fields. Energies 7 (7), 4221–4240.
30	Xu et al., 2011. An investigation of the solar powered absorption refrigeration system with advanced energy storage technology. Elsevier; Solar Energy 85 (9), 1794–1804.
History of UK refrigeration	Cooper, A.J., 1997. The World Below Zero. ACR Today and Battlepress, Aylesbury.

Papers recording research and development may be found in proceedings of international conferences:

- Compressors and their Systems, IMechE
- Natural working fluids, IIR
- The cold chain, IIR
- International congress of refrigeration, IIR

USEFUL WEB SITES

The following sites contain relevant and up-to-date information and should contain reliable links.

URL	Organisation	Relevance
www.ior.org.uk	Institute of Refrigeration (UK)	Latest news on all refrigeration events and legislation, download publications, search technical and conference papers
www.cibse.org	The Chartered Institution of Building Services Engineers	Air conditioning and heating in buildings
www.feta.co.uk	Federation of Environmental Trade Associations	UK trade associations including BRA and HPA.
www.iiar.org	International Institute of Ammonia Refrigeration (USA)	Updates on ammonia technology, technical publications
www.b-es.org	Building and Engineering Services Association	Ductwork publications
www.fsdf.org.uk	The Food Storage and Distribution Federation	All aspects of food and drink logistics
www.crtech.co.uk	Cambridge Refrigeration Technology	Transport refrigeration, environmental test chambers, technical library
http://www.journals.elsevier.com/international-journal-of-refrigeration	International Journal of Refrigeration	Abstracts of latest IJR refrigeration research papers
www.iifiir.org	International Institute of Refrigeration	All aspects
www.food.gov.uk/	Food standards agency	Food storage conditions
www.acrib.org.uk	Air Conditioning and Refrigeration Industry Board	Detailed updates on legislation, training requirements
www.r744.com	Carbon Dioxide Industry platform	All refrigerant CO_2 topics

Refrigeration, Air Conditioning and Heat Pumps
http://dx.doi.org/10.1016/B978-0-08-100647-4.00040-1

www.hydrocarbons21.com	Hydrocarbons Industry platform	All refrigerant hydrocarbon topics
www.r717.com	Ammonia Industry platform	All refrigerant NH_3 topics
www.r718.com	R718 (water) Industry platform	All refrigerant R718 topics

REFERENCES

Bajpai, V.K., 2012. Design of Solar Powered Vapour Absorption System, Proceedings of the World Congress on Engineering, London, vol III.

Bailey, C., Cox, R.P., 1976. The chilling of beef carcasses, IOR.

Berglöf, K., 2004. Methods and potential for performance validation of air conditioning, refrigeration and heat pump systems, IOR.

Blackhurst, D.R., 1999. Recent developments in thermosyphon cooling, IOR.

Boast M.F.G., 2003, Frost free operation of large and high rise cold storage areas, IOR.

Bostock, D., 2007. Designing to minimise the risk of refrigerant leakage, IOR conference 2007.

Bowers, C.D., et al., 2012. Refrigerant distribution effects on the performance of microchannel evaporators. International Refrigeration and Air Conditioning Conference, Purdue.

Brasz, J.J., September 2007. Capacity control and performance optimization of variable geometry variable speed centrifugal compressors. International Conference on Compressors and their Systems. IMechE.

British Standard BS1042, 1989. Measurement of fluid flow in closed conduits, 1989, and later standards PD ISO/TR 15377, 2007.

Brown, T., et al., 2003. Practical investigations of two-stage bacon tempering. Int. J. Refrig. 26 (6), 690–697.

Building Regulations, 2000. Approved document L1/2 Conservation of fuel and power in buildings other than dwellings, NBS/RIBA Enterprises, 2006.

Cavallini, A., 2004. Properties of CO2 as a refrigerant. European Seminar-CO2 as a refrigerant: theoretical and design aspects.

Chaudry, M., Abeysekera, M., Hosseini, S.H.R., Wu, J., Jenkins, N., 2014. Energy strategies under uncertainty: uncertainties in UK heat infrastructure development. UKERC Report UKERC/WP/FG/2014/005; UKERC.

CIBSE Guide A, 2015. Environmental Design.

CIBSE Guide B0 Applications and activities, 2016.

CIBSE Guide B2, 2016. Ventilation and Ductwork.

CIBSE Guide B3, 2016. Air Conditioning and Refrigeration.

CIBSE Guide B4, 2016. Noise and vibration control for building services systems.

CIBSE Guide C, 2007. Reference data.

CIBSE/IOR, 2016. Guide to Variable Refrigerant Flow Air Conditioning.

CIBSE TM44, 2012. Inspection of air conditioning systems.

Clark, J., Gillies, A., 2014. Comparison of evaporative and air cooled condensers in industrial applications, IOR.

Critoph, R.E., Zhong, Y., 2005. Review of trends in solid sorption refrigeration and heat pumping technology, Proceedings of the Institution of Mechanical Engineers, Part E. J. Process. Mech. Eng. 219, 310.

Critoph, R.E., Metcalf, S.J., 2012. Development of a domestic adsorption gas-fired heat pump, IOR.

Controlling noise at work, 2005. Health and Safety Executive.

Davies, T., et al., 2014. A novel low energy defrost process for the frozen food chain, IOR.

Drewry Shipping Consultants Ltd, 2013. Annual reefer shipping market review and forecast 2013/14.

Eames, I.W., 2005–6. Absorption refrigeration and jet pumps, IOR.

Elson J.P., et al., 1990–1. Scroll compressor design and application characteristics for air conditioning, heat pump and refrigeration applications, IOR.

Energy Technology List, 2015. Published by the Carbon Trust.

Eurovent Certification, On-line directory Paris.

Refrigeration, Air Conditioning and Heat Pumps
http://dx.doi.org/10.1016/B978-0-08-100647-4.00036-X

European Standard EN 378, 2008 (or later). Refrigerating systems and heat pumps – safety and environmental requirements, Parts 1–4, BSI, London.
European Standard EN12900, 2005. Refrigerant compressors – rating conditions, tolerances and presentation of manufacturer's performance data, BSI, London.
European Standard EN13215, 2000. Condensing units for refrigeration – rating conditions, tolerances and presentation of manufacturer's performance data, BSI, London.
European Standard EN13136, 2013. Refrigerating systems and heat pumps. Pressure relief devices and their associated piping. Methods for calculation, BSI, London.
European Standard EN14511, 2013. Parts 1–4 Air conditioners, liquid chilling packages and heat pumps with electrically driven compressors for space heating and cooling, BSI, London.
European Standard EN15240, 2008. Ventilation for buildings. Energy performance of buildings. Guidelines for inspection of air-conditioning systems, BSI, London.
Evans, J.A., Scarcelli, S., Swain, M.V.L., 2007. Temperature and energy performance of refrigerated retail display and commercial catering cabinets under test conditions. Int. J. Refrig. 30, 398–408.
Flucker, S., Tozer, R., 2011. Data centre cooling air performance metrics. CIBSE Technical Symposium.
Gosling, C.T., 1980. Applied Air Conditioning and Refrigeration. Elsevier, London.
Gosney, W.B., 1982. Principles of Refrigeration. Cambridge University Press, New York.
Guideline methods of calculating TEWI, Issue 2, 2006. British Refrigeration Association.
Hands, B., 1993. The Stirling cycle refrigerator programme at Oxford University, IOR.
Hoffmann, K., Pearson, D.F., 2011. Ammonia heat pumps for district heating in Norway – a case study, IOR.
HSE, 2014. Health and safety executive, safety of pressure systems – approved code of practice (ACOP) L122, (ISBN 0 7176 1767 X).
Hundy, G.F., 2001–2. Capacity control solutions with scroll compressors, IOR.
Institute of Refrigeration Cold Store Code of Practice, 2015, IOR.
James, S.J., Creed, P.G., Bailey, C., 1979. The determination of freezing time of boxed meat blocks, IOR.
James, C., Nicolson, M., James, S.J., 1999. Review of microbial contamination and control measures in abattoirs. Food Refrigeration and Process Engineering Research Centre. University of Bristol, ISBN 0-86292-498-7.
James, C., James, S.J., 1997. Meat decontamination – the state of the art. MAFF Advanced Fellowship in Food Process Engineering, University of Bristol, EU concerted action programme CT94 1881, ISBN 0-86292-460-X.
Jones, W.P., 2000. Air Conditioning Engineering. Butterworth Heinemann, Oxford.
Kühn, A. (Ed.), 2013. Thermally driven heat pumps for heating and cooling. IEA Heat Pump Annex, vol. 34.
Lane, A.-L., Benson, J., Hundy, G., Berglöf, K., 2014. System Efficiency Index, SEI, a key performance indicator for field measurements on refrigeration, air conditioning and heat pump systems. 3rd IIR International Conference on Sustainability and the Cold Chain, London.
Lester, C., Hundy, G., 2014. Early refrigeration at the Grimsby Ice Factory, IOR.
Macklin, R., 2006. Gloucester Police Headquarters heat pump project, IOR conference, "Smaller, Colder, Smarter."
Maidment, G.G., et al., 2007. Development of the system efficiency index (SEI) for refrigeration and air conditioning systems, IOR conference 2007.
Marques, C., et al., 2013. The use of phase change materials in domestic refrigerator applications, IOR.
NIST Refprop, (on-going). Reference fluid thermodynamic and transport properties database. National Institute of Standards and Technology, Boulder, USA.

Non-Domestic Heating, Cooling and Ventilation Compliance Guide, 2006. DCLG.
Oughton, R.J., 1987. Legionnaires' disease in refrigeration and associated equipment, IOR.
Paul, J., Jahn, E., 1997. Binary ice – application of liquid, pumpable ice slurries and status of the technology, IOR.
Paul, J., 2001–2. Innovative applications of pumpable ice slurry, IOR.
Pearson, S.F., 2006. A new look at evaporative condensers, IOR.
Pearson, S.F., 2004. Air conditioning for the future using carbon dioxide, IOR.
Pearson, S.F., 1990. Thermosyphon cooling, IOR.
Prakesh, J.A., 2006. Energy optimisation potential through improved onsite analysing methods in refrigeration. Masters thesis, Department of Energy Technology, Royal Institute of Technology (KTH), Stockholm.
SP Arbetsrapport, 2014. SP Technical Research Institute of Sweden, ISBN/ISSN 0284-5127.
Sulc, V., 2007. Characteristics of refrigerating systems – extension of the graphical-analytical method, International Conference on Compressors and their Systems, IMechE.
Süß, J., 2014–5. New ways to use water as a refrigerant, IOR.
Sundsten, S., Andersson, A., Tornberg, E., 2001. The effect of the freezing rate on the quality of hamburgers. Proceedings of the International Institute of Refrigeration Rapid Cooling – Above and below zero, Bristol.
Tozer, R.M., James, R.W., 1997. Fundamental thermodynamics of ideal absorption cycles. Int J. Refrig. 20 (2), 120–135.
VDMA Specification No. 24247, (on-going). Energy efficiency of refrigeration systems.
Wilson, N., Ozcan, S., Sandeman, K., 2007. Overview of magnetic refrigeration, IOR.

SUBJECT INDEX

B

D

F

G

Q

R

S

W

Y

Z

Printed in the United States
By Bookmasters